Lecture Notes in Mathematics

Edited by A. Dold and B. Eckmann

1339

T. Sunada (Ed.)

Geometry and Analysis on Manifolds

Proceedings of the 21st International Taniguchi Symposium held at Katata, Japan, Aug. 23–29 and the Conference held at Kyoto, Aug. 31 – Sept. 2, 1987

Springer-Verlag
Berlin Heidelberg New York London Paris Tokyo

Editor

Toshikazu Sunada
Department of Mathematics, Faculty of Science
Nagoya University, Nagoya 464, Japan

Mathematics Subject Classification (1980): 53-06, 58-06

ISBN 3-540-50113-4 Springer-Verlag Berlin Heidelberg New York
ISBN 0-387-50113-4 Springer-Verlag New York Berlin Heidelberg

Printing and binding: Druckhaus Beltz, Hemsbach/Bergstr.
2146/3140-543210

PREFACE

The twentyfirst Taniguchi International Symposium was held at Katata in Shiga prefecture, Japan from August 23rd through 29th, 1987 under the title

Geometry and Analysis on Manifolds.

The symposium was followed by a conference held at the Institute for Mathematical Science in Kyoto University from August 31st till September 2nd under the same title.

The symposium and conference were focused on various aspects of geometric analysis, including spectral analysis of the Laplacian on compact and noncompact Riemannian manifolds, harmonic analysis on manifolds, complex analysis and isospectral problems. The present volume contains expanded versions of most of the invited lectures in Katata and Kyoto.

We, the organizers and all the participants, would like to express our hearty thanks to Mr. Toyosaburo Taniguchi for his support. Thanks are due to Professor Shingo Murakami who, as the coordinator of the Taniguchi International Symposia, guided the organizing committee to the success of the symposium and conference.

Toshikazu Sunada

Participants in the Taniguchi International Symposium

Michael T. ANDERSON	Department of Mathematics, California Institute of Technology, Pasadena, CA 91125, U.S.A.
Shigetoshi BANDO	Department of Mathematics, Tohoku University, 980 Sendai, Japan
Gérard BESSON	Université de Grenoble I, Institut Fourier, Laboratoire de Mathématique, 38402 Saint-Martin-d'Hères Cedex, France
Peter BUSER	Department of Mathematics, Swiss Federal Institute of Technology Lausanne, CH-1015 Lausanne, Switzerland
Jean-Pierre DEMAILLY	Université de Grenoble I, Institut Fourier, Laboratoire de Mathématique, 38402 Saint-Martin-d'Hères Cedex, France
Harold DONNELLY	Department of Mathematics, Purdue University, W. Lafayette, IN 47909, U.S.A.
Jozef DODZIUK	Department of Mathematics, Queens Colleges, CUNY, Flushing, NY 11367, U.S.A.
Ichiro ENOKI	Department of Mathematics, College of General Education, Osaka University, 560 Toyonaka, Japan
Masahiko KANAI	Department of Mathematics, Keio University, 223 Yokohama, Japan
Toshiki MABUCHI	Department of Mathematics, College of General Education, Osaka University, 560 Toyonaka, Japan
Werner MÜLLER	Akademie der Wissenschaften der DDR, Karl-Weierstrass-Institut für Mathematik, DDR 1086, Berlin.
Shingo MURAKAMI	Department of Mathematics, Osaka University, 560 Toyonaka, Japan
Takushiro OCHIAI	Department of Mathematics, University of Tokyo, 113 Tokyo, Japan
Shin OZAWA	Department of Mathematics, Tokyo Institute of Technology, 152 Tokyo, Japan
Toshikazu SUNADA	Department of Mathematics, Nagoya University, 464 Nagoya, Japan

Japanese speakers in the Kyoto Conference

Kenji FUKAYA	Department of Mathematics, Faculty of General Education, University of Tokyo, Komaba, Tokyo, Japan
Atsushi KASUE	Department of Mathematics, Osaka University, 560 Toyonaka, Japan
Atsushi KATSUDA	Department of Mathematics, Nagoya University, 464 Nagoya, Japan
Hajime KAWAKAMI	Department of Mathematics, Kanazawa University, Kanazawa, Japan
Ryoichi KOBAYASHI	Department of Mathematics, Tohoku University, 980 Sendai, Japan
Ken-ichi SUGIYAMA	Department of Mathematics, University of Tokyo, 113 Tokyo, Japan

PROGRAM OF SYMPOSIUM (KATATA)

Monday, 24. 8.:

9:30 S. Ozawa : Surveys and open problems concerning eigenvalues of the Laplacian on a wildly perturbed domain.

14:00 S. Bando : Ricci flat Kähler metrics on non-compact Kähler manifolds.

15:30 I. Enoki : On compact Kähler manifolds with nonpositive Ricci curvature.

Tuesday, 25. 8.:

9:30 M.T. Anderson : Topology of complete manifolds of non-negative Ricci curvature.

11:00 M.T. Anderson : Space of positive Einstein metrics on compact manifolds.

14:00 G. Besson : On the multiplicity of eigenvalues of the Laplacian.

15:30 T. Mabuchi : Einstein Kähler metrics on toric varieties.

Thursday, 27. 8.:

9:30 H. Donnelly : Decay of eigenfunctions on Riemannian manifolds.

14:00 J. Dodziuk : Examples of Riemann surfaces of large genus with large λ_1.

15:30 T. Sunada : Fundamental groups and spectrum.

Friday, 28. 8.:

9:30 M. Kanai : Rough isometries between open manifolds.

11:00 M. Kanai : Geodesic flows of negatively curved manifolds with smooth stable foliations.

14:00 J.-P. Demailly : Characterization of affine algebraic manifolds by volume and curvature estimates.

15:30 W. Müller : Manifolds with corners and eta-invariants.

Saturday, 29. 8.:

9:30 P. Buser : An upper bound for the number of pairwise isospectral Riemann surfaces.

11:00 P. Buser : A finiteness theorem for the spectrum of Riemann surfaces.

PROGRAM OF CONFERENCE (KYOTO)

Monday, 31. 8.:

10:00 M.T. Anderson : Compactification of complete minimal submanifolds in R^n by Gauss map.

11:10 A. Kasue : Harmonic functions of finite growth on a manifold with asymptotically non-negative curvature.

13:30 J. Dodziuk : Lower bounds for the bottom of the spectrum of negatively curved manifolds.

14:40 A. Katsuda : Density theorem for closed geodesics.

15:50 H. Kawakami : On a construction of complete simply-connected Riemannian manifolds with negative curvature.

Tuesday, 1. 9.:

9:30 G. Besson : Number of bounded states and estimates on some geometric invariants.

10:30 R. Kobayashi : Kähler-Einstein metrics on algebraic varieties of general type.

11:30 J.-P. Demailly : Vanishing theorems and Morse inequalities for complex vector bundles.

14:30 H. Donnelly : Decay of eigenfunctions on Riemannian manifolds.

15:40 K. Sugiyama : Spectrum and a vanishing theorem.

Wednesday, 2. 9.:

9:30 P. Buser : Cayley graphs and planer isospectral domains.

10:30 K. Fukaya : Collapsing of Riemannian manifolds and eigenvalues of the Laplace operator.

11:30 W. Müller : On the generalized Hirzebruch conjecture.

CONTENTS

L^2 HARMONIC FORMS ON COMPLETE RIEMANNIAN MANIFOLDS

Michael T. Anderson*
Mathematics 253-37, California Institute of Technology
Pasadena, CA 91125

In this paper, we briefly survey selected recent developments and present some new results in the area of L^2 harmonic forms on complete Riemannian manifolds. In light of studies of L^2 cohomology relating to singular varieties, discrete series representations of Lie groups, arithmetic quotients of symmetric spaces among others, our discussion will be rather limited, focussing only on aspects of L^2 harmonic forms in global Riemannian geometry. This paper is partly intended as a completion of the announcement [1]. One may refer to [13] for a previous survey in this area.

Throughout the paper, all manifolds will be connected, complete, oriented Riemannian manifolds, of dimension n.

§1. L^2 cohomology and L^2 harmonic forms.

[1.1] Let Δ_M denote the Laplace-Beltrami operator acting on C^∞ p-forms $C^\infty(\Lambda^p(M))$ on the manifold M. The space of L^2 harmonic p-forms $\mathcal{H}^p_{(2)}(M)$ consists of those forms $\omega \in C^\infty(\Lambda^p(M))$ such that $\Delta_M\omega = 0$ and $\omega \in L^2$, i.e. $\|\omega\|^2 = \int_M \omega\wedge *\omega < \infty$, where $*: \Lambda^p(M) \to \Lambda^{n-p}(M)$ is the Hodge $*$ operator. The regularity theory for elliptic operators implies that $\mathcal{H}^p_{(2)}(M)$ is a Hilbert space with L^2 inner product.

If M is compact, the Hodge theorem implies that $\mathcal{H}^p_{(2)}(M) \cong H^p(M,\mathbb{R})$ so that these spaces are topological invariants of M. One guiding problem is to understand to what extent this remains true for non-compact manifolds. Note that since Δ and $*$ commute, $*$ induces on isomorphism $\mathcal{H}^p_{(2)}(M) \cong \mathcal{H}^{n-p}_{(2)}(M)$ (representing Poincaré duality for M compact). In particular, $*$ induces an automorphism of $\mathcal{H}^{n/2}_{(2)}(M)$ with $*^2 = (-1)^{n/2}$. Since $*$ is a conformal invariant on n/2-forms, one obtains the important fact that the Hilbert space structure on $\mathcal{H}^{n/2}_{(2)}(M)$ depends only on the conformal structure of M.

A well-known result of Andreotti-Vesentini [10] implies that

$$\mathcal{H}^p_{(2)}(M) = \{\omega\in C^\infty(\Lambda^p(M))\cap L^2 : d\omega=0 \text{ and } \delta w=0\},$$

where d is the exterior derivative and δ its formal adjoint. Thus one has a natural map

$$\mathcal{H}^p_{(2)}(M) \to H^p_{deR}(M).$$

We now relate the space of L^2 harmonic forms on M to its L^2 cohomology. The simplest definition of L^2 cohomology is

---

*Partially supported by NSF Grant DMS-8701137

$$H^p_{(2)}(M) = \frac{\ker d_p}{\operatorname{Im} d_{p-1}}, \tag{1.1}$$

where $\ker d_p = \{\omega \in C^\infty(\Lambda^p M) \cap L^2 : d\omega = 0\}$, $\operatorname{Im} d_{p-1} = \{\eta \in C^\infty(\Lambda^p M) \cap L^2 : d\alpha = \eta$, for some $\alpha \in \operatorname{dom}_{p-1}\}$ with $\operatorname{dom} d_{p-1} = \{\alpha \in C^\infty(\Lambda^{p-1} M) \cap L^2 : d\alpha \in L^2\}$. Clearly there is a natural map

$$i_\#: \mathcal{H}^p_{(2)}(M) \to H^p_{(2)}(M)$$

and one says the Strong Hodge Theorem holds if $i_\#$ is an isomorphism. Cheeger [6] has shown that $i_\#$ is always an injection (since M is assumed complete). However, in many cases $i_\#$ is not surjective. For example, it is easily calculated that $\mathcal{H}^1_{(2)}(\mathbb{R}) = \{0\}$, but $H^1_{(2)}(\mathbb{R})$ is infinite dimensional(c.f. [6]).

Define the reduced L^2 cohomology by

$$\overline{H}^p_{(2)}(M) = \frac{\ker \overline{d}_p}{\overline{\operatorname{Im} d_{p-1}}} \tag{1.2}$$

where the closure is taken in L^2 and $\overline{d}_p$ is the strong closure of d_p in L^2,, i.e., $\overline{d}_p\alpha = \beta$ if $\exists \alpha_i \in \operatorname{dom} d_p$ such that $\alpha_i \to \alpha$ and $d\alpha_i \to \beta$ in L^2. There is a natural surjection $H^p_{(2)}(M) \to \overline{H}^p_{(2)}(M)$ and we have the basic fact [6] that

$$\mathcal{H}^p_{(2)}(M) \cong \overline{H}^p_{(2)}(M) \tag{1.3}$$

for any complete Riemannian manifold M. (1.3) may be viewed as a non-compact Hodge theorem: the reduced L^2 de Rham cohomology of M is isomorphic to the space of L^2 harmonic forms. For a de Rham-type theorem relating $\overline{H}^p_{(2)}(M)$ to the simplicial L^2 cohomology of M, c.f. [12].

An immediate consequence of (1.3) is that $\mathcal{H}^p_{(2)}(M)$ (up to equivalence) depends only on the quasi-isometry class of the metric on M, since it is easily verified that the topology on $\overline{H}^p_{(2)}(M)$ is a quasi-isometry invariant. In particular, if M is a (non-compact) regular cover of a compact manifold N with metric lifted from N, then $\mathcal{H}^p_{(2)}(M)$ does not depend on the metric. One is led to expect in this case that $\mathcal{H}^p_{(2)}(M)$ is a topological invariant of the pair $(N = M/\Gamma, \Gamma)$ in this case. In fact, Dodziuk [11] has shown that the action of Γ on $\mathcal{H}^p_{(2)}(M)$ is a homotopy invarant of (N,Γ) (up to equivalence). In particular, the L^2 Betti numbers $b^p_\Gamma(N) = \dim_\Gamma \mathcal{H}^p_{(2)}(M)$, c.f. §2, are homotopy invariants of N.

In general, one is interested in understanding relations between the topology and geometry of M and the spaces $\mathcal{H}^p_{(2)}(M)$. However, in many cases $\dim \mathcal{H}^p_{(2)}(M)$ has been difficult to estimate, even whether it vanishes or not. Some examples and discussion follow.

[1.2] (i) $\mathcal{H}^0_{(2)}(M) = \mathcal{H}^n_{(2)}(M) = \begin{cases} 0 & \text{if } \operatorname{vol} M = \infty \\ \mathbb{R} & \text{if } \operatorname{vol} M < \infty \end{cases}$. Further, if M is simply connected, then $\mathcal{H}^1_{(2)}(M)$ is naturally identified with the space of harmonic functions $u: M \to \mathbb{R}$ with finite energy or Dirichlet integral $\int_M |du|^2 < \infty$.

(ii) $\mathcal{H}^p_{(2)}(\mathbb{R}^n, \text{flat}) = 0$, for $0 \le p \le n$.

(iii) Let $H^n(-1)$ be the hyperbolic space of constant curvature -1. Then

$$\dim \mathcal{H}^p_{(2)}(H^n(-1)) = \begin{cases} 0 & p \neq n/2. \\ \infty & p = n/2. \end{cases}$$

In fact, the same result holds for any irreducible symmetric space of non-compact type, c.f. [4].

[1.3] Let $(C^\infty_o(M), d)$ denote the de Rham complex of forms of compact support on M and let $H^p_o(M)$ be the corresponding cohomology with compact supports. There is a natural map

$$\Pi: H^p_o(M) \to \mathcal{H}^p_{(2)}(M) \tag{1.4}$$

given by $\Pi[\omega] = P(\omega)$, where P denotes orthogonal projection (in L^2) onto $\mathcal{H}^p_{(2)}(M)$. The weak Hodge theorem of Kodaira [20]

$$L^2 \cong \overline{\text{Im } \delta_{p+1,o}} \oplus \overline{\text{Im } d_{p-1,o}} \oplus \mathcal{H}^p_{(2)}(M) \tag{1.5}$$

implies that Π is well defined on cohomology. A result of Gaffney [16] that $\int_c \omega = \int_c P(\omega)$ for any compact p-cycle c in M and closed p-form $\omega \in C^\infty_o(\Lambda^p M)$ implies that $\ker \Pi \subset Z \equiv \{\omega \in C^\infty_o(\Lambda^p M): d\omega = 0, \int_c \omega = 0$ for all compactly supported p-cycles c$\}$.

For example, suppose $E \to X$ is a vector bundle over the compact n-manifold X with fibre $\mathbb{R}^m$. The Thom class $U \in H^m_{deR}(E)$ of E is represented by a closed m-form in $C^\infty_o(\Lambda^m E)$ and is non-zero (in $H^m_{deR}(E)$) if for instance the Euler class $e(E) \in H^m_{deR}(X)$ is non-zero, since $e(E) = z^*U$, where z is the 0-section. It follows in this case that $P[U] \in \mathcal{H}^m_{(2)}(E)$ represents a non-trivial L^2 harmonic n-form on E, for any complete metric on E.

[1.4] As an application of [1.3] we mention the following. Suppose M is a complete non-compact Riemannian manifold which admits a complete metric with non-negative curvature operator $R: \Lambda^2(TM) \to \Lambda^2(TM)$. The well-known Bochner-Weitzenbock formula implies that for any harmonic p-form ω on M

$$\tfrac{1}{2}\Delta|\omega|^2 = |\nabla\omega|^2 + \langle F_p\omega, \omega\rangle. \tag{1.6}$$

It is shown in [17] that $R \ge 0 \Rightarrow \langle F\omega, \omega\rangle \ge 0$, for any p. If $\omega \in L^2$, then one may integrate (1.6) by parts, and using a standard cutoff function argument it follows that

$$\int_M |\nabla\omega|^2 + \langle F_p\omega, \omega\rangle = 0.$$

Thus, $|\nabla\omega|^2 \equiv 0$, i.e. ω is a parallel p-form on M. It follows that $|\omega|^2$ = const and since $\omega \in L^2$ and $\mathrm{vol}(M) = \infty$, we see $\omega = 0$.

This shows that for any complete metric on M with $R \geq 0$, there are no L^2 harmonic p-forms, for any p. Since F_1 is basically the Ricci curvature Ric_M of M, we also see that if $\mathrm{Ric}_M \geq 0$ then there are no non-trivial harmonic 1-forms on M.

Corollary. Let X be a compact n-manifold and $E \to X$ a rank m vector bundle with non-zero Thom class $U \in H^m_{deR}(E)$. Then E admits no complete metric with non-negative curvature operator R.

Proof. By [1.3] above, $\Pi[U]$ is a non-trivial L^2 harmonic m-form on E. If E has a complete metric with $R \geq 0$, we contradict the discussion above.

For example, TS^n admits no complete metric with $R \geqslant 0$ for n even, even though S^n admits metrics with $R > 0$. Note that TS^n admits complete metrics with sectional curvature $K \geq 0$. An open question of Gromoll asks whether every vector bundle over a compact manifold of non-negative sectional curvature admits a complete metric of non-negative sectional curvature.

[1.5] By [1.3], one may produce L^2 harmonic forms on M if M has compactly supported cohomology and homology. In fact, it is easily verified that $H_o(M)$ and Z are topological invariants of M so that $Y \equiv H_o(M)/Z \subset \mathrm{Im}\,\Pi \subset \mathcal{H}_{(2)}(M)$ depends only on the topology of M.

Note further that one may use the * operator to produce harmonic forms not in $\mathrm{Im}\,\Pi$. For instance, if $n = 4k+2$, then $*^2 = -1$ on 2k+1 forms so that $\mathcal{H}^{2k+1}_{(2)}(M)$ is even dimensional. If $E \to X$ is a rank 2k+1 bundle over X^{2k+1} with non-zero Thom class, then $\mathrm{Im}\,\Pi_{2k+1} = \langle \Pi[U] \rangle$ so that $*\Pi(U) \notin \mathrm{Im}\,\Pi$.

The space $*Y$ is also seen to be a topological invariant so if we set $\mathcal{H}^p_c = \overline{Y_p \cup *Y_{n-p}}$ we may form $\mathcal{H}^p_\infty \equiv \mathcal{H}^p_{(2)}(M)/\mathcal{H}^p_c$. A basic problem is to determine whether, and if so how, $\mathcal{H}^p_\infty$ is characterized by the geometry of M at infinity (up to quasi-isometry).

Based on examples, one expects to a certain extent that if M is "large" at infinity, e.g. strictly negative curvature, exponential volume growth, etc., then M should possess L^2 harmonic p-forms in $\mathcal{H}^p_\infty$, for some p. If M is "small" at infinity, e.g. non-negative curvature, polynomial volume growth, etc., then M may possess no L^2 harmonic p-forms in $\mathcal{H}^p_\infty$.

In this respect, we raise the following questions.

1) If M is a complete, non-compact manifold of positive sectional curvature, is it true that M admits no non-trivial L^2 harmonic p-forms? Of course, the Gromoll-Meyer theorem [19] implies that M is diffeomorphic to $\mathbb{R}^n$ so there is no compactly supported homology. By the remarks above, the question is true if M has non-negative curvature operator or M is quasi-isometric to $\mathbb{R}^n$.

One may ask if the curvature condition above can be weakened. For example, if $H_1(M) = H_{n-1}(M) = 0$ and $M^n, n\geq 3$, is of polynomial volume growth, i.e. $\mathrm{vol}B(r) \leq c\cdot r^k$ for $r\geq 1$, some k, where B(r) is the geodesic r-ball in M, is every L^2 harmonic 1-form on M zero? Similarly, if $H_p(M) = H_{n-p}(M) = 0$, is every L^2 harmonic p-form on M^n, $p\neq \frac{n}{2}$ zero? One must exclude the middle dimensions because of conformal invariance. However, we will see in [3.7] below that these questions are false in every dimension.

2) If M^{2n} is simply connected with curvature $K_M \leq -1$, does M admit a non-trivial L^2 harmonic p-form? One expects the answer should be yes and that one can weaken the hypothesis to $K_M \leq -1$ outside a compact set.

§2. L^2 Betti numbers.

[2.1] The L^2 Betti numbers $b^k_{(2)}(M)$, introduced by Atiyah in [2], are homotopy invariants attached to a compact manifold M defined as follows. Let $\tilde{M}$ be the universal cover of M equipped with a lifted metric so that $\Gamma = \pi_1(M)$ acts by isometries. The Hilbert space $\mathcal{H}^k_{(2)}(\tilde{M})$ thus becomes a Γ-module. Let $P: L^2 \to \mathcal{H}^k_{(2)}(\tilde{M})$ be the orthogonal projection with associated smooth kernel $p(x,y) \in \mathrm{Hom}(\Lambda^k_y(\tilde{M}),\Lambda^k_x(\tilde{M}))$. Note that Γ commutes with Δ and preserves inner products so that it commutes with P. In particular,

$$p(\gamma x,\gamma y) = p(x,y), \ \forall\gamma\in\Gamma. \tag{2.1}$$

If $\{\phi_n\}$ is an orthonormal basis of $\mathcal{H}^k_{(2)}(\tilde{M})$, then one has the sequence of partial sums

$$p_N(x,y) = \sum_1^N \phi_n(x) \otimes \phi_n^*(y),$$

where ϕ_n^* is the dual of ϕ_n defined by the metric in $(\Lambda^p(M))^*$. P_N defines a projection of finite rank on L^2 and the sequence P_N converges strongly in L^2 to P. In particular,

$$p(x,y) = \sum_1^\infty \phi_n(x) \otimes \phi_n^*(y),$$

where the convergence is uniform on compact sets in $\tilde{M}$. In particular,

$$\mathrm{tr}\ p(x,x) = \sum_1^\infty |\phi_n(x)|^2 \geq 0. \tag{2.2}$$

Since p(x,y) is Γ-invariant, tr p(x,x) is a Γ-invariant function on $\tilde{M}$ and thus defines a function on M. Define the L^2 Betti number $b^k_{(2)}(M)$ by

$$b^k_{(2)}(M) = \dim \mathcal{H}^k_{(2)}(\tilde{M}) = \int_M \mathrm{tr}\ p(x,x)dV_x. \tag{2.3}$$

Note that one may define the L^2 Betti numbers $b^k_2(M_1\Gamma)$ of a Galois cover $\bar{M} \to M$ with group Γ in exactly the same way. We list below several properties and results for L^2 Betti numbers. Many of these carry over to the case $b^k_{(2)}(M,\Gamma)$, but for simplicity we assume $\Gamma = \pi_1(M)$.

[2.2] It is clear that the L^2 Betti numbers are largely dependent on the structure of $\pi_1(M)$. Of course, if $\pi_1(M) = 0$, then $b^k_{(2)}(M) = \dim H^k(M,\mathbb{R})$ by the Hodge theorem. Thus, one is really only interested in the case $|\pi_1(M)| = \infty$ and we will always assume this below. The group $\pi_1(M)$ enters in two ways: (i) in determining the basic features of the geometry and topology of $\tilde{M}$ and thus of the space $\mathcal{H}^k_{(2)}(\tilde{M})$, and (ii) via the action of Γ on $\mathcal{H}^k_{(2)}(\tilde{M})$. Roughly speaking, (i) determines whether $b^k_{(2)}(\tilde{M})$ is zero or not, while (ii) leads to the exact value of $b^k_{(2)}(M)$ (assumed positive).

As an example of the π_1-dependence, note that if M_1 and M_2 are compact manifolds with $\tilde{M}_1$ isometric to $\tilde{M}_2$ (or quasi-isometric), then $b^k_{(2)}(M_1) > 0 \Longleftrightarrow b^k_{(2)}(M_2) > 0$. For example, L^2 Betti numbers of all flat manifolds are zero.

[2.3] Since we assume $|\pi_1(M)| = \infty$, the following statements are equivalent:

(i) $b^k_{(2)}(M) > 0$

(ii) $\dim \mathcal{H}^k_{(2)}(\tilde{M}) > 0$, i.e. there is a non-trivial L^2 harmonic k-form on $\tilde{M}$.

(iii) $\dim \mathcal{H}^k_{(2)}(\tilde{M}) = \infty$.

The L^2 Betti numbers behave multiplicatively under finite covers, i.e. if $\tilde{M} \to M$ is an ℓ-sheeted cover, then

$$b^k_{(2)}(\tilde{M}) = \ell \cdot b^k_{(2)}(M).$$

Also, the L^2 Betti numbers satisfy a Poincaré duality $b^k_{(2)}(M) = b^{n-k}_{(2)}(M)$.

[2.4] The L^2 index theorem of Atiyah [2] implies that a number of topological invariants of M can be computed in terms of the L^2 Betti numbers. In particular, for the Euler characteristic $\chi(M)$ one has

$$\chi(M) = \sum_{k=1}^{n} (-1)^k b^k_{(2)}(M). \tag{2.4}$$

For the signature $\sigma(M)$, assuming $n \equiv 0(4)$, one has

$$\sigma(M) = \beta_+^{n/2} - \beta_-^{n/2}, \tag{2.5}$$

where $\beta_\pm^{n/2}$ denotes the Γ-dimensions of the ± 1 eigenspaces of the $*$ operator acting on $\mathcal{H}^{n/2}_{(2)}(\tilde{M})$.

[2.5] It follows easily from standard elliptic theory that one has an estimate of the form

$$\operatorname{tr} P(x,x) \leq c(\operatorname{geo}(\tilde{M})), \tag{2.6}$$

where c is a constant depending on $\sup|K_M|$ and $\inf \operatorname{Inj}(x,\tilde{M})$. Thus, if M collapses with bounded covering geometry, i.e. if M admits a sequence of metrics g_i such that $geo(\tilde{M},g_i) \leq c$ and $vol(M,g_i) \to 0$, then

$$b^k_{(2)}(M) = \int_M tr\, P_i(x,x)dv_i \leq c\cdot vol(M,g_i) \to 0.$$

This observation of Cheeger-Gromov [7] shows that the L^2 Betti numbers are obstructions to the collapse of M with bounded covering geometry. For instance, any manifold M of the form $M = N\times S^1$ collapses with bounded covering geometry. In general, of course, (2.6) leads to the upper bound

$$b^k_{(2)}(M) \leq c(n,k,geo(\tilde{M}))\cdot volM.$$

[2.6] Suppose $\pi_1(M)$ is an amenable group. This may be characterized geometrically by the condition $h_{Ch}(\tilde{M}) = 0$, where $h_{Ch}(\tilde{M}) = \inf_{U\subset\subset M} \frac{vol\partial U}{volU}$ is the Cheeger isoperimetric constant on $\tilde{M}$ (c.f. [5]). Examples of amenable discrete groups include all nilpotent and solvable groups, as well as groups of subexponential growth with respect to the word metric. On the other hand, fundamental groups (and subgroups) of compact negatively curved manifolds are non-amenable (unless infinite cyclic).

Cheeger-Gromov [9] prove the interesting result that if $\pi_1(M)$ is amenable, then the natural map

$$\rho\colon H^p_{(2)}(\tilde{M}) \to H^p_{deR}(\tilde{M}) \qquad (2.7)$$

is injective, for any p. This has the following immediate consequence.

(i) $b^1_{(2)}(\tilde{M}) = 0$, i.e. M has no L^2 harmonic 1-forms. Using a result of Brooks [5], this implies that if X is any compact manifold, then $\tilde{X}$ has L^2 harmonic 1-forms only if $\lambda_0 > 0$, where λ_0 is the infinum of the L^2-spectrum of Δ on functions. Similarly, by a result of Lyons-Sullivan [21], $\tilde{X}$ has L^2 harmonic 1-forms only if $\tilde{X}$ carries a non-constant bounded harmonic function.

(ii) If M is a $K(\pi,1)$, then $\tilde{M}$ has no L^2 harmonic p-forms, for any p.

(iii) If $\tilde{M}$ has a non-zero L^2 harmonic p-form, then $\dim H^p_{deR}(M) = \infty$.

For general $\pi_1(M)$, the method of Cheeger-Gromov can easily be shown to imply that

$$b^p_{(2)}(M,\rho) \leq c(geo(M))\cdot h_{ch}(\tilde{M}),$$

where $b^p_{(2)}(M,\rho) = \dim_\Gamma \ker \rho$. It would be interesting to bound c in terms of weaker invariants, e.g. $\inf Ric_M$ and $diam_M$.

Note that the converse of the result above is false, i.e., there exist compact manifolds with non-amenable $\pi_1(M)$ with all L^2 Betti numbers zero, c.f. [2.5] for example.

Also, (2.7) is not valid for general non-compact manifolds with say, polynomial volume growth, c.f. [3.7].

[2.7] Suppose M is a non-compact homogeneous manifold with transitive group of isometries G. It follows that the projection $P: L^2 \to \mathcal{H}^k_{(2)}(M)$ is G-invariant, so that the trace of the projection kernel $\mathrm{tr}\,p(x,x)$ is a constant function α^k on M. In particular, $\alpha^k > 0$ if and only if $\mathcal{H}^k_{(2)}(M)$ is infinite dimensional. Note that if M admits a discrete, cocompact subgroup of isometries, then the L^2 Betti numbers of M/Γ are given by simply

$$b^k_{(2)}(M/\Gamma) = \alpha^k \mathrm{vol}(M/\Gamma).$$

Is it true for an arbitrary (contractible) homogeneous space that $\alpha^k = 0$ for $k \neq n/2$? The argument of Cheeger-Gromov [2.6] does not apply to amenable Lie groups, since the characterization of amenability in [2.6] is false. There are amenable, in fact solvable, Lie groups with left-invariant metrics with $h_{Ch} > 0$, c.f. [3.6]. (Of course, these do not have compact quotients.)

We refer to the works of Cheeger-Gromov listed in the references for a number of further and deeper results on L^2 Betti numbers and L^2 harmonic forms.

§3. L^2 harmonic forms on negatively curved manifolds

[3.1] Throughout this subsection, M denotes a complete simply connected n-manifold of negative sectional curvature K_M and we will study L^2 harmonic forms on such manifolds. To a large extent, this section completes the announcement [1].

In the standard model case $M = H^n(-1)$, it is easy to compute the dimensions of $\mathcal{H}^p_{(2)}(M)$ and one finds

$$\dim \mathcal{H}^p_{(2)}(H^n(-1)) = \begin{cases} 0 & p \neq n/2. \\ \infty & p = n/2. \end{cases} \tag{3.1}$$

In particular, if N is a compact hyperbolic manifold, then the L^2 Betti numbers satisfy $b^p_{(2)}(N) = 0$, $p \neq n/2$ and $b^{n/2}_{(2)}(N) = c(n)\cdot\mathrm{vol}(N) > 0$.

Dodziuk-Singer [13], [25] conjectured that (3.1) may hold for any complete simply connected manifold M such that $-b^2 \leq K_M \leq -a^2$, for some constants a,b. This L^2 form conjecture was partially motivated by the well-known Hopf conjecture: If N^{2m} is a compact manifold of negative sectional curvature, then $(-1)^m\chi(N) > 0$. By means of the L^2 index theorem [2.4], we see that the L^2 form conjecture (3.1) immediately implies the Hopf conjecture.

Some positive evidence for the L^2 form conjecture was obtained by Donnelly-Xavier [14] in case the curvature of M is sufficiently pinched. If $-a^2 \leq K_M \leq -1$, they show that $\mathcal{H}^p_{(2)}(M) = 0$ if $0 < p < \frac{n-1}{2}$ and $a < \frac{n-1}{2p}$. Using the asymptotics of the Bergmann metric, Donnelly-Fefferman [15] showed that the L^2 form conjecture is true for smooth strictly pseudo convex domains in $\mathbb{C}^N$ with the Bergmann metric.

This subsection will be concerned with the proof of the following theorem, announced in [1].

<u>Theorem 3.1</u>. For any $n \geq 2$, $0 < p < n$ and $a > |n-2p|$, there exist complete simply connected manifolds M^n with

$$-a^2 \leq K_M \leq -1,$$

such that $\dim \mathcal{H}^p_{(2)}(M) = \infty$.

The Theorem gives counterexamples to the L^2 form conjecture above. The examples constructed, although having relatively large isometry groups, do not have discrete, cocompact subgroups and thus cannot be used to construct counterexamples to the Hopf conjecture.

<u>Proof</u>. I. The metric.

Let L be a fixed totally geodesic hyperplane $H^{2p-1}(-a^2) \subset H^{2p}(-a^2)$ and let H be one of the components of $H^{2p}(-a^2) - L$. We first consider Riemannian manifolds of the form

$$N^n = H \times_f S^{n-2p}(1),$$

where $S^{n-2p}(1)$ is the space form of curvature +1 and $f: H \to \mathbb{R}$ is a smooth positive function to be determined below. Although not necessary, for simplicity in the computations, we will assume $f(x) = h(s(x))$, where $s(x)$ is the distance to L and $h: [0,\infty) \to [0,\infty)$. The metric on N is a warped product of the form

$$ds^2 = ds^2_{H^{2p}(-a^2)} + f^2 ds^2_{S^{n-2p}(1)} .$$

Note that N^n is diffeomorphic to $B^n - L$, where B^n is an n-ball. We first require that $h(s)$ is a smooth, strictly convex function such that $h(0) = 0$, $h'(0) = 1$. A standard example is $h_0(s) = \sinh s$. Clearly f extends to a smooth function on $H \cup L$ with $L = f^{-1}(0)$. Thus

$$M^n \equiv N^n \cup L = (H \cup L) \times_f S^{n-2p}(1)$$

is a C^∞ manifold diffeomorphic to $\mathbb{R}^n$. Note that this decomposition of $\mathbb{R}^n$ is just an open book decomposition with leaves H and binding L. It is also easily verified that the conditions above on h guarantee that the metric ds^2 on N extends to a complete C^∞ metric, also called ds^2, on M. The isometry group of ds^2 is $\mathrm{Isom}(H^{2p-1}(-a^2)) \times \mathrm{Isom}(S^{n-2p}(1))$.

We now compute the curvature of M. Let $\{X_i\}$ be a local orthonormal framing of H by eigenvalues of $D^2 f$ and $\{V_j\}$ be a local orthonormal framing of $S^{n-2p}(1)$. Using the formulas [3] for the curvature of a warped product, one sees that the set of 2-forms $\{X_i \wedge X_j\}$, $\{X_i \wedge V_j\}$, $\{V_i \wedge V_j\}$ diagonalizes the curvature operator $R: \Lambda^2(TM) \to \Lambda^2(TM)$ with corresponding sectional curvature

$$-a^2, \ \frac{-\lambda_i}{f}, \ -1, \tag{3.2}$$

where $\{\lambda_i\}$ are the eigenvalues of the Hessian $D^2 f$ on H. Further, under the appropriate labelling

$$\frac{-\lambda_i}{f} = \frac{-h'}{f} \text{ atanhas}, \ i < 2p; \ -\frac{\lambda_{2p}}{f} = \frac{-h''}{f}. \tag{3.3}$$

In particular, for the standard example h_0, the sectional curvatures lie in the range $[-a^2,-1]$.

II. Harmonic forms on M.

We consider harmonic p-forms on M which are invariant under a large symmetry group. Thus let ω be a p-form on $H^{2p}(-a^2)$ invariant under the $\mathbb{Z}_2$ action on $H^{2p}(-a^2)$ given by isometric reflection through L. Then ω defines a smooth p-form on H∪L and one may extend ω to a smooth p-form on M by using the isometric action of SO(n-2p+1). Let Ω_S^p be the space of p-forms on M so constructed.

One computes (c.f. for example [24]) that

$$\Delta_M \omega = \Delta_{H^{2p}} \omega + (-1)^p [d \circ i_F - i_F \circ d]\omega, \tag{3.4}$$

where $F = (n-2p)\nabla \log f$ is the negative of the mean curvature vector field of $S^{n-2p} \subset M^n$ and i_F denotes interior multiplication. Note of course that F is tangent to the H factor, i.e., F defines a vector field on H. (Although i_F is not defined on L, $\omega \in \Omega_S^p$ satisfies a Neumann condition at L so that the forms $i_F \circ d\omega$ and $d \circ i_F \omega$ are well-defined and smooth on M.)

First we give a simple procedure reducing the case of general p to p = 1. $H^{2p}(-a^2)$ may be written as a warped product

$$H^{2p}(-a^2) = H^2(-a^2) \times_g H^{2p-2}(-a^2), \tag{3.5}$$

where $g\colon H^2(-a^2) \to \mathbb{R}$ is given by $g(x) = \cosh ar(x)$, where r is the distance function to a fixed point $0 \in H^2(-a^2)$. Here we assume $H^{2p-2}(-a^2) \subset L$ and $0 \in \pi_1(L)$, where π_1 is projection on the first factor in (3.5). Under this decomposition, one sees that F is tangent to the $H^2(-a^2)$ factor. Now set

$$\omega = \phi \wedge \eta, \ \phi \in \Lambda^1(H^2(-a^2)), \ \eta \in \Lambda^{p-1}(H^{2p-2}(-a^2)). \tag{3.6}$$

If η is any harmonic (p-1)-form on $H^{2p-2}(-a^2)$, then ω satisfies the equation $\Delta_M \omega = 0$ if and only if

$$\Delta\phi - [d \circ i_F - i_F \circ d]\phi = 0 \quad \text{on } \Lambda^1(H^2(-a^2)). \tag{3.7}$$

Of course, as above, we require ϕ to be invariant under isometric reflection through the geodesic $\gamma = \pi_1(L) \subset H^2(-a^2)$. We may assume the

1-form ϕ = du for some function u so that (3.7) reduces to the scalar equation

$$\Delta_{H^2} u - i_F \circ du = 0; \tag{3.8}$$

$i_F \circ du$ is just the (Lie) derivative of u in the direction F.

We will show explicitly that (3.8) has a large space of solutions. To do this, it is convenient to use the conformal equivalence F of the half-disc $\pi_1(H) \subset H^2(-a^2)$ with the strip $\Omega = \{(x,\sigma) \in \mathbb{R}^2 : x \in \mathbb{R},\ \sigma \in (0,\pi/2)\}$ sending γ to $\{\sigma=0\}$. Then (3.8) takes the form

$$Lu = \frac{\partial^2 u}{\partial x^2} + \frac{\partial^2 u}{\partial \sigma^2} + (n-2p)\mu(\sigma)\frac{\partial u}{\partial \sigma} = 0, \tag{3.9}$$

where $\mu(\sigma) = \frac{\partial \log f}{\partial \sigma}$ and $f = f\Big|_{H^2(-a^2)}$ considered as a function on Ω.

More explicitly, one computes that

$$f = f(\sigma) = h(\ln(\tfrac{\alpha}{\beta})^{\frac{1}{2a}}) \tag{3.10}$$

where $\alpha = 1+\sin\sigma$, $\beta = 1-\sin\sigma$. In the standard case $h_0 = \sinh s$, one finds

$$f_0(\sigma) = \frac{1}{2}\,\frac{\alpha^{1/a}-\beta^{1/a}}{\cos^{1/a}\sigma}\ . \tag{3.11}$$

It is important to note that μ degenerates on $\partial\Omega$, so that (3.9) is not a uniformly elliptic equation (although the leading term is). In particular, one cannot apply standard elliptic theory to assert that (3.9) has many solutions smooth up to $\partial\Omega$. This is why we will solve (3.9) explicitly.

In order to obtain solutions to (3.9) whose behavior on $\partial\Omega$ is controlled, we will show that the mixed boundary value problem

$$\begin{aligned} &L(u) = 0 \\ &\frac{\partial u}{\partial \sigma}(x,0) = 0 \\ &u(x,\tfrac{\pi}{2}) = \nu \in C^\infty(\mathbb{R}) \end{aligned} \tag{3.12}$$

has a unique solution which is C^1 on $\Omega \cup \{\sigma=0\}$ and C^0 on $\bar{\Omega}$. It is easily seen that the Neumann condition at $\sigma = 0$ implies that u extends (by reflection), to a smooth function on $\tilde{\Omega} = \{(x,\sigma): x \in \mathbb{R},\ \sigma \in (-\frac{\pi}{2},\frac{\pi}{2})$ and thus via F to a smooth function on $H^2(-a^2)$. In particular, $\omega = du \wedge \eta$ belongs to Ω^p_S. We will only be interested in the case where ν is uniformly bounded in C^∞norm. To solve (3.12), let $\Omega_i \subset\subset \Omega$ be an exhaustion of Ω by smooth domains and extend ν to a C^∞ function on $\bar{\Omega}$ with bounded C^∞ norm. The Dirichlet problems

$$\begin{aligned} &L(u_i)=0 \\ &u\,|_{\partial\Omega_i} = \nu \end{aligned} \tag{3.13}$$

on Ω_i have unique smooth solutions in $\overline{\Omega}_i$ by standard elliptic theory since L is uniformly elliptic on Ω_i and L satisfies maximum principle.

We claim that as $i\to\infty$, a subsequence of $\{u_i\}$ converges to a solution of (3.12). To do this, we first show that the C^1 norm of $\{u_i\}$ is uniformly bounded. Since L satisfies the maximum principle,

$$|u_i| \le c(\sup|\nu|). \tag{3.14}$$

Note further that if u solves $L(u) = 0$, then so do the derivatives $(u_i)_x$, $(u_i)_{xx}$, etc., so that

$$|u_i|+|(u_i)_x|+|(u_i)_{xx}| \le c(C^2\text{norm of } \nu). \tag{3.15}$$

It remains to obtain bounds on the σ-derivatives of u_i. For this, write (3.9) in the form

$$(u_i)_{\sigma\sigma} + \mu(\sigma)(u_i)_\sigma = -(u_i)_{xx}. \tag{3.16}$$

View (3.16) as on O.D.E. along the curves $x = x_0$ fixed, $\sigma\in[0,\pi/2)$. Set $g_i(\sigma) = -(u_i)_{xx}(x_0,\sigma)$ (and note that $|g_i(\sigma)|\le c$), $v_i(\sigma) = (u_i)_\sigma(x_0,\sigma)$ so that one has

$$v_i' + u(\sigma)v_i = g_i \tag{3.17}$$

Since $u(\sigma) = (\log f)_\sigma$, (3.17) has general solutions of the form

$$v_i(\sigma) = \tfrac{1}{f}[C + \textstyle\int_0^\sigma g_i f]. \tag{3.18}$$

One may use the fact that $\frac{d\sigma}{ds} = a\cos\sigma \sim e^{-as}$ as $\sigma \to \frac{\pi}{2}(s\to\infty)$ to estimate $\frac{1}{f}\int_0^\sigma |g_i f| \le \frac{c}{f(s)}\int_0^s f(s)e^{-as}ds$. For now, we assume f is chosen so that this latter integral is bounded, although this will follow from (3.27) below. We claim that C = 0. To see this, integrate (3.18) from σ_0 to σ and use (3.14) with the preceding bound to obtain

$$C\cdot\int_0^\sigma \frac{1}{f(\sigma)}d\sigma \le A < \infty. \tag{3.19}$$

But as $\sigma \to 0$, $f(\sigma)\sim\sigma$ to first order so the integral in (3.19) diverges as $\sigma\to 0$. Thus, C=0 as claimed. We now have

$$v_i(\sigma) = \frac{1}{f}\int_0^\sigma g_i f \tag{3.20}$$

and

$$|v_i(\sigma)| \le M. \tag{3.21}$$

Since $(u_i)_x$ also solves $L(u) = 0$, exactly the same argument shows $|(u_i)_{x\sigma}|$ is uniformly bounded. To bound $|(u_i)_{\sigma\sigma}|=|v_i'|$ near $\sigma = 0$, by (3.16) it suffices to bound $|\mu(\sigma)v_i|$ or equivalently show

$$\frac{f\sigma}{f}\frac{1}{f}\int_0^\sigma f \le B < \infty \tag{3.22}$$

as $\sigma\to 0$. As above, this follows from the fact that $f(\sigma)\sim\sigma$ as $\sigma\to 0$.

It follows that $\{u_i\}$ has uniformly bounded C^1 norm in $\bar{\Omega}$, so that $\{u_i\}$ has a subsequence, call it $\{u_i\}$, converging uniformly on $\bar{\Omega}$ to a solution u of $L(u) = 0$ which achieves boundary values ν on $\{\sigma=\pi/2\}$. Since the C^2 norm of $\{u_i\}$ is uniformly bounded near $\{\sigma=0\}$, u also satisfies the Neumann condition of $\sigma=0$ since (3.20) holds for u_σ. This establishes (3.12).

Next, in order to carry out III below, we will need to find solutions to (3.12) whose gradient decays as $x \to \pm\infty$; for example

$$|du|(x,\sigma) \le c\cdot e^{-\delta|x|} \tag{3.23}$$

for some constant $\delta>0$ is more than sufficient. Let $h = h(\sigma)$ be a smooth positive function and compute

$$L(e^{-\delta|x|}h(\sigma)) = e^{-\delta|x|}(\delta^2 h + h_{\sigma\sigma} + \mu h_\sigma).$$

Thus $w(x,\sigma) = h(\sigma)e^{-\delta|x|}$ is a positive supersolution if

$$\delta^2 h + h_{\sigma\sigma} + \mu h_\sigma \le 0. \tag{3.24}$$

One easily verifies that $h(\sigma) = -c_1\sigma^2 + c_2$, for appropriate constants c_1, c_2 satisfies (3.24) if $\delta>0$ is small. Note that w satisfies the boundary conditions

$$w(x,\tfrac{\pi}{2}) = ce^{-\delta|x|} > 0$$

$$\left.\frac{\partial w}{\partial\sigma}\right|_{\sigma=0} = 0.$$

Thus if u is any solution to (3.12) with $u(x,\pi/2) \le w(x,\pi/2)$, it follows by the maximum principle [18] that $u \le w$ in Ω. Of course the same argument may be applied to $-u$ to bound u from below.

Recall that if u solves $L(u) = 0$, then so do u_x, u_{xx} with Dirichlet boundary data ν_x, ν_{xx}. Thus, if one chooses $\nu\colon \mathbb{R} \to \mathbb{R}$ such that $|\nu|+|\nu_x|+|\nu_{xx}|<h(\pi/2)e^{-\delta|x|}$, it follows that

$$|u|+|u_x|+|u_{xx}| < w$$

in Ω. In particular u_x satisfies (3.23). To obtain the same bound on $|u_\sigma|$, just return to (3.16) and use the estimate (3.20).

III. L^2 estimate

We estimate the L^2 norm of the p-forms ω constructed in II. We have

$$\int_M |\omega|^2 dV = \int_{H\times S^{n-2p}} |\omega|^2 f^{n-2p} dV_H \wedge dV_S = \mathrm{vol}\, S^{n-2p}(1) \int_{\pi_1(H)\times H^{2p-2}(-a^2)} |du|^2 |\eta|^2 f^{n-2p} dV_{H^2} \wedge dV_{H^{2p-2}}$$

$$= \mathrm{vol}\, S^{n-2p}(1) \int_{H^{2p-1}(-a^2)} |\eta|^2 dV \cdot \int_{\pi_1(H)} |du|^2 f^{n-2p} dV. \tag{3.25}$$

Note that both integrals on the right hand side of (3.25) are conformally invariant. Conformally identify $H^{2p-2}(-a^2)$ with $B^{2p-2}(1)$ with flat metric and set for instance $\eta = dx_1 \wedge \ldots \wedge dx_{p-1}$. Then one has

$$\int_{H^{2p-2}(-a^2)} |\eta|^2 \leq C_1 < \infty.$$

Next, conformally identify $\pi_1(H)$ with (Ω, flat metric) so that if u satisfies (3.12) and (3.23), we have

$$\int_{\pi_1(H)} |du|^2 f^{n-2p} dV = \int_\Omega |du|^2_{\mathrm{Eucl}} f^{n-2p} dx \wedge d\sigma \leq C_2 \int_0^{\pi/2} f^{n-2p}(\sigma) d\sigma. \tag{3.26}$$

Finally, we note

$$\int_0^{\pi/2} f^{n-2p}(\sigma) d\sigma \leq C_3 \cdot \int_0^\infty f^{n-2p}(s) e^{-as} ds, \tag{3.27}$$

so that if the latter integral is finite, one has an infinite dimensional space of L^2 harmonic p-forms on M. The Theorem follows by taking for instance $f = f_0$, since in this case, $f_0(s) \sim e^s$. ∎

We have not determined the full L^2 cohomology of the spaces M constructed above, but expect there should be L^2 harmonic ℓ-forms, for $\ell \neq p$.

[3.2] Basically what has been done in the proof of Theorem 3.1 is to use the compact isometry group SO(n-2p+1) to shift the domain of conformal invariance from n/2 to p. For clarity, we will briefly express the idea in greater generality.

Let M be a complete, non-compact Riemannian manifold and G a compact group of isometries with dim G $\geq$ 1. Let Π: M $\to$ M/G be the projection onto the orbit space and suppose M/G is reasonably well-behaved, e.g. assume M/G is a manifold with corners. We choose the natural metric on the interior int(M/G) such that Π is a Riemannian submersion on $\Pi^{-1}(\mathrm{int}(M/G))$. Note that int(M/G) is the space of principal orbits. We also assume dim(M/G) = 2p and let f: M/G $\to$ $\mathbb{R}$ be $f(x) = \mathrm{vol}_{n-2p}(\mathrm{orbit}(x))$.

The problem of finding G-invariant harmonic p-forms φ on M is equivalent to finding p-forms ω on M/G which satisfy an elliptic system $L(\omega) = 0$ on int(M/G) and certain (typically degenerate) boundary conditions on $\partial(M/G)$. The operator L is basically similar to that given by (3.4). Now because we are in the middle dimension on M/G, it can be seen that L is a conformally invariant operator. Assume that one can prove the existence of solutions to $L(\omega) = 0$ with boundary conditions, so that $\varphi = \pi^*\omega$, is a smooth harmonic p-form on M. One has

$$\int_M |\varphi|^2 dV = \int_{M/G} f\cdot|\omega|^2 dV. \tag{3.28}$$

The main point is that the right hand side of (3.28) is conformally invariant. Suppose for instance $(M/G, ds^2)$ is conformally equivalent to a Riemannian manifold with corners $(M/G, ds_0^2)$ with bounded diameter and that $\|\omega\|_0 < C$ in this metric. Then

$$\int_{M/G} |\omega|^2 f dV \leq C\cdot\int_{M/G} f dV_0$$

so that one has L^2 harmonic p-forms on M if the manifold $(\mathrm{int}(M/G), f ds_0^2)$ has finite volume. This latter condition is often not hard to arrange.

[3.3] The pinching condition in Theorem 3.1 can be considerably improved is a certain sense. Set for instance n=3, p=1 and a=1. Then the manifold $M^3 = H^2(-1)\times_f S'(1)$ has L^2 harmonic 1-forms if

$$\int_0^\infty f(s)e^{-s} ds < \infty$$

with $f(0) = 0$, $f'(0) = 1$. Choose for instance a smooth function f(s) with $f(s) = \frac{\sinh s}{1+s^2}$ for $s \geq s_0$ and $f'' > 0$ on $(0, s_0]$; it is easy to see (draw a picture) that such f exist. The curvatures K_M of M^3 are given by -1, $-\frac{f''}{f}$, $-\frac{f'}{f}\tanh s$. Thus, we have $K_M<0$ and $K_M \to -1$ as $s\to\infty$, in fact $|K_M+1| \leq \frac{C}{1+s}$. Note further that given any $\varepsilon>0$, one may choose s_0 sufficiently large so that $|K_M+1| \leq \varepsilon$ and $|K_M+1| \leq \frac{c}{1+s}$ as $s\to\infty$. In particular, the curvature of M converges to -1 in neighborhoods of every point on the sphere at infinity $S(\infty)$ of M except the "poles" $L\cap S(\infty)$.

Similar results hold for n-2p arbitrary.

[3.4] Of course, it would be interesting to find conditions under which the L^2 form conjecture remains true. [3.3] indicates it may be difficult to find a condition involving only the structure of $S(\infty)$. One result in this direction, due to Mazzeo [22] states that if M conformally compactifies, i.e., if ds_0^2 is a smooth metric of bounded diameter on $\overline{M} = M\cup\partial M$ and the metric on M is a complete metric of negative curvature of the form $ds^2 = \rho^2 ds_0^2$, $\rho: M \to \mathbb{R}$, then the L^2 form conjecture is true.

The metric in [3.3] does not conformally compactify on $S(\infty) - (L \cap S(\infty))$.

[3.5] Although Theorem 3.1 may be viewed as some evidence against the Hopf conjecture, we believe the following question should be true.

Question: Let M^{2n} be a compact manifold of strictly negative curvature. Is it true that $b^n_{(2)}(M) > 0$?

[3.6] Consider again the case $n = 3$, $p = 1$ for the metric constructed in the proof of Theorem 3.1. The binding L is a complete geodesic in $H^2(-a^2)$. Suppose, following a suggestion of D. Sullivan, we consider a sequence of (isometric) metrics g_i with bindings L_i converging to a point p_∞ in the ideal boundary $S(\infty)$ of $H^2(-a^2)$. The sequence of metrics $\{g_i\}$ converges to a metric g_∞ which is homogeneous. In fact, the S^1 action now becomes an $\mathbb{R}$ action and isometries of $H^2(-a^2)$ which leave the point p_∞ fixed extend to isometries of g_∞. This is just the subgroup of $PSL(2,\mathbb{R})$ of the form $\begin{pmatrix} \alpha & \beta \\ 0 & 1/\alpha \end{pmatrix}$--the real translations and dilations of the upper half plane. In fact, it is easily verified that $H = \mathrm{Isom}(g_\infty)$ is the semi-direct product $(\mathbb{R}\oplus\mathbb{R}) \times \mathbb{R}$, where $\mathbb{R}$ acts on $\mathbb{R}\oplus\mathbb{R}$ by $t \to \begin{pmatrix} e^{at} & 0 \\ 0 & e^{t} \end{pmatrix}$. In particular, G is solvable.

It is an open question whether g_∞ has non-zero L^2 harmonic 1-forms for $a>1$, although we suspect the answer is no. Note that we may do the same process with the metrics constructed in [3.3]. In this case, the limiting metric g_∞ is of constant curvature -1 and G is embedded as a subgroup of $\mathrm{Isom}(H^3(-1))$. G is just the group above with $a=1$, endowed with a left-invariant metric of curvature -1. In particular, $G = H^3(-1)$ has no L^2 harmonic 1-forms. Consequently, the dimension of the space of L^2 harmonic forms may drop when passing to geometric limits of a given metric.

[3.7] The method used in [3.1] can be used to construct harmonic forms in other situations, for instance on contractible manifolds with polynomial volume growth.

As before, let $M^n = (H \cup L) \times_f S^{n-2p}(1)$, where H is a half-ball in $B^{2p}(1)$, $\partial H = L = B^{2p} \cap P$, where P is a hyperplane in $\mathbb{R}^{2p}$ through the origin. Now instead of taking a hyperbolic metric on H, we choose a complete conformally flat metric ds^2 on $B^{2p}(1)$ (invariant under reflection through P), which has polynomial volume growth, e.g.

$$ds^2 = \frac{1}{(1-r)^{2k}}\, ds_0^2, \ k>1, \tag{3.29}$$

where ds_0^2 is the Euclidean metric, r the Euclidean radius. If ρ is the distance to the origin in the ds^2 metric, one finds $\rho(r) = (1-r)^{1-k}-1$

and vol $B(\rho) \leq c\cdot\rho^{1+(2p-1)\frac{k}{k-1}}$ for $\rho>1$ and some constant $c>0$. Here $B(\rho)$ is the geodesic ball of radius ρ in the metric ds^2.

One may decompose the standard Euclidean ball $B^{2p}(1)$ as

$$B^{2p}(1) = B^2(1)\times_g B^{2p-2}(1) \tag{3.30}$$

for some warping function g, where $B^2(1)$ is a flat disc through the origin and we assume $B^{2p-2} \subset L$. Let $\gamma = \pi_1(L) \subset B^2(1)$, where π_1 is projection on the first factor in (3.30). Choose f of the form $f = y\circ\pi_1$ where $y: B^2(1) \to \mathbb{R}$ is a non-negative function with $y^{-1}(0) = \gamma$, y invariant under reflection through γ and $\nabla y = \nabla s$ on γ, where s is the distance to γ on the half disc $\pi_1(H)$.

Consider p-forms ω invariant under the $\mathbb{Z}_2\times SO(n-2p+1)$ action as before of the form

$$\omega = du\wedge\eta,$$

where du is a 1-form on $B^2(1)$ and η a (p-1)-form on $B^{2p-2}(1)$. Since du and η are in the middle dimension, using the conformal invariance one sees $\Delta_M\omega = 0$ if $\Delta\eta = 0$ on B^{2p-2} and

$$\Delta u - i_F\circ du = 0, \tag{3.31}$$

where the equation is with respect to the metric ds^2 induced on B^2, $F = (n-2p)\nabla\log y$.

Now choose f (or y) for instance to be a bounded function. Following the procedure in [3.1] it is straightforward to verify that (3.31) has an infinite dimensional space of solutions, satisfying the appropriate boundary condition, such that the corresponding p-forms ω on M are in L^2. Since f is bounded, M has volume growth

$$\mathrm{vol}_M B_M(r) \leq C\cdot r^{1+(2p-1)\frac{k}{k-1}}$$

equivalent to the volume growth of (B^{2p},ds^2).

Note also that in case p = 1, the harmonic 1-forms du on M are, by construction, differentials of bounded harmonic functions u on M. In particular, in any dimension, there are bounded harmonic functions on manifolds of polynomial volume growth, in fact of growth $v(r) \leq cr^{2+\varepsilon}$, any $\varepsilon>0$ (compare [21]).

[3.8] The L^2 harmonic 1-forms constructed in [3.1] (or [3.3], [3.7]) correspond to harmonic functions $u: M^n \to \mathbb{R}$ of finite energy. In particular one obtains non-trivial harmonic maps

$$F: M^n \to \mathbb{R}^m \tag{3.32}$$

of finite energy. Also, by construction, the functions u are bounded harmonic functions so that F maps into a bounded set in $\mathbb{R}^m$.

It then appears likely that non-constant harmonic maps F of finite energy should exist mapping M into any Riemannian manifold N. It is shown in [23] that harmonic maps F as in (3.32) never exist if $M^n = H^n(-1)$.

References

[1] M. T. Anderson, L^2harmonic forms and a conjecture of Dodziuk-Singer, Bulletin Amer. Math. Soc. 13 (1985), 163-165.

[2] M. F. Atiyah, Elliptic operators, discrete groups and Von Neumann algebras, Asterisque, 32-33 (1976), 43-72.

[3] R. Bishop and B. O'Neill, Manifolds of negative curvature. Trans. Amer. Math. Soc. 145 (1969), 1-49.

[4] A. Borel and N. Wallach, Continuous cohomology, discrete subgroups and representations of reductive groups, Ann. of Math. Studies, vol. 104 (1980).

[5] R. Brooks, The fundamental group and spectrum of the Laplacian, Comm. Math. Helv. 36 (1981), 581-598.

[6] J. Cheeger, M. Goresky and R. Macpherson, L^2cohomology and intersection homology of singular algebraic varieties, Ann. of Math. Studies, vol. 102 (1982), 303.

[7] J. Cheeger and M. Gromov, On the characteristic numbers of complete manifolds of bounded curvature and finite volume, Rauch Memorial Volume, I. Chavel and H. Farkas, Eds., Springer Verlag, Berlin (1985), 115-154.

[8] J. Cheeger and M. Gromov, Bounds on the Von Neumann dimension of L^2cohomology and the Gauss-Bonnet Theorem for open manifolds, Jour. Diff. Geo. 21 (1985), 1-34.

[9] J. Cheeger and M. Gromov, L^2cohomology and group cohomology, Topology 25 (1986), 189-215.

[10] G. deRham, Varietes differentiables, Hermann, Paris (1960).

[11] J. Dodziuk, DeRham-Hodge theory for L^2cohomology of infinite coverings, Topology, 16 (1977), 157-165.

[12] J. Dodziuk, Sobolev spaces of differential forms and deRham-Hodge isomorphism, J. Diff. Geo. 16 (1981), 63-73.

[13] J. Dodziuk, L^2harmonic forms on complete manifolds, Ann. of Math. Studies, vol. 102 (1982), 191-302.

[14] H. Donnelly and F. Xavier, On the differential form spectrum of negatively curved Riemannian manifolds, Amer. J. Math 108 (1984), 169-185.

[15] H. Donnelly and C. Fefferman, L^2cohomology and index theorem for the Bergmann metric, Ann. of Math. 118 (1983), 593-618.

[16] M. Gaffney, The heat equation method of Milgram and Rosenblum for open Riemannian manifolds, Ann. of Math. 60 (1954), 458-466.

[17] S. Gallot and D. Meyer, Operateur de courbure et Laplacien des formes differentielles d'une variete riemannienne, J. Math. Pure Appl. 54 (1975), 259-284.

[18] D. Gilbarg and N. Trudinger, Elliptic partial differential equations of second order, Springer Verlag, New York (1977).

[19] D. Gromoll and W. Meyer, On complete open manifolds of positive curvature, Ann. of Math. 90 (1979), 74-90.

[20] K. Kodaira, Harmonic fields in Riemannian manifolds, Ann. of Math. 50 (1949), 587-665.

[21] T. Lyons and D. Sullivan, Function theory, random paths and covering spaces, Jour. Diff. Geo., 19 (1984), 299-323.

[22] R. Mazzeo, MIT Thesis (1986).

[23] H. Sealey, Some conditions ensuring the vanishing of harmonic differential forms..., Math. Proc. Camb. Phil. Soc. 91 (1982), 441-452.

[24] S. Zucker, L^2cohomology of warped products and arithmetic groups, Inv. Math. 70 (1982), 169-218.

[25] S.-T. Yau, Problem section, Seminar on Differential Geometry, Ann. of Math. Studies vol. 102 (1982).

RICCI-FLAT KÄHLER METRICS ON AFFINE ALGEBRAIC MANIFOLDS

Dedicated to Professor Ichiro Satake
on his 60th Birthday

Shigetoshi Bando and Ryoichi Kobayashi

Althougn rather satisfying results on the existence and uniquness of Einstein-Kähler metrics of negative Ricci curvature on noncompact manifolds were obtained in [9] and [14], it seems that not much is known about Ricci flat Kähler metrics. To the authors' knowledge only methods of construction which appeared in the literature are Calabi's construction (cf. [4], [5], [6] and [7]) and ones to make hyperkähler metrics, for instance, twistor construction and hyperkähler reduction (finite or infinite dimensional) (cf. [6], [11], [12], [13], [16], [17], and [18]). Both of them can deal with only very restricted situations and can not produce one on $P^n - D$ whose existence was anounced by Yau in [4]. In this paper we shall give a simple proof of it under the assumption of the existence of an Einstein-Kähler metric of positive Ricci curvature on D.

Let us first note why the existence of an Einstein-Kähler metric is a natural assumption. In the special case : $C^n = P^n - P^{n-1}$, we get the flat metric $|dz|^2$. If we delete the J-rotated direction of the normal from the induced metric on the sphere of redius $r = |z|$, we get

$$|dz|^2 - \frac{|\langle\bar{z}, dz\rangle|^2}{|z|^2} .$$

After normalization

$$\frac{1}{|z|^2}\{ |dz|^2 - \frac{|\langle\bar{z}, dz\rangle|^2}{|z|^2} \} = \sqrt{-1}\,\partial\bar{\partial}\, \mathrm{Log}\, |z|^2 ,$$

which is the Fubini-Study metric on P^{n-1}. This suggests some relation of Ricci-flat metrics and Einstein-Kähler metrics at the infinity.

We consider a little bit more general situation than $P^n - D$. Let X be a compact Kähler manifold of $\dim X = n \geqslant 3$ with positive first Chern class $c_1(X) > 0$, and D a smooth reduced divisor such that $c_1(X) = \alpha[D]$, where [D] is the Poincaré dual of D and $\alpha > 1$. Our theorem is :

Theorem 1.

If D admits an Einstein-Kähler metric and the condition (*) stated below holds, then there exists a complete Ricci-flat Kähler metric on $X - D$.

Recently Siu and Tian proved the existence of an Einstein-Kähler metric on the Fermat hypersurface F of degree $n - 1$ and n in P^n ([21], [22]).

Corollary.

$P^n - F$ admits a complete Ricci-flat Kähler metric.

For the proof of Theorem 1 we need some lemmas, which one can prove modifying the arguments in [19]. Let K_X, K_D and L_D be the canonical bundles of X, D and the line bundle on X defined by D, respectively. Since $K_D = (K_X + L_D)|_D$, $c_1(D) = (c_1(X) - [D])|_D = (\alpha - 1)[D]|_D > 0$, X and D are simply connected (cf. [15], [26]). For a volume form V we define its Ricci form by $-\sqrt{-1}\,\partial\bar{\partial} \log V$. Then we have

Lemma 1.

X has a Ricci-flat volume form V with a pole of order 2α along D.

Assume the condition (*) $H^1(D, T_X \otimes L_D^{-k}|_D) = 0$ for all $k \geq 2$, which is satisfied in the case $X = P^n$, then by [29] [30] we obtain

Lemma 2.

There exists a neighborhood of D in X which is biholomorphic to a neighborhood of the o-section of L_D in the jet level along D up to any order, by a diffeomorphism ϕ.

Lemma 3.

There exists a neighborhood of the 0-section of L_D which is biholomorphic to a neighborhood of the ∞-section of K_D up to covering and quotient.

Calabi showed in [5]

Lemma 4.

The total space K_D admits a complete Ricci-flat Kähler metric.

For the completeness we include here the construction. Identify a Kähler metric with its Kähler form. Let θ be the Einstein-Kähler metric on D such that $\mathrm{Ric}(\theta) = -\sqrt{-1}\,\partial\bar{\partial}\log\theta^{n-1} = \theta$. The volume form θ^{n-1} defines a fiber metric $\|\cdot\|$ on K_D, and gives a function $t = \log\|\eta\|$, for $\eta \in K_D$. Then $\theta = \sqrt{-1}\,\partial\bar{\partial}\,t$. We seek a Ricci-flat Kähler metric $\bar{\omega}_0$ on K_D in the form $\sqrt{-1}\,\partial\bar{\partial}\,\bar{F}(t)$. Putting $\bar{F}' = f$, we have

$$\bar{\omega}_0 = f\,\theta + f'\sqrt{-1}\,\partial t \wedge \bar{\partial} t,$$

$$\bar{\omega}_0 = f^{n-1} f'\,\theta^{n-1}\sqrt{-1}\,\partial t \wedge \bar{\partial} t,$$

$$\mathrm{Ric}(\bar{\omega}_0) = -\sqrt{-1}\,\partial\bar{\partial}\log(f^{n-1} f') + \sqrt{-1}\,\partial\bar{\partial} t.$$

If we take f to be $(1 + e^t)^{\frac{1}{n}}$, then $\mathrm{Ric}(\bar{\omega}_0) = 0$.

Clearly $\bar{\omega}_0$ is compatible with the covering and quotient operation in Lemma 3. Thus from Lemma 2 we get a Kähler metric $\omega_0 = \sqrt{-1}\,\partial\bar{\partial} F$, where $F = \phi^*\bar{F}$, near the infinity of $Y = X - D$, which is complete toward D and is almost Ricci-flat in the following sense:

Lemma 5.

$\log\frac{\omega_0^n}{V}$ is a constant in the jet level up to any order along D.

Proof.

The above construction shows that ω_0 also has a pole of order 2α along D. Thus the function $h = \log\frac{\omega_0^n}{V}$ is bounded and extends smoothly to a function which is pluriharmonic in the jet level up to any order (as one wants) along D. Because D is simply connected and $H^0(D, -kL_D|_D) = 0$, for $k \geqslant 1$, h must be a constant in the jet level up to any order along D.

Since $Y = X - D$ is affine algebraic and the Kähler potential F of ω_0 tends to ∞ toward the infinity of Y, we can extend ω_0 to a complete Kähler metric defined on the whole Y, which we still call ω_0. Then ω_0 has $C^{k,\alpha}$- bounded geometry.

Definition.

A Riemannian metric g on a manifold M is called of $C^{k,\alpha}$-bounded geometry if for each point $p \in M$ there exists a coordinate chart $x = (x^1, x^2, \cdots, x^m)$ centered at p such that

i) x runs over a unit ball B^m in R^m.

ii) If we write $g = g_{ij}(x)\, dx^i dx^j$, then the matrix (g_{ij}) is bounded from below by a constant positive matrix independent of p.

iii) $C^{k,\alpha}$ norms of g_{ij} are uniformly bounded.

On such a manifold we can define the Banach space $C^{k,\alpha}$ of uniformly $C^{k,\alpha}$-bounded functions.

All the above implies that ω_0 satisfies the assumption of the following Theorem 2, and we get Theorem 1.

Theorem 2.

Let ω_0 be a Kähler metric on an n-dimensional $(n \geqslant 3)$ open complex manifold Y of $C^{k,\alpha}$-bounded geometry with $k \geqslant 4$, $0 < \alpha < 1$. Assume that ω_0 satisfies the Sobolev inequality and that the volume of the geodesic ball grows like r^{2n}. If Y admits a Ricci-flat volume form V such that $\omega_0{}^n = e^f V$ with $f \in C^{k,\alpha} \cap L^2$, then Y admits a complete Ricci-flat Kähler metric.

Proof.

We seek a Ricci-flat Kähler metric in the form $\omega = \omega_0 + \sqrt{-1}\,\partial\bar{\partial} u$ by the continuity method. Namely we solve the following equation with $t \in [0, 1]$,

$$\omega^n = e^{(1-t)f} V = e^{-tf} \omega_0{}^n .$$

To set the stage, we define two Banach spaces B_1, B_2 as follows.

$$B_1 = \{ u \in C^{k+2,\alpha} \cap L^{p_0} \mid \sqrt{-1}\,\partial\bar{\partial} u \in L^2 \},$$

$$B_2 = C^{k,\alpha} \cap L^2,$$

where the L^p space, with the norm $\|\cdot\|_p$, means one with respect to the metric ω_0 and $p_0 = \dfrac{2n}{n-2}$. Set

$$\mathcal{O} = \{ u \in B_1 \mid \exists a > 0 \ \text{ s.t. } \ a\omega_0 < \omega = \omega_0 + \sqrt{-1}\,\partial\bar\partial u < a^{-1}\omega_0 \}.$$

We consider the following operator E from $\mathcal{O}$ to B_2,

$$E(u) = \log \frac{\omega^n}{\omega_0^n} .$$

Then the equation (1) become

$$E(u) = - tf , \tag{2}$$

which has the solution $u = 0$ at $t = 0$, and the equation we want to solve is (2) at $t = 1$. Thus we have to show the openness and the closedness of the interval in which (2) has a solution.

First we consider the openness. We show the linearization of E has the bounded inverse. The linearization at ω is the Laplacian Δ with respect to the metric ω. Note that ω is equivalent to ω_0. So ω also satisfies the Sobolev inequality and the volume growth estimate. Moreover L^p spaces with respect to ω_0 and ω are the same. Since it is easy to see the equation $\Delta v = g$ is always solvable for any $g \in B_2$ as the limit of the solutions with the Dirichlet boundary condition on the relatively compact smooth domains, we have only to show that $v \in B_1$. Multiplying the equation by $-|v|^{p-2}v$ and integration by parts give

$$\int |\partial |v|^{\frac{p}{2}}|^2 = - \frac{p^2}{4(p-1)} \int f|v|^{p-2}v .$$

The Sobolev inequality gives that for $p \geq p_{-1} = \dfrac{2(n-1)}{n-2} > 2$

$$\left(\int |v|^{p\gamma}\right)^{\frac{1}{\gamma}} \leq Cp \int |f|\ |v|^{p-1} , \tag{3}$$

with $\gamma = \frac{n}{n-1}$. Here and hereafter C means a positive constant which may differ in a different context. By the Hölder inequality we have

$$\left(\int |v|^{p\gamma}\right)^{\frac{1}{\gamma}} \leq Cp \left(\int f^2\right)^{\frac{1}{2}} \left(\int |v|^{2(p-1)}\right)^{\frac{1}{2}} .$$

Taking $p = p_{-1}$, we have $p_{-1}\gamma = 2(p_{-1} - 1) = \frac{2n}{n-2} = p_0$, and

$$\left(\int |v|^{p_0}\right)^{\frac{1}{p_0}} \leq C\left(\int f^2\right)^{\frac{1}{2}} .$$

Thus $v \in L^{p_0}$. Again applying the Hölder inequality to (3), we have

$$\left(\int |v|^{p\gamma}\right)^{\frac{1}{\gamma}} \leq Cp\left(\int f^p\right)^{\frac{1}{p}}\left(\int |v|^p\right)^{\frac{p-1}{p}}$$

$$\leq Cp\ \|f\|_2^{\frac{1}{p}}\ \|f\|_\infty^{\frac{p-2}{p}}\ \|v\|_p^{p-1} .$$

Putting $p_i = p_{-1}\gamma^{i+1} = p_0\gamma^i$, we have

$$\|v\|_{p_{i+1}} \leq (Cp_i)^{\frac{1}{p_i}} \|f\|_2^{\frac{1}{p_i^2}}\ \|f\|_\infty^{\frac{p_i - 2}{p_i^2}}\ \|v\|_{p_i}^{\frac{p_i - 1}{p_i}}$$

The iteration argument shows that with $\theta = \frac{2}{n}$

$$\|v\|_\infty \leq C\ \|f\|_2^\theta\ \|f\|_\infty^{1-\theta} .$$

Then the interior Schauder estimates gives $C^{k+2,\alpha}$-bound of v.
For the L^2 estimate of $\sqrt{-1}\,\partial\bar\partial v$, we use the following lemma which finish the proof of the openness.

<u>Lemma 6</u>.

$$\int |\sqrt{-1}\,\partial\bar\partial v|^2 = \int f^2 .$$

<u>Proof</u>.

Let ρ be a general cut off function. Then by integration by parts we have that

$$\int \rho^2 |\sqrt{-1}\,\partial\bar\partial v|^2 - \int \rho^2 |\Delta v|^2$$

$$= C\left(\int \rho\ |\partial\rho|\ |\partial v|\ |\sqrt{-1}\,\partial\bar\partial v|\right)$$

$$= O\left(\left(\int \rho^2 \, |\sqrt{-1}\,\partial\bar\partial\, v|^2\right)^{\frac{1}{2}} \left(\int |\partial\rho|^2 \, |\partial v|^2\right)^{\frac{1}{2}}\right).$$

Thus the proof is reduced to the following :

$$\frac{1}{R^2}\int_{T(R)} |\partial v|^2 \longrightarrow 0 \ , \ \text{as} \ \ R \longrightarrow \infty \ , \tag{4}$$

where $T(R) = B(3R) - B(2R)$. To prove this we choose ρ as

$$\rho = \begin{cases} 1 & \text{on } T(R) \\ 0 & \text{outside of } B(4R) - B(R) \ , \end{cases}$$

$$|\partial\rho| = O\left(\frac{1}{R}\right) .$$

Then we have

$$\int \rho^2 \, |\partial v|^2 \le \int \rho^2 \, |v| \, |\Delta v| + 2\int \rho \, |\partial v| \, |v| \, |\partial\rho|$$

$$\le \int \rho^2 \, |v| \, |\Delta v| + 2\left(\int \rho^2 |\partial v|^2\right)^{\frac{1}{2}} \left(\int |v|^2 |\partial\rho|^2\right)^{\frac{1}{2}} ,$$

$$\int_{T(R)} |v| \, |\Delta v| \le \left(\int_{T(R)} |v|^{p_0}\right)^{\frac{1}{p_0}} \left(\int_{T(R)} |\Delta v|^2\right)^{\frac{1}{2}} (\mathrm{vol}(B(T(R))))^{\frac{1}{2n}}$$

$$= o(R^2) ,$$

and

$$\left(\int_{T(R)} |v|^2\right)^{\frac{1}{2}} \le \left(\int_{T(R)} |v|^{p_0}\right)^{\frac{1}{p_0}} (\mathrm{vol}(T(R)))^{\frac{1}{2n}} = o(R^2) .$$

The assertion (4) follows from these inequalities.

Now we proceed to the closedness.

$$(1 - e^{-tf}) \, \omega_0{}^n = \omega_0{}^n - \omega^n$$

$$= (- \sqrt{-1}\,\partial\bar\partial\, u) \, (\omega_0{}^{n-1} + \omega^{n-2}{}_0 \, \omega + \cdots + \omega^n).$$

Multiplying the above equation by $|u|^{p-2}u$ with $p \ge p_{-1}$, and applying integration by parts, we see that (to be precise, with a similar reasoning as in the proof of lemma 6)

$$\int \sqrt{-1}\, \partial |u|^{\frac{p}{2}} \wedge \bar{\partial} |u|^{\frac{p}{2}} \left(\omega_0{}^{n-1} + \omega^{n-2}{}_0\, \omega + \cdots + \omega^n \right)$$

$$= \int (1 - e^{-tf})\, |u|^{p-2} u\, \omega_0{}^n .$$

By the Sobolev inequality with respect to ω_0, we get the following inequality with respect to the volume form $\omega_0{}^n$:

$$\left(\int |u|^{p\gamma} \right)^{\frac{1}{\gamma}} \leqslant Cp \int |f|\, |u|^{p-1} .$$

By the same argument as before we get the a priori estimates for $\|u\|_{p_0}$ and $\|u\|_\infty$. To get a C^2-estimate for u, we use Yau's infinitesimal Schwarz lemma ([27]).

Lemma 7.

Let (M_i, θ_i), $i = 1, 2$ be Kähler manifolds and $\varphi : M_1 \longrightarrow M_2$ a holomorphic map, then

$$\Delta_{\theta_1} \log |\partial\varphi|^2 \geqslant \frac{\mathrm{Ric}_{\theta_1}(\partial\varphi, \overline{\partial\varphi})}{|\partial\varphi|^2} - \frac{\mathrm{Bisect}_{\theta_2}(\partial\varphi, \overline{\partial\varphi}, \partial\varphi, \overline{\partial\varphi})}{|\partial\varphi|^2} .$$

We apply this lemma to $\mathrm{id} : (Y, \omega) \longrightarrow (Y, \omega_0)$. Since $|\mathrm{Ric}(\omega)|_{\omega_0} = (1 - t)\, |\mathrm{Ric}(\omega_0)|_{\omega_0}$ and $|\mathrm{Bisect}_{\omega_0}|_{\omega_0}$ are bounded,

$$\Delta \log \mathrm{tr}_\omega\, \omega_0 \geqslant - C\, \mathrm{tr}_\omega\, \omega_0 ,$$

with some constant $C > 0$, and $\Delta = \Delta_\omega$. Then

$$\Delta\, (\log \mathrm{tr}_\omega\, \omega_0 - Au) \geqslant - An + (A - C)\, \mathrm{tr}_\omega\, \omega_0 .$$

Choosing $A > C$ and applying Yau's meximum principle ([8], [25]), we have an upper bound for $\mathrm{tr}_\omega\, \omega_0$. Together with the equation (1) we obtain an a priori constant $a > 0$ such that

$$a\omega_0 < \omega < a^{-1}\omega_0 .$$

As to the $C^{k+2,\alpha}$ estimate we refer to [1], [10]. For the L estimate of $\sqrt{-1}\,\partial\bar{\partial}\,u$, we remark that it can be derived from the proof of the openness because the estimate in the linearized equation depends only on a.

To prove Theorem 1 we chose to work on the noncompact manifold Y but it is also possible to work on the compact manifold X. The idea is as follows. Using the Einstein-Kähler metric on D we suitably construct a good fiber metric $\|\cdot\|$ on L_D, whose curvature form θ is in paticular positive. For each $0 < \varepsilon < 1$ we solve the following equation on X.

$$\mathrm{Ric}(\omega_\varepsilon) = \alpha\,(\,\theta + \sqrt{-1}\,\partial\bar{\partial}\,\log\,(\,\|\sigma\|^2 + \varepsilon\,))\ \geqslant\ 0\ ,$$

where σ is a section of L_D whose zero locus in D. This is solvable by the resolution of Calabi's conjecture by Yau ([26]), if ω_ε belongs to some scalar multiple of the anti-canonical class. Letting $\varepsilon \longrightarrow 0$, we see that ω_ε blows up along D and we obtain the desired Ricci-flat metric on Y.

Finally we remark that we did not prove the uniqueness of Ricci-flat Kähler metrics. Although we have the satisfactory uniqueness results in the compact case with any sign of first Chern class ([1], [3], and [26]), it is very difficult problem in the noncompact Ricci-flat case. Even on C^n we can not prove the uniqueness. We think some compactification of Ricci-flat manifolds like one in [20] may have some relation (cf. [28]).

Note. The authors would like to thank Professors Hajime Tsuji and Ngaiming Mok for pointing out an error in the first version of Lemma 2 and Professor Hajime Sato for informing them the reference [29].

Appendix.

In this appendix, we give an outline of the proof of the following

Theorem A.

Let X be a compact Kähler manifold with positive first Chern class and D a smooth hypersurface in X such that $c_1(X) = \alpha[D]$ for $1 < \alpha < 2$. Assume that D admits an Einstein-Kähler metric. Then $X - D$ admits a complete Ricci-flat Kähler metric.

We reduce this Theorem to Theorem 2. Let L_D be a holomorphic line bundle determined by D. From the assumption, we can find a Hermitian metric $\|\cdot\|$ for L_D suth that the restriction $\tilde{\theta}$ of the curvature form θ for L_D to D is an Einstein-Kähler metric with $\mathrm{Ric}(\tilde{\theta}) = (\alpha - 1)\tilde{\theta}$. Pick a holomorphic section $\sigma \in H^0(X, \mathcal{O}_X(D))$ so that the zero divisor of σ is D. Put $t = \log\|\sigma\|^{-2}$. Then $\theta = \sqrt{-1}\partial\bar{\partial}\log\|\sigma\|^{-2}$. We consider a function $F(t) = \int f(t)dt$, where $f(t) = (1 + \exp(\alpha - 1)t)^{1/n}$. The function $F(t)$ is defined near the infinity of $X - D$ and we can extend $F(t)$ to a strictly plurisubharmonic function F defined on the whole $X - D$. Set $\omega_o = \sqrt{-1}\partial\bar{\partial}F$. Then ω_o is a complete Kähler metric on $X - D$ and $\omega_o^n = \exp(-g)V$, where $g = 0$ on D and V is a Ricci-flat volume form on $X - D$. A direct calculation shows that if $1 < \alpha < 2$ then the function g belongs to $L_2(X, \omega_o)$. We thus use Theorem 2 to solve

$$(\omega_o + \sqrt{-1}\partial\bar{\partial}u)^n = V = \exp(g)\omega_o^n$$

and we get a complete Ricci-flat Kähler metric $\omega = \omega_o + \sqrt{-1}\partial\bar{\partial}u$ on $X - D$. If $X = P^n$ and D is a smooth hypersurface of degree d, then $1 < \alpha < 2$ is equivalent to $(n+1)/2 < d \leqq n$.

References.

[1] T. Aubin, Nonlinear analysis on manifolds, Monge-Ampère equations, Springer-Verlag, Berlin/New York, 1982.

[2] ——, Réduction du cas positif de l'équation de Monge-Ampère sur les variétés Kählériennes compactes à la demonstration d'une inégalité, J. Funct. Anal., 57 (1984), 143-153.

[3] S. Bando and T. Mabuchi, Uniqueness of Einstein Kähler metrics modulo connected group actions, in "Algebraic Geometry, Sendai, 1985", ed. T. Oda, Adv. Stud. Pure Math., 10, Kinokuniya, Tokyo, and North-Holland, Amsterdam, 1987, 11-40.

[4] J. P. Bourguignon et al., Première classe de Chern et courbure de Ricci : preuve de la conjecture de Calabi, Astérisque 58, Soc. Math. France, 1978.

[5] E. Calabi, Métriques Kählériennes et fibrés holomorphes, Ann. Sci. Ec. Norm. Sup. Paris, 4me Sér. 12 (1979), 269-294.

[6] ——, Isometric families of Kähler structures, in "Chern Sump. 1979", ed. W.-Y. Hsiang et al., Springer-Verlag, Berlin/New York, 1980, 23-39.

[7] ——, Extremal Kähler metrics, in "Seminar on Differential Geometry", ed. S.-T. Yau, Princeton Univ. Press, Ann. Math. Stud., 102 (1982), 259-290.

[8] S.-Y. Cheng and S.-T. Yau, Differential equations on Riemannian manifolds and their geometric applications, Comm. Pure Appl. Math., 28 (1975), 333-354.

[9] ——, On the existence of a complete Kähler-Einstein metric on noncompact complex manifolds and regularity of Fefferman's equation, Comm. Pure Appl. Math., 32 (1980), 507-544.

[10] D. Gilbarg and N. S. Trudinger, Elliptic partial differential equations of second order, 2nd ed., Springer-Verlag, Berlin/New York, 1983.

[11] N. J. Hitchin, Polygons and gravitons, Math. Proc. Camb. Phil. Soc., 83 (1979), 465-476.

[12] N. J. Hitchin, A. Karlehede, U. Lindstrom, and M. Rocek, Hyperkähler metrics and supersymmetry, Comm. Math. Phys., 108 (1987), 535-589.

[13] M. Itoh, Quaternion structure on the moduli space of Yang-Mills connections, Math. Ann., 276 (1987), 581-593.

[14] R. Kobayashi, Einstein-Kähler V-metrics on open Satake V-surfaces with isolated quatient singularities, Math. Ann., 272 (1985), 385-398.

[15] S. Kobayashi, On compact Kähler manifolds with positive definite Ricci tensor, Ann. Math., 7 (1961), 570-574.

[16] ——, Simple vector bundles over symplectic Kähler manifolds, Proc. Japan Acad., 62, Ser. A, (1986), 21-24.

[17] P. B. Kronheimer, ALE gravitational instantons, thesis, Oxford Univ., (1986).

[18] ——, Instantons gravitationals et singularités de Klein, C. R. Acad. Sci. Paris, 303, Ser. I, (1986), 53-55.

[19] J. Morrow and K. Kodaira, Complex manifolds, Holt, Rinehart and Winston, Inc., New York, 1971.

[20] A. Nadel and H. Tsuji, Compactification of complete Kähler manifolds of negative Ricci curvature, (preprint, Harvard Univ. and Tokyo Metropolitan Univ.).

[21] Y.-T. Siu, The existence of Kähler-Einstein metrics on manifolds with positive anticanonical line bundle and finite symmetry group, (preprint, Harvard Univ.).

[22] G. Tian, On Kähler-Einstein metrics on certain Kähler manifolds with $c_1(M) > 0$, Invent. Math., 89 (1987), 225-246.

[23] G. Tian and S.-T. Yau, Kähler-Einstein metrics on complex surfaces with $c_1 > 0$, Comm. Math. Phys., 112 (1987), 175-203.

[24] H. Tsuji, Complete negatively pinched Kähler surfaces of finite volume, (preprint, Harvard Univ. and Tokyo Metropolitan Univ.).

[25] S.-T. Yau, Harmonic functions on complete Riemannian manifolds, Comm. Pure Appl. Math., 28 (1975), 201-228.

[26] ——, On the Ricci curvature of a compact Kähler manifold and complex Monge-Ampère equation, I, Comm. Pure Appl. Math., 31 (1978), 339-411.

[27] ——, A general Schwarz lemma for Kähler manifolds, Amer. J. Math., 100 (1978), 197-203.

[28] ——, Nonlinear analysis in geometry, L'Enseig. Math., 33 (1987), 109-158.

[29] J. Cerf, Topologie de certains espaces de plongements, Bull. Soc. math. France, 89 (1961), 227-380.

[30] H. Grauert, Über Modifikationen und exzeptionelle analytische Mengen, Math. Ann., 146 (1962), 331-368.

Authors' Address:

Mathematical Institute
Tohoku University
Sendai, 980 Japan

ON THE MULTIPLICY OF THE EIGENVALUES OF THE LAPLACIAN

Gérard BESSON

Institut Fourier
UNIVERSITÉ DE GRENOBLE I
Département de Mathématiques
BP 74
38402 ST MARTIN D'HÈRES CEDEX - (France)

0. Introduction

The study of the eigenvalues of the Laplacian pertains to the realm ofspectral geometry (see [Bér]), an attempt to recognize a Riemannian manifold by the "sound" it produces (see [Kac] and [Pro]). If (X, g) is a closed (compact, connected without boundary) Riemannian manifold, the collection of eigenvalues of the Laplace Beltrami operator Δ is an increasing sequence of numbers

$$0 = \lambda_0 < \lambda_1 \leqslant \lambda_2 \leqslant \cdots \lambda_n \nearrow +\infty .$$

Most of the works that have been done concentrate on obtaining informations on the first eigenvalue λ_1 or on "the last ones" (asymptotic expansion of λ_n as n goes to infinity).

Among the interesting invariants attached to the eigenvalues are their multiplicities,

$$m_i = \text{multiplicity of } \lambda_i .$$

Let us recall that $m_0 = 1$, since there are no non constant harmonic functions on a closed, connected Riemannian manifold.

Then we ask ourselves the

0.1. QUESTION. *What are the possible values of m_i ?*

As usual it is very difficult to obtain relevent informations (such as sharp inequalities) on m_i for $i > 1$, so that we shall concentrate in the sequel on the study of m_1 .

The common belief is that the more symmetric the manifold, the higher the multiplicity. In particular as in the finite dimensional case (for symmetric matrices in $\mathbf{R}^N$) one hopes that the generic situation is that all m_i's are equal to one; this is indeed true, due to the following.

0.2. THEOREM (K. Uhlenbeck [Uhl]). — *Generically in the space of metrics all eigenvalues are simple.*

Here generic means for a metric in acountable intersection of open dense set in the space of Riemannian metrics. (For a precise statement and more results the reader is referred to [Uhl], [Alb], [Ble-Wil] and [Ba-Ura]).

On the other hand there are examples of Riemannian manifolds such that m_1 is larger than one, such as the canonical spheres and the flat tori. In these cases the metrics are so particular that the spectrum can be explicitly described (see [B-G-M]). Then the 1st question can be modified in

0.3. QUESTION.*How large can m_1 be?*

A first answer to this was given in 1977 by S.Y. Cheng, more precisely one has the

0.4. THEOREM ([Che]). — *Let X be a closed orientable surface of genus γ then for any Riemannian metric,*

$$m_1 \leqslant (1/2)(2\gamma + 2)(2\gamma + 3)\ . \tag{$*$}$$

In other words, in dimension 2 (orientability is not really a restriction here) the multiplicity of the first eigenvalue is limited by the topology.

0.5. EXAMPLES.

i) When $X = S^2$ Theorem 0.4 asserts that for any Riemannian metric

$$m_1 \leqslant 3\ .$$

On the other hand for the canonical metric, denoted by can, on S^2 it is a well known fact (see [B-G-M]) that

$$m_1(\text{can}) = 3$$

so that inequality $(*)$ is sharp in this case.

ii) When $X = \mathbf{T}^2$ Cheng's theorem gives

$$m_1 \leqslant 10$$

but the only known examples of multiplicity for tori are given by the flat metrics, and the maximal value for m_1 among them is 6, achieved for the "equilateral torus", *i.e.* the flat metric (denoted equi. in the sequel) induced by the Euclidean one on $\mathbf{R}^2$ when $\mathbf{T}^2$ is thought of as $\mathbf{R}^2/\mathcal{L}$ where $\mathcal{L}$ is the lattice generated by the vectors $(1,0)$ and $(1/2, \sqrt{3}/2)$.

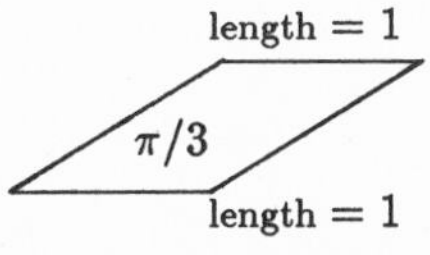

$$m_1(\text{flat metric}) \leqslant m_1(\text{equi}) = 6\ .$$

This fact leads us to

0.6. QUESTION.*Do we have $m_1 \leqslant 6$ for any metric on $\mathbf{T}^2$?*

iii) Nothing could be said about the sharpness of inequality $(*)$ in the case of higher genus.

In 1980, the author was able to improve the result of Theorem 0.4. More precisely

0.7. THEOREM ([Bes 1]). — *Under the same assumptions as in 0.4 one has*

$$m_1 \leqslant 4\gamma + 3 \,.$$

0.8. THEOREM ([Bes 1]). — *If $X = \mathbf{T}^2$ one has*

$$m_1 \leqslant 6 \,.$$

for any Riemannian metric on X .

The case of the real projective plane was also studied in [Bes 1] and a sharp inequality was obtained in this case, namely

$$m_1 \leqslant 5 = m_1(\text{can}) \,.$$

These results were only concerned by the two dimensional case; nevertheless they strengthened the common belief mentionned above. The first crack in the building came in 1982 when H. Urakawa ([Ura]) exhibited a metric on S^3 such that

$$m_1 = 7$$

when $m_1(\text{can}) = 4$ (!!) (with an obvious definition of the metric can) (see also [Béber-Bou]).

This text aim at describing some recent advances in the study of the number m_1 which in particular completely answer question 0.3 in the case of dimension greater than 2 (and thus explain Urakawa's result); furthermore one is led to a natural conjecture on what should be the sharp upper bound in dimension 2.

The author has the feeling that this text should be more descriptive than giving the full technical details. It is intended to be a guide for the reading of the articles that will be mentioned later by pointing out the basic principles and the general schemes.

It seems interesting to recall the ideas used in the proof of Cheng's theorem (§1) before describing the more recent results (§2). This will lead us to the above mentioned conjecture and to what can be thought of as a first step towards its proof (§3). We shall conclude by some related open questions (§4).

The author wishes to thank Professor T. Sunada and the Taniguchi Foundation for giving him the opportunity of describing the following ideas, and for their exceptional hospitality.

Special thanks are due to C. Anné, P. Bérard and N. Torki whose valuable comments helped me improve the first version of this text.

1. The ideas in Cheng's Theorem and some improvements

In order to describe the recent developments in the study of the multiplicity of the first eigenvalue, it is useful to recall the basic ideas that lead to Cheng's result.

It relies on the study of the so-called nodal set of an eigenfunction, *i.e.* the zero set of an eigenfunction in X, both from the local point of view and from the global one.

A. Local study.

Let us assume that the Riemannian surface X has a multiple first eigenvalue ($m_1 > 1$) and let E be the corresponding eigenspace.

i) If $\dim E$ is large, then one can construct at any point $0 \in X$, by linear combination, a non-zero eigenfunction φ such that φ vanishes at high order at 0, namely, with obvious notations,

$$\varphi(0) = \varphi'(0) = \cdots = \varphi^{(N-1)}(0) = 0 \ ;$$

more precisely, since X is two dimensional if $\dim E = \frac{N(N+1)}{2} + 1$ then there exists a function $\varphi \in E \setminus \{0\}$ vanishing at order at least $(N-1)$ at 0.

ii) An eigenfunction has a very special nodal set. Indeed, in normal coordinates around 0, the Taylor expansion of φ is

$$\varphi(x) = p_N(x) + O(|x|^{N+1})$$

$p_N =$ homogeneous polynomial of total degree N and $\Delta\varphi = \lambda\varphi \Rightarrow p_N$ is a *harmonic* polynomial.

Now in $\mathbf{R}^2$ the zero set of an homogeneous harmonic polynomial of degree N is an equiangular system of N straight lines passing though the origine

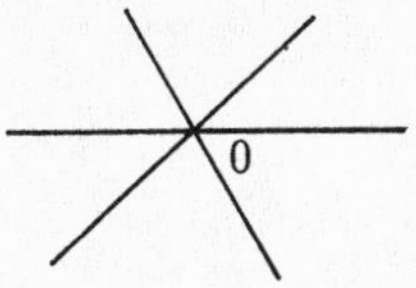

zero set of p_3 in $\mathbf{R}^2$.

An easy analytic argument shows that the nodal set of φ is C^2-diffeomorphic to the above picture near 0

zero set of φ near a point where it vanishes at order 2.

B. Global study.

A global picture of the nodal set of φ is then

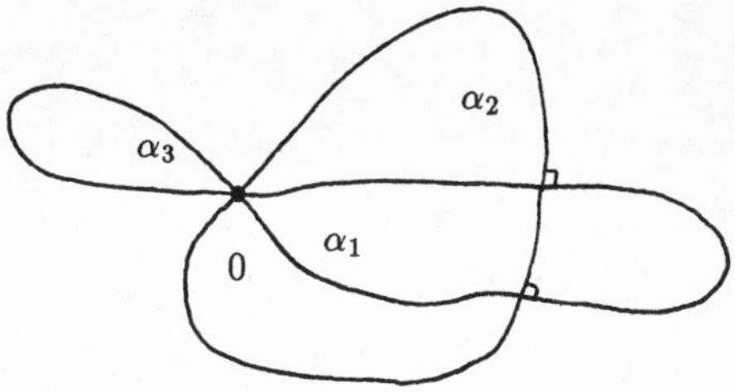

It is in fact a ***graph*** embedded in the manifold X with vertices at the points where φ and φ' vanish (φ' is $d\varphi$ of course).

It is clear now that $\varphi^{-1}(0)$ contains at least N (three on the picture) piecewise smooth embedded circle $\alpha_1, \ldots, \alpha_N$.

Obviously "many" of these "circles" will divide X in a lot of connected components : the following lemma is straightforward

1.1. Lemma ([Che]). — *If $N \geqslant 2\gamma + k$ then $X \setminus \cup_{i=1}^{N} \alpha_i$ has at least $k+1$ connected components.*

So if N is too large this contradicts the celebrated,

1.2. THEOREM (R. Courant [Cou-Hil]). — *If φ is a first eigenfunction of the Laplacian then*

$$\sharp\{\text{ connected components of } X \setminus \varphi^{-1}(0)\} = 2 .$$

Putting all these ideas and numbers together leads to Cheng's bound.

C. Improving the local study.

In order to force the order of vanishing of an eigenfunction one has to annihilate the corresponding derivatives, for example in normal coordinates $x = (x_1, x_2)$.

$$\varphi(0) \bullet$$

$$\frac{\partial\varphi}{\partial x_1}(0) \bullet \qquad \bullet \frac{\partial\varphi}{\partial x_2}(0)$$

$$\frac{\partial^2\varphi}{\partial x_1^2}(0) \bullet \qquad \bullet \frac{\partial^2\varphi}{\partial x_1 \partial x_2}(0) \qquad \boxed{\bullet \frac{\partial^2\varphi}{\partial x_2^2}(0)}$$

but

$$\Delta\varphi = \lambda\varphi \Rightarrow \frac{\partial^2\varphi}{\partial x_2^2}(0) + \frac{\partial^2\varphi}{\partial x_1^2}(0) = \lambda\varphi(0)$$

so that the vanishing of $\frac{\partial^2\varphi}{\partial x_2^2}(0)$ is a consequence of the vanishing of $\frac{\partial^2\varphi}{\partial x_1^2}(0)$ and $\varphi(0)$. It is then sufficient that $\dim E \geqslant 6$ (instead of 7) in order to have

$$\varphi(0) = \varphi'(0) = \varphi''(0) = 0$$

differentiating the equation

$$\Delta\varphi = \lambda\varphi$$

at 0 gives similar improvements at each step. This leads to the bound (see [Bes 1]),

$$(**) \qquad m_1 \leqslant 4\gamma + 3 .$$

1.3. EXAMPLES.

i) If $X = S^2$ we again obtain the (sharp!) bound

$$m_1 \leqslant 3 .$$

ii) If $X = \mathbf{T}^2$ (**) reads $m_1 \leqslant 7$ for any Riemannian metric, and it is not yet the upper bound that ought to be sharp (see the introduction).

D. Improving the global study when $X = \mathbf{T}^2$.

If X is a two torus we are in a precise topological situation so that we can try to improve the topological lemma used in section **B**.

i) First, Lemma 1.1 has the consequence that one cannot have an eigenfunction on X vanishing at order three at some point (recall that we are working with eigenfunctions for the first eigenvalue only), *i.e.* $N \leqslant 3$.

ii) But an eigenfunction vanishing at order 2 does exist on the equilateral torus

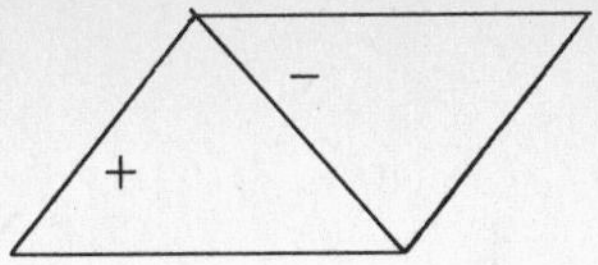

φ vanishes at order two at 0

—— = nodal set of φ

where

$$\varphi(x_1, x_2) = \sin\left[2\pi(x_1 - x_2\sqrt{3}/3)\right] + \sin\left[2\pi((2\sqrt{3}/3)x_2)\right] - \cdots$$
$$- \sin\left[2\pi(x_1 + x_2\sqrt{3}/3)\right] .$$

iii) For such a function it is not difficult to show that the three circles $\alpha_1, \alpha_2, \alpha_3$ satisfy

a) No two of them are homotopic.

b) They intersect only at 0.

Indeed if a) or b) is not true, then $X \setminus \{\alpha_1 \cup \alpha_2 \cup \alpha_3\}$ has at least three connected components.

iv) Now if $m_1 = 7$ then one has a subspace of E of dimension 2 constituted of eigenfunctions vanishing at order 2 at 0. Choosing a direction, say θ at 0 in X , it is then possible to find a normalized eigenfunction (the L^2-norm is 1) f_θ in this subspace so that one of the tangent to the branches of the nodal set at 0 of f_θ is given by θ .

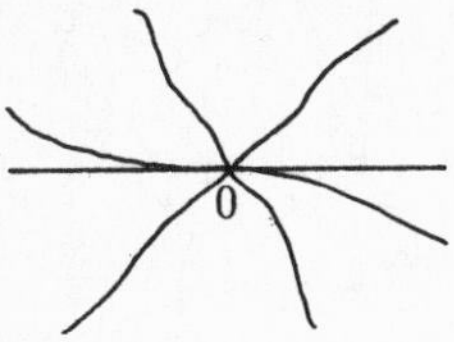

θ–direction

nodal set of f_θ near 0 .

Only f_θ and $-f_\theta$ have this property for otherwise there is an eigenfunction vanishing at order 3.

The "circles" $\alpha_1(\theta)$, $\alpha_2(\theta)$, $\alpha_3(\theta)$, $\theta \in S^1$ corresponding to f_θ then give an homotopy between $\alpha_1(1)$,$\alpha_2(1)$, $\alpha_3(1)$ contradicting the above property a); so finally

$$m_1 \leqslant 6 \qquad \text{on } X = \mathbf{T}^2 .$$

E. Concluding remarks.

1) When the maximal multiplicity is achieved for a nice metric (*i.e.* canonical with a lot of isometries), then one can deform it and preserve the action of a finite subgroup of the isometry group which acts irreducibly in the first eigenspace : this gives a one parameter family of metrics with m_1 constantly equal to the maximal value. We conclude that

m_1 maximal is not characteristic of a nice metric .

2) It is worth while pointing out that the proof relies on the following property of the eigenspaces E; any $f \in E$ satisfies Courant's theorem. In fact Cheng's bound would be true for the dimension of any subspace of functions satisfying this property (one also need that they have a "nice" nodal set).

3) In particular this is valid for the second eigenspace of any Schrödinger operator on X

$$\Delta + V$$

where V is a smooth function.

2. Some recent developments

The study of the multiplicity has recently considerably progressed. An important break through has been made by Y. Colin de Verdière ([ColVer 1-4]). In this paragraph we describe the basic principles and applications developed in the above references, the technicality being left aside.

Certainly the central idea is the one developed in the celebrated paper by V.I. Arnold ([Arn]) "modes and quasimodes" for the finite dimensional case that

enough parameters allow to force the multiplicity

but let us make precise what is meant here.

A. The basic principle.

Let us assume that we are given a family $\mathcal{F}$ of operators which is, for example, a (Banach) manifold near one of them called $\mathcal{O}$, see fig.1

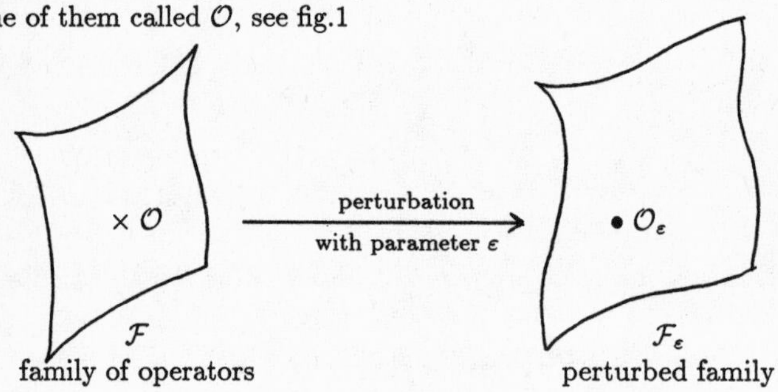

fig.1

For the sake of simplicity the reader may think of $\mathcal{F}$ as being the family of Laplace-Beltrami operators corresponding to the various metric on a manifold X or the family of Schrödinger operators with fixed metric and varying potentials. Another important example will be described later.

Let us then assume that we have a perturbation of the whole family with a parameter called ε (small!) into another family $\mathcal{F}_\varepsilon$.

Finally we assume that the operator in $\mathcal{F}$ denoted by $\mathcal{O}$ has a multiple eigenvalue λ of multiplicity N.

2.1. QUESTION. *What does happen to eigenvalues of operators in $\mathcal{F}$ under the perturbation ?*

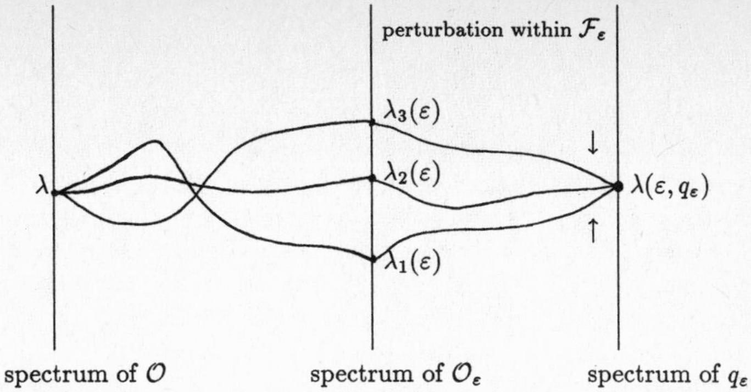

fig.2

As usual if the perturbation is smooth enough the multiple eigenvalue λ of $\mathcal{O}$ splits into N branches $\lambda_i(\varepsilon)$, which are eigenvalues of $\mathcal{O}_\varepsilon$ and which are not likely to be multiple in general (see [Kat] for the perturbation theory).

We could also perturb $\mathcal{O}_\varepsilon$ within the family $\mathcal{F}_\varepsilon$ (see fig.1) so that we may think of the branches $\lambda_i(\varepsilon, q)$ depending also on $q \in \mathcal{F}_\varepsilon$. Now if we have enough parameters defining the family $\mathcal{F}_\varepsilon$ such that the branches vary with ***enough independence*** then by slightly moving the operator around $\mathcal{O}_\varepsilon$ we may force the various numbers $\lambda_i(\varepsilon, q)$ to coïncide again for a given ε; in other words we may find $q_\varepsilon \in \mathcal{F}_\varepsilon$ such that

$$\lambda_1(\varepsilon, q_\varepsilon) = \cdots = \lambda_N(\varepsilon, q_\varepsilon) \ .$$

Practically $\mathcal{F}$ and $\mathcal{F}_\varepsilon$ are often infinite dimensional manifold and we want finitely many eigenvalues to coïncide, so that it is reasonable to think that this procedure works. This is not quite true in general, one needs to make sure that the functions $\lambda_i(\varepsilon, q)$ on $\mathcal{F}_\varepsilon$ (or equivalently on $\mathcal{F}$) are independent; this is translated into the so called

stability conditions

(see [ColVer 2-3-4] and [Bes 2]).

There are various different ways of stating a good stability condition. For the above process to be efficient only a weak form is required, called Weak Arnold Hypothesis (W.A.H.) in [ColVer 2]. In section 3 we shall state a stronger condition called Strong Arnold Hypothesis (S.A.H.) which (implying W.A.H.) is more suitable to what is done in section 3 and is easier to test on examples.

The stability conditions will be conditions on the family $\mathcal{F}_\varepsilon$ near $\mathcal{O}_\varepsilon$ or equivalently on $\mathcal{F}$ near $\mathcal{O}$ and since we shall use an infinitesimal version we shall say that it is a condition on $\mathcal{O} \in \mathcal{F}$.

In the examples we shall use the procedure the reverse way, namely we shall be given a situation that we shall perturb, using most of the time singular perturbation, with a parameter ε going to zero, so that we start with a family $\mathcal{F}_\varepsilon$ and end with $\mathcal{F}$ on which we know the existence of an operator $\mathcal{O}$ with a multiple (first) eigenvalue; the two following steps are needed :

1) compute an asymptotic expansion of $\lambda_i(\varepsilon, q)$ for all $q \in \mathcal{F}_\varepsilon$ when ε goes to zero, and identify $\mathcal{F}$ and $\mathcal{O} \in \mathcal{F}$,

2) verify the stability condition for $\mathcal{O} \in \mathcal{F}$.

The applications of the basic principle that we shall describe briefly in paragraph **C** need the use of some operators on graphs.

B. Laplacians on graphs.

Let Γ be a finite graph, *i.e.* a set of $v = N+1$ vertices linked by unoriented edges, such that there is at most one edge between any two vertices. Let e be the total number of edges and n_i the valence of the vertex i , *i.e.* the number of edges between i and another vertex. We shall mainly use the complete graph on $N+1 = v$ vertices K_v , which has the maximal valence for each vertex,

$$\text{for all } i \text{ vertex of } K_v \quad n_i = v-1 = N$$

or equivalently

$$e = \frac{(N+1)N}{2}$$

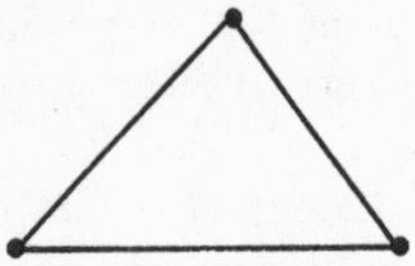

K_3

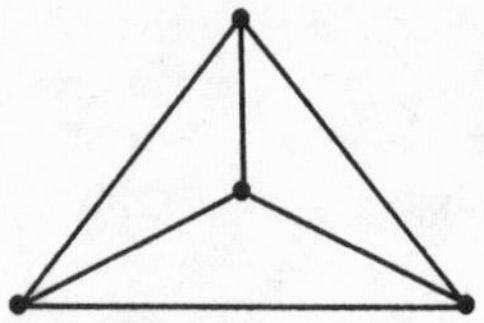

K_4

We shall consider the space $\mathcal{L} \simeq \mathbf{R}^v$ of functions on the vertices of Γ and study the following operators for $x \in \mathcal{L}$, $x = (x_i)$

$$(Ax)_i = \left(\frac{1}{n_i} \sum_j c_{ij}(x_i - x_j) \right) + V_i$$

$c_{ij}, V_i \in \mathbf{R}$ and $c_{ij} > 0$ if (i,j) is an edge and $c_{ij} = 0$ otherwise.

These are Schrödinger operators with potential function $V = (V_i)$. The matrix (c_{ij}) has to be interpreted has the inverse of a Riemannian metric (or weights) supported on the set of edges.

In particular if $\Gamma = K_v$ there is a preferred operator defined by

$$(\mathcal{O}x)_i = \frac{1}{N} \sum_j (x_i - x_j)$$

with eigenvalues

$$\begin{aligned} \lambda_0 &= 0 \quad \text{of multiplicity } 1 \\ \lambda_1 &= \tfrac{N+1}{N} \quad \text{of multiplicity } N \end{aligned}$$

(see [ColVer 3] and [ColVer 4] for more details on operators on graph).

These operators have a zeroth eigenvalue, called λ_0 which is simple. We shall call λ_1 , the first eigenvalue, the next one. In the sequel $\Gamma = K_v$, $\mathcal{F}$ will be this family of operators and $\mathcal{O}$ the above operator which has a first eigenvalue of multiplicity N .

C. Applications 1 : embedding graphs in X.

Recall that X is a closed Riemannian manifold of dimension n . Let us assume that we can embed the complete graph K_v into X ,

$$K_v \hookrightarrow X$$

and let us perturb the metric on X in the following way.

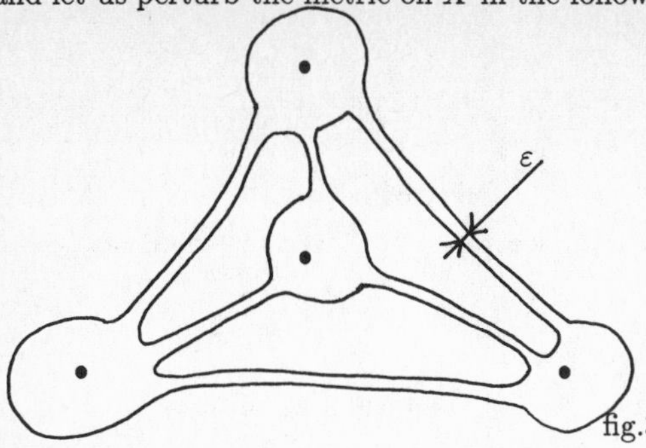

fig.3

Let W_ε be a neighbourhood of the graph in X which is a tubular neighbourhood of the edges and balls around the vertices of fixed size. Outside W_ε we multiply the Riemannian metric by ε . Now

i) We let ε go to zero.

ii) Since the perturbed metric is not continuous any more a regularization process has to take place near the boundary of W_ε .

A difficult and very technical study of the asymptotics of the eigenvalues of the perturbed metric shows that the $v = N+1$ first eigenvalues asymptotically behave (up to a power of ε) as the eigenvalues of an operator on the graph of the type described in paragraph **B**.

Applying the basic principle we get the

2.2. THEOREM (Y. Colin de Verdière [ColVer 1,3]). —

1) If $n \geqslant 3$ then for any integer N there exists a metric on X such that $m_1 = N$.

2) If $n = 2$ and X orientable there exist a metric on X and a smooth function V such that the multiplicity of the second eigenvalue of the corresponding Schrödinger operator is,

$$m_1 = \mathrm{Int}\left[\frac{5+\sqrt{48\gamma+1}}{2}\right] .$$

(where $\mathrm{Int}(x)$ is the integer part of x).

2.3. REMARKS.

i) If $n \geqslant 3$ for any N the complete graph in $N+1$ vertices can be embedded in X .

ii) If $n = 2$ the maximal number of vertices of a complete graph which can be embedded in orientable surface of genus γ is bounded above by

$$v = N+1 \leqslant \mathrm{Int}\left[\frac{7+\sqrt{48\gamma+1}}{2}\right] \quad \text{([Rin])} .$$

iii) It is more convenient for the study of the spectrum of a Riemannian manifold to use the quadratic form

$$q(f) = \int_X |df|^2 = \text{Dirichlet integral for } f \in H^1(X)$$

instead of the operator, and Theorem 2.2 is obtained by studying the effect of the perturbation described above on this quadratic form.

In dimension 2, q is conformally invariant so that multiplying the metric by ε outside W_ε has no effect on q when evaluated on function with support outside W_ε . In that case one has to use another perturbation (see [ColVer 3]).

This roughly explains why in dimension 2 a Schrödinger operator appear instead of a single Laplacian. It is not known whether one can take V to be zero.

iv) A complete graph always satisfies the weak stability condition.

v) The problem raised in Question 0.4 is thus completely solved in dimension greater than 2.

D. Applications 2 : modelling a surface on a graph.

In this paragraph we assume that X is a surface endowed with a metric of constant curvature -1 (thus $\gamma \geqslant 2$).

Let S be a finite union of closed geodesics such that $X \backslash S$ has $N+1$ connected components, then we can modify the conformal structure by multiplying the length of each geodesic in S by a number ε .

It is now known (see [Do-Pi-Ra-Su] and [Col 1]) that the $N+1$ first eigenvalues ($\lambda_0 = 0 < \lambda_1 \leqslant \lambda_2 \cdots \leqslant \lambda_N$) behave asymptotically (see the above reference for a precise statement) as those of a graph whose vertices correspond to the thick parts of the manifold and the edges to the tubular neighbourhood of the geodesics in S .

We are thus in a situation where we can apply the basic principle. One can arrange that the limiting graph is complete; the following theorem has been obtained by B. Colbois and Y. Colin de Verdière,

2.4. THEOREM ([Col-ColVer]). — *If X is an orientable surface of genus $\gamma \geqslant 3$ there exists a metric with constant curvature -1 on X such that*

$$m_1 = \mathrm{Int}\left[\frac{1+\sqrt{8\gamma+1}}{2}\right]$$

2.5. REMARK.

i) The asymptotics of the eigenvalues used in the theorem is the one obtained in [Col]. See also [Bur] for a refinement.

ii) In [Col-Bur] it is shown using group theory and coverings that for any prime number p there exists a compact Riemann surface with genus $\gamma = p(p-1)+1$ and $m_1 = p-1$.

iii) It is to be noticed that these are the first examples of Riemann surface the multiplicity of the first eigenvalue of which satisfies $m_1 \underset{\gamma \to +\infty}{\sim} c\sqrt{\gamma}$.

iv) Any Riemann surface can be embedded in any manifold of dimension greater than 2 thus by a process similar to the one describes in paragraph **C** one can get Theorem 2.2 in a somewhat easier (technically) way (see [Bes 2] for the description).

v) In [Col-ColVer] the theorem also concerns non compact Riemann surface.

D. Applications 3 : Adding handles and blowing up points on surfaces.

By cutting off small balls (of radius ε) in a Riemannian manifold one does not perturb very much the eigenvalues (see [Oza] for a review on this problem).

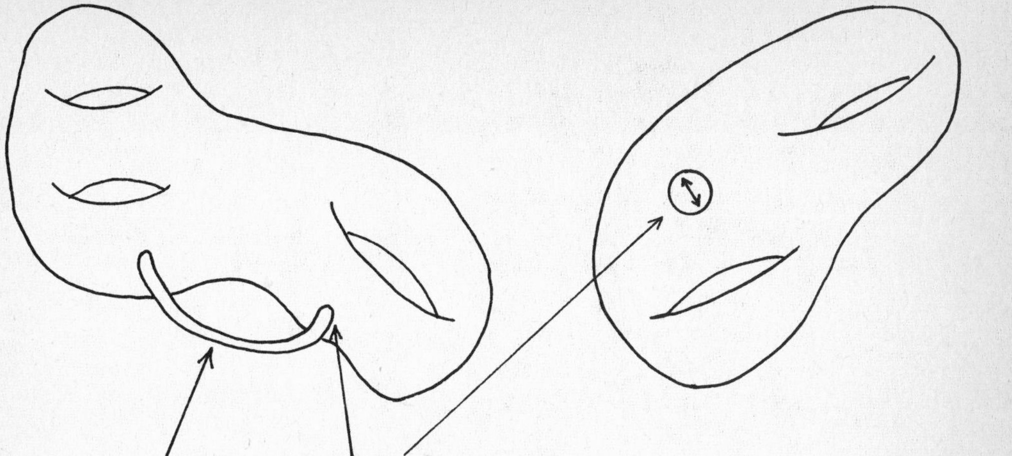

fig.4

Then one can add to a given surface a small handle which degenerates. It has been shown in that situation that the eigenvalues converge ([Cha-Fel]). Similarly one can identify opposite points on the boundary of a geodesic ball (blowing up) of radius ε , again the spectrum converges when ε goes to zero (see [ColVer 3]).

The basic principle is applied with these asymptotics and $\mathcal{F}$ defined to be the family of Laplacians for the various metrics on the limiting surface.

2.6. THEOREM (Y. Colin de Verdière [ColVer 3]). —

1) If X is an orientable Riemann surface then there exists a Riemannian metric on X with $m_1 = 6$.

2) If X is a non orientable surface then there exist a Riemannian metric with $m_1 = 5$.

2.7. REMARK.

i) If $X \# Y$ denote the connected sum of the two surfaces X and Y . Then

$$X \text{ orientable of genus } \gamma \geqslant 1 \Rightarrow X = \underbrace{\mathbf{T}^2 \# \mathbf{T}^2 \# \mathbf{T}^2 \cdots \# \mathbf{T}^2}_{\gamma \text{ times}}$$

and we add handles to a torus on which a metric with $m_1 = 6$ exists (and it satisfies the appropriate stability condition, see next section).

If X is non orientable then X is obtained from the projective space by adding handles or blowing up points and on $P^2(\mathbf{R})$ the canonical metric has a (stable) first eigenvalue with $m_1 = 5$.

ii) In particular if $X = \mathbf{K}^2 = P^2(\mathbf{R}) \# P^2(\mathbf{R})$ is the Klein bottle then there is a metric with $m_1 = 5$ and it is proved in [ColVer 3] that this is the maximal value, *i.e.*

$$\text{for any metric on } \mathbf{K}^2 , \qquad m_1 \leqslant 5 .$$

Thus the inequality is sharp.

The proof of the upper bound is a refinement of the topological lemma used in Cheng's theorem for the case of $\mathbf{K}^2$.

3. A conjecture and a first step towards its proof

In this section we shall finally give a stability condition which suits our purpose, namely studying "generic properties of eigenfunctions under multiplicity". The study is motivated by a conjecture due to Y. Colin de Verdière and it is intended to be a first step towards its proof (in the direction chosen).

A. A conjecture for m_1 on surfaces.

As we mentioned before the problem of finding the possible values of m_1 on a Riemannian manifold of dimension greater than 2 is completely solved by Theorem 2.2.

On the other hand, the case of surfaces is still open; we know by Theorem 0.4 that there is an upper bound and we have examples of large multiplicity on surfaces, but the bound

$$m_1 \leqslant 4\gamma + 3$$

is certainly not sharp (it is not sharp already for $\gamma = 1$) so the best upper bound is still to be discovered.

The method described briefly in the preceeding section relies heavily on graph theory and more precisely on embeddings of a complete graph in $N+1$ vertices in the manifold. If X is a surface of genus γ , the maximal value for such an N is

$$N_{\max} = \mathrm{Int}\left[\frac{5+\sqrt{48\gamma+1}}{2}\right] \quad (\mathrm{Int} = \text{Interger part}) ,$$

it is then reasonable to

3.1. Conjecture. — *For any Riemannian metric on X (or Schrödinger operator)*

$$m_1 \leqslant N_{\max} .$$

3.2. Remarks.

i) The conjecture is true when $\gamma = 0$ or 1 .

ii) If the above conjecture is true for Schrödinger operator then the inequality is sharp as shown by Theorem 2.2.

iii) In [Bur] the upper bound

$$m_1 \leqslant \sqrt{(6(\gamma-1))}$$

for metrics is proved when X is close to a graph in the sense of Application **D** of the previous section.

In order to make a step towards the proof of the conjecture we need to find a good property of a multiple eigenspace; Courant's theorem was one but it is certainly not enough in order to get the bound 3.1.

B. A stability condition.

In the following four paragraphs X is not limited to be a surface. If we are given a multiple eigenvalue λ of multiplicity N corresponding to a Riemannian metric g_0 , we wish to study the following set

$$W = \{\text{metric } g \text{ close to } g_0 \setminus g \text{ has } \lambda \text{ as eigenvalue with multiplicity } N\} ,$$

if E_0 is the eigenspace corresponding to g_0 , by small perturbations of the metric the eigenvalue λ split in N eigenvalues $\mu_1, \ldots, \mu_N$

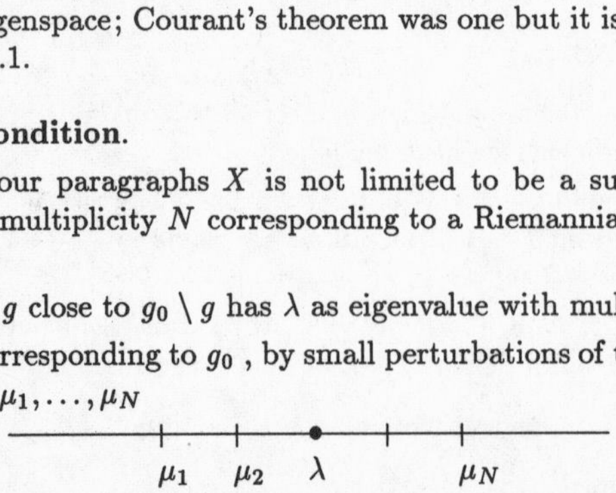

and the eigenspace E_0 in the sum E of the eigenspaces corresponding to the eigenvalues $\mu_1, \ldots, \mu_N$.

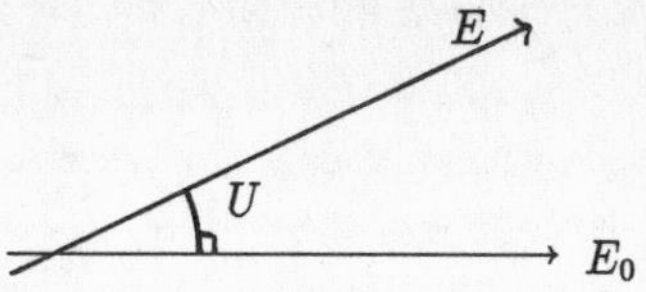

In particular in the Sobolev space $H^1(X)$, E is close to E_0 , so that we can easily construct a canonical L^2-isometry, say U , between them (see [ColVer 1]).

Let q_0(resp. q') be the Dirichlet integral corresponding to the metric g_0(resp. g) , then

$$q_0 \restriction E_0 = \lambda \langle \cdot , \cdot \rangle$$

where $\langle \cdot , \cdot \rangle$ is the L^2-scalar product; define q by,

$$q(x,y) = q'(Ux, Uy) \text{ for } x, y \in E_0$$

q is the quadratic form q' transported on the *fixed* vector space E_0 , we thus have a map Φ ,

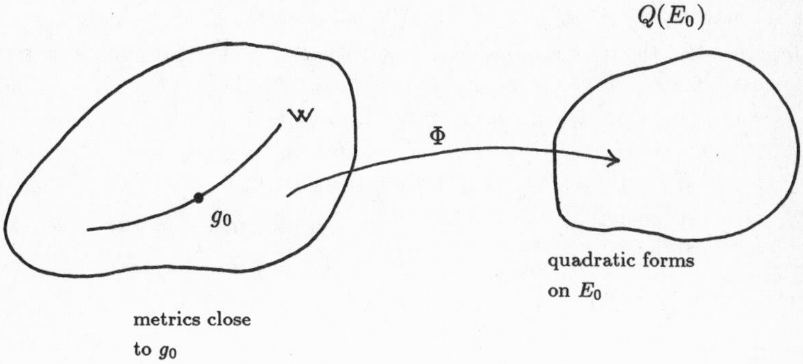

such that $\Phi(g_0) = \lambda \langle \cdot , \cdot \rangle$.

We then give the

3.3. DEFINITION. — *We shall say that the eigenvalue λ is **stable** if Φ is a submersion at g_0* .

3.4. REMARKS.

i) This is the S.A.H. property of [ColVer 2] .

ii) The stability of λ implies that W is being a submanifold near g_0 , and this will allow us to do some differential geometry on W .

iii) One has to be careful with the topology on the space of metrics, in particular it is more convenient to work with Banach manifolds so that we shall use the C^k topology on this space for k large enough. These details will not be discussed here, the reader is referred to [Bes 2].

At first glance it does not seem easy to verify that a given eigenvalue is stable in this sense; in [ColVer 2] it is proved using representation theory that

3.5. THEOREM (Y. Colin de Verdière). —

i) All the eigenvalues of the canonical metric on $X = S^2$ *are stable.*

ii) An eigenvalue of a flat two-torus is stable if and only if it has multiplicity not bigger than 6 .

In particular unstable eigenvalues do exist. In order to decide upon the stability of an eigenvalue we shall give a criterion in a general setting.

C. A criterion for stability.

The notion of stability for a given eigenvalue relies on perturbation theory. In Definition 3.4 we used general perturbations of the metric g_0; we could as well restrict ourselves to metrics which are pointwise conformal to g_0 , or to perturbations of the Laplacian associated to g_0 by smooth potentials. A notion of stability is attached to each of them.

We shall say that an eigenvalue is either stable, conformally stable or stable for potentials, depending on the type of perturbations under consideration.

Recall that we denoted by E_0 the multiple eigenspace and let $(u_i)_{1\leqslant i\leqslant N}$ be an orthonormal basis of E_0 , then we have the

3.6. PROPOSITION (see [Bes 2]). — *The eigenvalue* λ_0 *is*

i) not conformally stable or not stable for potentials iff there is a non-trivial relation of the type

$$\sum_{i\leqslant j}\alpha_{ij}u_i(x)u_j(x)=0 \quad \text{for all } x\in X$$

with $\alpha_{ij}\in\mathbf{R}$, $(\alpha_{ij})_{i,j}\neq 0$.

ii) not stable iff the following non-trivial relations are satisfied simultaneously

$$\begin{cases} \sum_{i\leqslant j}\alpha_{ij}u_i(x)u_j(x)=0 & \text{for all } x\in X \\ \text{and} & \\ \sum_{i\leqslant j}\alpha_{ij}(d_xu_i\circ d_xu_j)\equiv 0 & \end{cases}$$

where $a\circ b$ *denote the symmetric product of* a *and* b, *and* $\alpha_{ij}\in\mathbf{R}$.

3.7. Remarks.

i) The proof is easy since one knows exactly the derivative of the eigenvalues with respect to a one-parameter perturbation (see [Bes 2] for the details) of the type considered.

ii) The property of being stable is a property of the *eigenspace* E_0 , indeed if one define

$$E_0^2=\{\text{functions } u\cdot v, \text{ with } u\in E_0 \text{ and } v\in E_0\}$$

then in the case i) (for example) the stability is equivalent to

$$\dim E_0^2=\frac{N(N+1)}{2},$$

in the sequel we may write E_0 stable instead of λ_0 stable.

3.8. Examples.

i) If $X=S^n$ endowed with the canonical metric, the first eigenspace is spanned by the restriction to S^n of the coordinates of $\mathbf{R}^{n+1}$, $x_1,x_2,\dots,x_{n+1}$. It is then obvious that no non trivial relation

$$\sum_{i\leqslant j}\alpha_{ij}x_ix_j\equiv 0$$

holds. Thus the first eigenvalue of S^n is stable in any sense. But other eigenvalues may not be stable, for example on S^3 ([Tor]).

ii) If $X = \mathbf{T}^2$ thought of as

$$X = \mathbf{R}^2/\mathcal{L} \qquad \mathcal{L} \text{ spanned by } (1,0) \text{ and } (0,1)$$

then the flat quotient metric has a first eigenspace of multiplicity 4 spanned by the functions

$$\cos(2\pi x)\ ,\ \sin(2\pi x)\ ,\ \cos(2\pi y)\ ,\ \sin(2\pi y)\ ,$$

and of course

$$\cos^2(2\pi x) + \sin^2(2\pi x) \equiv 1 \equiv \cos^2(2\pi y) + \sin^2(2\pi y)\ ,$$

so that the corresponding eigenvalue is *not* conformally stable. However due to Theorem 3.5 (and it is easy to verify with the criterion) it is stable.

D. Generic properties of eigenfunctions.

With the notation of the previous paragraph let us define the following map

$$\begin{aligned} X &\xrightarrow{F_g} \mathbf{R}^N \\ x &\longmapsto (u_1(x), \ldots, u_N(x)) \end{aligned}$$

where $(u_i)_{i=1,2,\ldots,N}$ is a basis of the eigenspace of dimension N relative to the eigenvalue λ of the metric $g \in W$ (near g_0). The reference to g is omitted in (u_i) for the sake of the simplicity.

It is clear that the basis (u_i) can be taken depending smoothly (for example C^1) on the metric (see [Bes 2]) using the canonical isometry U introduced in paragraph C .

In order to find a generic property of the map Φ it is useful to recall what we know about maps.

$$X^n \xrightarrow{G} \mathbf{R}^N\ , \qquad n = \dim X\ .$$

Let G be an arbitrary C^k map of X into $\mathbf{R}^N$ (k is big enough) then generically

G is a Morse function if $N = 1$ [Mor]
G is an immersion if $N \geqslant 2n$ [Whi]
G is an imbedding if $N \geqslant 2n + 1$ [Whi]

the last two results are the celebrated Whitney's theorem. (The precise definition of the topology and of genericity is omitted,for the sake of simplicity).

What can we say when G is replaced by F_g? The first step has been made by K. Uhlenbeck [Uhl] : when $N = 1$ generically F_g is a Morse function. (if $N = 1$ the eigenvalue λ_0 is simple and clearly W is an open neighbourhood of g_0 , in that case since X is compact generic means for an open dense set).

The two other steps are true also, provided that λ_0 is stable, precisely

3.8. PROPOSITION. — *If λ_0 is a stable multiple eigenvalue of multiplicity N then for an open dense set of metric in W .*

i) F_g is an immersion if $N \geqslant 2n$.

ii) F_g is an embedding if $N \geqslant 2n + 1$.

3.9. REMARKS.

i) The proof is done by making use of a transversality theorem. It is a little bit technical since one has to study the behaviour of the eigenfunctions with respect to a metric in W . It boils down to showing that if $G(x_0, y)$ is a Green function and v is an eigenfunction then

$$G(x_0, \cdot)v(\cdot)$$

cannot be C^∞ .

In order to prove that, we need the metric g_0 to be smooth (C^∞) but in order to apply the transversality theorem we need W be a Banach manifold, thus allowing C^k perturbations of g_0 (k large). The reader is referred to [Bes 2] for the details.

The author wishes to emphasize that Proposition 3.8 relies on a *local study* (first step in the proof of Cheng's theorem) of the Green function.

ii) In the statement of Proposition 3.8 stability is meant to be one of the three cases presented above and then W is a submanifold of the corresponding space (all metrics, metrics that are conformal to a given one, functions considered as potentials).

iii) Proposition 3.8 would be empty if there wereno stable eigenvalue.

E. More stable eigenvalues.

What are the stable multiple eigenvalues besides the canonical examples?

Roughly speaking all the examples described by Y. Colin de Verdière, B. Colbois and Y. Colin de Verdière are stable.

Indeed they are obtained through the following procedure

1) a subspace of L^2 is given by the situation. It is called a test space E_{test} . In the case of the embedding of a complete graph in a manifold X . E_{test} is an embedding in $L^2(X)$ of the eigenfunctions of the complete graph (corresponding to the multiple eigenvalue) by associating to each of them a function constant on the thick parts (see fig.3), and harmonic on the thin parts.

2) Then a geometric construction exhibits a metric with an eigenvalue of multiplicity $\dim(E_{\text{test}})$ and an eigenspace E close to E_{test} in the Sobolev space $H^1(X)$.

Now the conformal stability (for example) can be restated in the following way :

$$\mathbf{R}^{\frac{N(N+1)}{2}} \xrightarrow{L} L^2(X)$$

$$\alpha = (\alpha_{ij}) \longmapsto \sum_{i \leqslant j} \alpha_{ij} u_i u_j$$

E stable $\iff$ L injective

$\iff$ L open i.e. $\exists C \;\; L(\alpha) \geqslant C|\alpha|$

where $|\alpha| = \left(\sum_{i \leqslant j} \alpha_{ij}^2\right)^{1/2}$

It is then clear that the proximity of E and E_{test} in $H^1(X)$ implies that,

$$E \text{ stable} \iff E_{\text{test}} \text{ stable} \quad .$$

In each particular cases one is then led to prove the stability for the explicit space E_{test} here of course the conformal stability of E_{test} is understood as

$$\dim(E_{\text{test}}^2) = \frac{N(N+1)}{2}$$

and a similar definition for the stability.

3.10. REMARKS.

i) In the previous section it was written that a complete graph always satisfy a stability condition. It is true with the weak form of it (W.A.H. of [ColVer 2]), the only one necessary to construct the multiple eigenvalue; it is also true with the strong version of the stability condition used in this section but it is not true for the conformal stability or the stability for potentials. Indeed the multiple eigenspace E_0 of the canonical operator on a complete graph is of dimension N , if $v = N+1$, and the space $L^2(K_v)$ is of dimension $N+1$ so that

$$N+1 < \frac{N(N+1)}{2} = \dim(E_0^2)$$

if $N \geqslant 3$.

So that the conformal stability is *never* satisfied if $N \geqslant 3$.

However the embedding E_{test} of E_0 in a manifold used in Application 1 of section 2 do satisfy this strong condition.

The reader must keep in mind these two versions of the stability condition.

ii) When graphs are used to construct multiple eigenvalues we already said that the strong stability condition is satisfied for E_{test} (see [Bes 2] for the details), this takes care of Applications 1 and 2 for section 2.

In Application 3 the space E_{test} is just the eigenspace corresponding to the first eigenvalue for the canonical metric on the projective plane or for a flat metric on the torus (and thus has dimension $\leqslant 6$) . Both are known to be stable (maybe not conformally stable on the torus).

iii) The explicit examples of high multiplicity allows to state the

3.11. THEOREM. — *On any closed manifold X^n there exist Riemannian metrics such that X^n can be embedded in $\mathbf{R}^{2n+1}$ with eigenfunctions for the first eigenvalue as coordinates.*

F. What about the conjecture?.

Let us come back to the case of surfaces; throughout this paragraph X will be of *dimension 2*. To prove the conjecture for a general situation is certainly very difficult since pathologically high multiplicity might appear. The strong stability condition ensures that such a phenomenon does not occur, indeed it is a condition which allows to preserve the multiplicity under well chosen perturbations and as such deserve the name of "stability condition".

The coming idea is an attempt to prove the conjecture for stable eigenvalues.

Assume that X possesses a metric such that the first eigenvalue is stable and that $m_1 = N$ is larger than 4 , (*i.e.* , $N \geqslant 5 = 2 \times 2 + 1$). Then there are metrics close to this one such that the map,

$$\begin{aligned} X &\xrightarrow{F_g} \mathbf{R}^N \\ x &\longmapsto (u_1(x), \ldots, u_N(x)) \end{aligned}$$

is an embedding (with the notations of the previous paragraph).

This result was obtained by a *local study* of the eigenfunctions and if we want to follow the scheme of Cheng's theorem we need to find a global property of this embedding.

The translation to Courant's nodal theorem on F is the following that we shall call the TPO property :

- *any hyperplane in $\mathbf{R}^N$ passing through the origin separates $F(X)$ in exactly two connected components.*

and this reminds us of a well-known property of an embedding of a surface

$$X \xrightarrow{G} \mathbf{R}^N$$

called the two piece property (TPP) which reads :

- *for any half space H of $\mathbf{R}^N$ (the boundary hyperplane does not necessarily pass through the origin) $H \cap G(X)$ is connected.*

Th. Banchoff proved that for ***closed surface*** the TPP is equivalent to the tightness of the embedding (see [Kui 1]). Certainly the embedding F is not tight since one has the

3.12. THEOREM (N. Kuiper [Kui 2]). — *If G is a tight substantial (i.e. $G(X)$ is not included in a subspace) C^2 embedding of the surface X in $\mathbf{R}^N$ then*

$$N \leqslant 5 .$$

And we know metrics for which $m_1 > 5$ (of course F is smooth).

But on the other hand one has the beautiful theorem due to Th. Banchoff ([Ban]).

3.13. THEOREM ([Ban]). — *With the same assumptions as above and if G is only assumed to be a polyhedral embedded (i.e. $G(X)$ is a polyhedron in $\mathbf{R}^N$) then*

$$N \leqslant \operatorname{Int}\left[\frac{5+\sqrt{48\gamma+1}}{2}\right] .$$

The idea would be then to deform the TPO smooth embeddings F into a TPP polyhedral embedding G and apply Banchoff's bound thus proving the conjecture.

3.14. REMARKS.

i) The bound which appears in 3.13 has the same origin as the one appearing in the conjecture, since, for a polyhedron, being tight is a property of the 1-skeleton (see [Küh]) which is a graph.

Furthermore the proof is very similar to the one leading to the upper bound for the number of vertices of an embedded complete graph in X .

ii) It is easy to verify that the first eigenspace of a flat equilateral torus which is of multiplicity 6 gives an embedding in $\mathbf{R}^6$ which is not tight. Indeed it suffices to construct a function which is a linear combination of the corresponding eigenfunctions and the constants that has a non strict local minima at some points and this can be done explicitely in this situation.

iii) Needless to say this is only one possible approach towards the proof of the multiplicity conjecture and many others could be envisaged.

4. Conclusion

The recent progress in the study of multiplicities of the eigenvalues of the Laplacian has brought through new ideas, in particular the link between the operator theory on graphs and operator theory on manifolds. This has lead to a complete solution of the original problem in dimension greater than 2. The case of Dimension 2 is still open, but the conjecture mentioned above is an aim to reach.

Many other problems are left in connection to what has been presented here. Let us briefly describe some of them.

Problem 1 : Operators on graph.

In [ColVer 4], Y. Colin de Verdière defines a new invariant of graphs using the multiplicity of the second eigenvalue of some operators on graph (the first one being always simple). In particular he obtained, using the known bounds on m_1, criterion for the non-embeddability of a graph in a surface of genus γ. A good question is

What is the link between this invariant and the chromatic number of the graph?

One of the possible answer could give a new proof of the four colour theorem. (see [ColVer 4] for the details, and also [ColVer-Mar]).

Almost nothing is known about this invariant.

Problem 2 : Surfaces of constant negative curvature.

On a two-torus there are flat metrics which have eigenvalues of arbitrary large multiplicities (these are not λ_1 of course).

It is reasonable to think that this does not occur for a surface endowed with a metric of constant curvature -1, so a question is

Does there exist an upper bound for m_i on a given fixed surface X of genus $\geqslant 2$, independent of i?

Problem 3 : A technical question.

In [Bes 2] we study linear combination of eigenfunctions (u_i) in the same eigenspace.

Does there exist an analytic continuation principle for functions of the type $\sum_{i\leqslant j} \alpha_{ij} u_i u_j$, $\alpha_{ij} \in \mathbf{R}$?

i.e. if $\sum_{i\leqslant j} \alpha_{ij} u_i(x) u_j(x) = 0$ for x in an open set is it true that the above function vanishes identically?

Certainly this is true for a product $u_1 u_2$ or $u_1^2 - u_2^2$. This is true also if the metric is real analytic since the functions are then real analytic.

The first problem is by and large the most interesting one.

Other questions about the spectrum of the Laplacian can be found in [T-S-G].

5. Bibliography

[Alb] J. ALBERT. — *Genericity of simple eigenvalues for elliptic PDE's*, Proc. Amer. Math. Soc., **48** (1975), 413-418.

[Arn] V.I. ARNOLD. — *Modes and quasimodes*, Functional Anal. Appl., **6** (1972), 94-101.

[Ba-Ura] S. BANDO, H. URAKAWA. — *Generic properties of eigenvalues of the Laplacian for compact Riemannian manifolds*, Tôhoku Math. J., **35** (2) (1983), 155-172.

[Ban] TH. BANCHOFF. — *Thightly embedded 2-dimensional polyhedral manifolds*, Amer. J. Math., **87** (1965), 245-256.

[Bér] P. BÉRARD. — *Spectral geometry : direct and inverse problems*, Lecture notes in Math., n° 1207, Springer Verlag, 1986.

[Béber-Bou] L. BÉRARD BERGERY, J.P. BOURGUIGNON. — *Laplacians and Riemannian submersions with totally geodesic fibers*, Illinois J. Math., **26** (1982), 181-200.

[Bes 1] G. BESSON. — *Sur la multiplicité de la première valeur propre des surfaces riemanniennes*, Ann. Inst. Fourier, **30** (1980), 109-128.

[Bes 2] G. BESSON. — *Propriétés génériques des fonctions propres et multiplicité*, Prépublication de l'Institut Fourier n° 81, Grenoble , 1987.

[B-G-M] M. BERGER, P. GAUDUCHON et E. MAZET. — *Le spectre d'une variété riemannienne*, Lecture Notes, n° 194, Springer N.Y., 1971.

[Ble-Wil] D. BLEECKER, L. WILSON. — *Splitting the spectrum of a Riemannian manifold*, Siam J. Math. Analysis, **11** (1980), 813-818.

[Bur] M. BURGER. — Dégénérescence de surfaces de Riemann et petites valeurs propres,Preprint.

[Bur-Col] M. BURGER, B. COLBOIS, M. BURGER. — *A propos de la multiplicité de la première valeur propre d'une surface de Riemann*, C. R. Acad. Sci. Sér. I Math., **300** (1985), 247-250.

[Cha-Fel] I. CHAVEL, E. FELDMANN. — *Spectra of manifolds with small handles*, Comment. Math. Helv., **56** (1981), 83-102.

[Che] S.Y. CHENG. — *Eigenfunctions and nodal sets*, Commentarii Math. Helv., **51** (1976), 43-55.

[Col 1] B. COLBOIS. — *Sur la multiplicité de la première valeur propre non nulle du Laplacien des surfaces à courbure -1*, Thèse Université de Lausanne, 1987.

[Col-ColVer] B. COLBOIS, Y. COLIN DE VERDIÈRE. — Multiplicité de la première valeur propre positive du Laplacien d'une surface à courbure constante,To appear in Commentarii Math. Helv..

[ColVer 1] Y. COLIN DE VERDIÈRE. — *Sur la multiplicité de la première valeur propre non nulle du Laplacien*, Comm. Math. Helvetici, **61** (1986), 254-270.

[ColVer 2] Y. COLIN DE VERDIÈRE. — Sur une hypothèse de transversalité d'Arnold, To appear in Comm. Math. Helvetici.

[ColVer 3] Y. COLIN DE VERDIÈRE. — Construction de Laplaciens dont une partie du spectre est donnée,To appear in Annales Scient. E.N.S..

[ColVer 4] Y. COLIN DE VERDIÈRE. — *Sur un nouvel invariant des graphes et un critère de planarité*, Prépublication de l'Institut Fourier n° 71, Grenoble , 1987.

[ColVer-Mar] Y. COLIN DE VERDIÈRE, A. MARIN. — *Triangulations presque-équilatérales d'une surface*, Prépublication de l'Institut Fourier n° 88, Grenoble , 1987.

[Cou-Hil] R. COURANT, D. HILBERT. — *Methods of Mathematical Physics*, Wiley-Interscience I, 1953, II 1962.

[Do-Pi-Ra-Su] J. DODZIUK, T. PIGNATARO, D. SULLIVAN, B. RANDOL. — Estimating small eigenvalues of Riemann surfaces,preprint.

[Kac] M. KAC. — *Can one hear the shape of a drum?*, Amer. Math. Monthly, **73** (1966), 1-23.

[Kat] T. KATO. — *Perturbation theory for linear operators*, Grundlehren der mathematischen Wissenschaften 132, Springer-Verlag, 1973.

[Küh] W. KÜHNEL. — *Tight and 0-right polyhedral embeddings of surfaces*, im, **58** (1980), 161-177.

[Kui 1] N. KUIPER. — *Tight embeddings and maps, submanifolds of geometrical class three in E^n*, The Chern symposium, Proc. Int. Symp. Calif. Springer-Verlag, (1979), 97-145.

[Kui 2] N. KUIPER. — *Immersions with minimal total absolute curvature*, Coll. de Géométrie Diff. Bruxelles CBRM, (1958), 75-88.

[Mor] M. MORSE. — *The critical points of a function of n variables*, Transactions of the A.M.S., **33** (1931), 71-91.

[Oza] S. OZAWA. — These proceedings,.

[Pro] M.H. PROTTER. — *Can one hear the shape of a drum? revisited*, Siam Review, Vol. 29, n° 2, june, 1987.

[Rin] G. RINGEL. — *Map color Theorem*, Grundlehren der mathematischen Wissenschaften in Einzeldarstellungen Band 209, Springer-Verlag, 1974.

[Tor] N. TORKI. — Private communications,.

[T-S-G] SÉMINAIRE DE THÉORIE SPECTRALE ET GÉOMÉTRIE. — *Institut Fourier de Grenoble*,Année 1986-87.

[Uhl] K. UHLENBECK. — *Generic properties of eigenfunctions*, Amer. J. Math., **98** (1976), 1059-1078.

[Ura] H. URAKAWA. — *On the least eigenvalue of the Laplacian for compact group manifold*, J. Math. Soc. Japan, **31** (1982), 181-200.

[Whi] H. WHITNEY. — *Geometric Integration theory*, Princeton Math. Series 21, 1957.

– ◊ –

Institut Fourier
B.P.74
38402 ST MARTIN D'HÈRES Cedex
(France)

(2 décembre 1987)

RIEMANN SURFACES OF LARGE GENUS AND LARGE λ_1.

Peter Buser
Departement de Mathématiques
EPF - Lausanne
CH-1015 Lausanne, Switzerland

Marc Burger
Math. Institut der Universität
Rheinsprung 21
CH-4051 Basel, Switzerland

Jozef Dodziuk*
Department of Mathematics
Queens College of CUNY
Flushing, NY 11367, USA

0. Introduction.

Let $\Lambda_g = \sup \{\lambda_1(S) \mid S$ - a compact Riemann surface of genus $g\}$, where $\lambda_1(S)$ is the smallest positive element of the spectrum of the Riemann surface S. P. Buser [B1] posed a problem of determining whether the limit $L = \limsup_{g \to \infty} \Lambda_g$ is positive. He observed later [B2] that deep results of Selberg [S] and Jacquet-Langlands [JL] imply that this limit is positive and in fact is greater than or equal to 3/16. It follows easily from [C] that $L \leq 1/4$ and it is natural to conjecture that $L = 1/4$. In this paper we give a more concrete, geometric construction of Riemann surfaces of arbitrarily large genus and $\lambda_1 \geq c$, with c arbitrarily close to 3/16. As in [B2] we begin by considering the principal congruence subgroups Γ_N of $SL(2,Z)$ consisting of 2×2 matrices with integer entries congruent to the identity modulo N. If $N > 2$, Γ_N has no torsion and acts on the upper half-plane U freely so that the quotient $S_0 = U/\Gamma_N$ is a Riemann surface of finite area with cusps. The number of cusps ν_∞ and the genus g of S_0 are given by [Sh]

$$\nu_\infty = \frac{N^2}{2} \prod_{p|N} \frac{p^2 - 1}{p^2}, \qquad g = 1 + \frac{N^2(N-6)}{24} \prod_{p|N} \frac{p^2 - 1}{p^2}$$

where the product is taken over all primes p dividing N. For our purpose it is important to note that the number of cusps is always even and that the genus tends to infinity when N grows. It is known that for every Riemann surface of finite area, the intersection of the spectrum of the Laplace operator Δ with $[0,1/4)$ consists of finitely many eigenvalues of finite multiplicity. 1/4 is always in the continuous spectrum. Selberg [S] showed that for the surfaces S_0 the smallest positive element of the spectrum $\lambda_1(S_0)$ satisfies $\lambda_1(S_0) \geq 3/16$. He conjectured that $\lambda_1(S_0) = 1/4$. Given a fixed surface S_0 as above, we will show how to construct compact surfaces S_t "approximating" S_0 as t tends to zero so that $\limsup_{t \to 0} \lambda_1(S_t) \geq \lambda_1(S_0) \geq 3/16$. The idea for constructing the surfaces S_t is as follows. Consider a maximal set of simple, closed disjoint geodesics on S_0. Cutting along these geodesics we obtain a decomposition of S_0 into three-holed spheres. The boundary components of the three-holed spheres are either geodesics or punctures. Replace every three-holed sphere which contains one or two punctures with a three-holed sphere with geodesic boundaries of lengths defined as follows. If the boundary component was a geodesic we keep its length unchanged. If it was a puncture give it length $t > 0$. We use the same length t for all punctures and treat the number t as a parameter. Now reassemble the pieces using old identifications of the boundary components for the components which came from the geodesics of S_0. We obtain a surface with boundary consisting of an even number of geodesics of length t. Group these geodesics in pairs and identify each pair to form a compact surface S_t. Note that the new surface S_t has genus larger than the genus of S_0. We remark that our strategy has a chance of succeeding, which can be seen as follows. The surfaces S_t come equipped with a maximal set of disjoint, simple, closed geodesics. Some of these geodesics have lengths independent of t, while others have lengths equal to t. However, the number $L_1(t)$ equal to the minimum of the sum of lengths of geodesics belonging to our family and forming a chain separating S_t is bounded away from zero by a constant which depends only on the choice of the dissection of S_0. By the theorem of [SWY] (cf. also [DR], [DPRS]) $\lambda_1(S_t)$ is bounded from below by a positive constant independent of t. We remark that in the actual construction we shall choose the parameter t in a different but equivalent way.

The paper is organized as follows. Section 1 contains a detailed construction of the surfaces S_t. In Section 2 we show that $\limsup_{t \to 0} \lambda_1(S_t) \geq \lambda_1(S_0)$ whenever surfaces S_t are obtained as above from a given surface S_0 with an even number of cusps.

1. Description of surfaces.

In this section we give a more precise description of the geometry of a family of compact surfaces S_t, $t>0$, which approximate a given surface with cusps S_0 when $t \longrightarrow 0$. The surface S_0 has an even number of cusps and large genus, but to simplify the notation, we will consider the case of two cusps. Let γ' be a simple closed curve enclosing the two cusps, i. e. such that one of the components of $S_0 - \gamma'$ is a twice punctured disk. The free homotopy class of γ' contains a unique simple closed geodesic γ, and by [E], one of the components of $S_0 - \gamma$ is a twice punctured disk. Denote the closure this component G_0. We observe that G_0 can be dissected into four congruent quadrilaterals one of which is drawn in Figure 1.

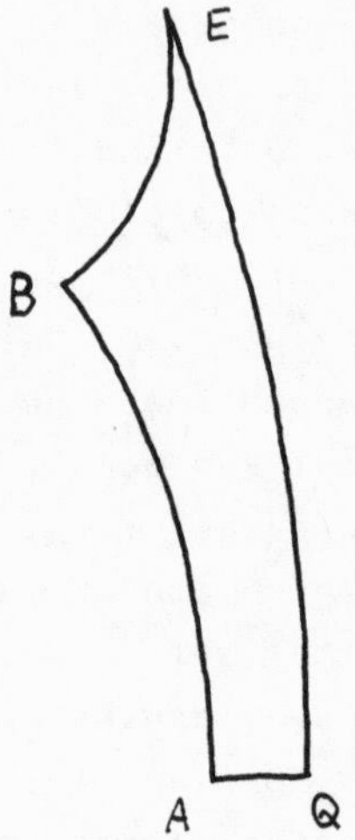

Figure 1

These quadrilaterals have angles $\pi/2$, $\pi/2$, $\pi/2$, and 0, and are uniquely determined by the length of the side AQ (cf. [Be], p. 156). To see that such a dissection exist, observe that G_0 is conformally equivalent to the unit disk in the complex plane punctured at -1/2 and 1/2. The four components of the complement of the real and imaginary axes are clearly conformally (or anti-conformally) equivalent, hence isometric with respect to the metric of G_0. The dissection of G_0 is accomplished by cutting the unit disk along the interval [-1/2,1/2] of the real axis and along the interval [-i,0] of the imaginary axis. Unrolling the resulting simply connected figure in the hyperbolic plane we obtain the following heptagon AA′B′E′B″EB. The figure also shows the four congruent quadrilaterals. G_0 is obtained by identifying AB with A′B′, BE with B″E, and B′E′ with B″E′.

Our next task is to "thicken" G_0 (identifying E and E′ we may interpret G_0 as an infinite handle) to form a finite handle. The parameter in this construction, a small real number $t > 0$, is different from the parameter in the introduction.

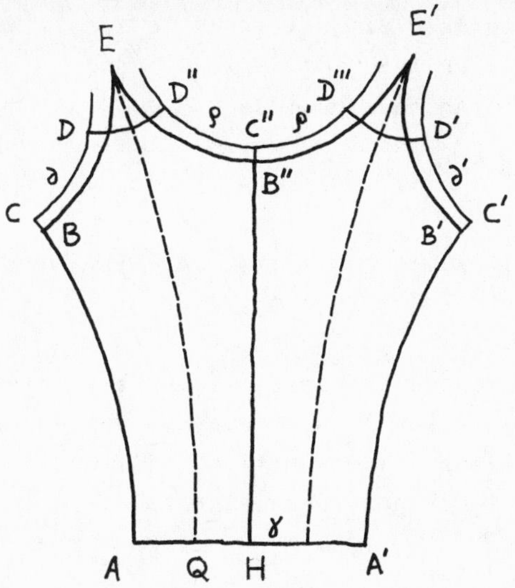

Figure 2

Let H be the midpoint of the segment AA′. Continue the geodesic segments AB, A′B′, and HB″ for distance t past B, B′, and B″ respectively. From the endpoints C, C′, C″ draw perpendiculars ∂, ∂', ρ, ρ' as shown in Figure 2. It follows from elementary hyperbolic geometry and continuity that for small $t > 0$ the common perpendicular DD″ of geodesics determined by ∂ and ρ has length approaching 0. Moreover, by symmetry the lengths of DD″ and D′D‴ are equal. The handle G_t is obtained now by identifying AC with A′C′, CD with C″D″, C′D′ with C″D‴, and finally DD′ with D″D‴ (see Figure 2). The image of DD″ in G_t will be the disappearing geodesic κ_t, i.e. the length of κ_t tends to zero as t approaches zero. Note that the handle G_t is isometric to G_0 near the geodesic γ, so that G_t can be attached to $S_0 - G_0$ along γ by the same identification as G_0. The resulting surface will be denoted by S_t. As a consequence of the construction, we see that the handles G_t are thicker than the cusps. More precisely, we have the following lemma.

<u>Lemma 1.1</u>. For every $R > 0$ there exist $T > 0$ and $\iota > 0$ so that $\mathrm{inj}(x) \geqslant \iota$ whenever $x \in S_t$, $0 \leqslant t \leqslant T$, and $\mathrm{dist}(x, S_0 - G_0) \leqslant R$.

2. Eigenvalues and eigenfunctions.

In this section we investigate the behavior of the first positive eigenvalue of the surface S_t as t tends to zero. We introduce the following notation. For $t > 0$, $\lambda_t = \lambda_1(S_t)$ and φ_t is a normalized eigenfunction belonging to λ_t. Thus

$$\Delta\varphi_t + \lambda_t \varphi_t = 0,$$

$$\int_{S_t} \varphi_t \, dA = 0, \qquad \int_{S_t} \varphi_t^2 \, dA = 1.$$

For $t = 0$, λ_0 will denote the smallest positive element of the spectrum of S_0. If $\lambda_0 < 1/4$, then it is an eigenvalue and we write φ_0 for the corresponding normalized eigenfunction (cf. [DPRS], p. 106).
If $\lambda_0 = 1/4$, then it may not be an eigenvalue.

<u>Theorem 2.1</u>. $\limsup_{t\to 0} \lambda_t \geq \lambda_0$.

<u>Proof</u>: By the theorem of [SWY] (see also [DR] and [DPRS]) $\liminf_{t\to 0} \lambda_t > 0$. Therefore $\lambda = \limsup_{t\to 0} \lambda_t$ is positive. By construction the diameter of S_t tends to infinity as t approaches 0. Hence, by a theorem of Cheng [C], $\lambda \leq 1/4$. Let $F = S - G_0 = S_t - G_t$. Fix a point $x_0 \in F$. For $t \geq 0$ let $p_t : U \to S_t$ be the universal covering map of the upper half-plane U onto S_t normalized so that

a) $p_t(z_0) = x_0$ for a fixed point $z_0 \in U$ and all $t \geq 0$.

b) $p_t(z) = p_{t'}(z)$ for all t, t' and all z in a neighborhood of z_0.

We define $\psi_t = p_t \circ \varphi_t$. By the inequalities of Sobolev and Gårding [BJS] for every integer $k \geq 0$

$$|\nabla^k \psi_t(x)| \leq c(r,k) \sum_{\ell=0}^{N(k)} |\Delta^\ell \psi_t|_{L^2(B_r(x))},$$

Where the constant depends only on the radius r and the number k of derivatives. It follows from Lemma 1.1, that for x in a compact set there exists $r > 0$, so that the covering maps p_t are injective on the balls $B_r(x)$. Therefore, the inequality above implies that

$$|\nabla^K \psi_t(x)| \leqslant c(r,K) \sum_{\ell=0}^{N(K)} \lambda_t^{\ell} .$$

It follows that the derivatives of all orders of the functions ψ_t are bounded uniformly in t on compact subsets of the upper half-plane. Thus we can choose a subsequence $\psi_{t(i)}$, $t(i) \longrightarrow 0$ as $i \longrightarrow \infty$ which converges to a limiting function ψ on U with derivatives of all orders uniformly on compacta. Furthermore, we can arrange that ψ satisfies $\Delta\psi + \lambda\psi = 0$. From now on we shall write only t as a subscript when we mean t(i). Observe that there is nothing to prove if $\lambda = 1/4$. Thus we only consider the case $\lambda < 1/4$. To prove the theorem, it will suffice to show that $\psi = \varphi \circ p_0$ for a smooth function φ in $L^2(S_0)$ which does not vanish identically. Indeed such a function φ would be an eigenfunction belonging to the eigenvalue $\lambda > 0$ so that $\lambda \geqslant \lambda_0$.

We first show that ψ is the pull-back of a function defined on S_0. Choose a fundamental domain K for the action of $\pi_1(S_0)$ with the property that the inverse image $\tilde{\gamma}$ of the geodesic γ disconnects K in such a way that one of the components is congruent to the heptagon in Figure 2. Call this component $\tilde{G}_0$. Let $\tilde{F}$ be the other component of $K - \tilde{\gamma}$. It is clear that the functions φ_t converge uniformly with all derivatives to a limiting function on F. Consider two boundary points of $\tilde{G}_0$, say x and x′, which lie over the same point in G_0. If x lies on AB and x′ lies on A′B′ (see Figure 3) then it is clear that

$$(2.1) \qquad \psi_t(x) = \psi_t(x'), \qquad \frac{\partial \psi_t}{\partial n}(x) = -\frac{\partial \psi_t}{\partial n}(x').$$

Therefore the same equalities hold for the function ψ. We can show that the same is true if x, x′ lie on BE, B″E respectively as follows. Let s be the distance between x and B (which is the same as the distance from x′ to B″). If $\tilde{G}_t$ is the lift of G_t containing $\tilde{G}_0$ and congruent to the octagon ACDD″D‴D′C′A′ we define x_t and x_t' as follows. x_t is chosen as the point on CD at distance s from C, x_t' is on C″D″ at distance s from C″. This is shown in Figure 3.
Clearly,

$$\psi_t(x_t) = \psi_t(x_t'), \qquad \frac{\partial \psi_t}{\partial n}(x_t) = -\frac{\partial \psi_t}{\partial n}(x_t').$$

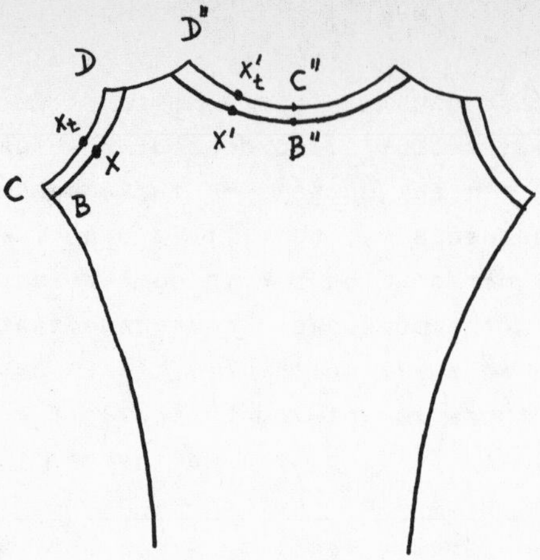

Figure 3

As t approaches zero x_t and x_t' converge to x and x′ respectively. It follows that (2.1) holds in this situation as well. Thus $\psi|K$ is the pullback of a function φ on S_0. A straightforward application of Green's formula and (2.1) shows that

$$\int (\Delta u + \lambda u)\varphi dA = 0$$

for every compactly supported smooth function u. By elliptic regularity [BJS] φ is real analytic on S_0. Similarly ψ is analytic on U. Since ψ and $\varphi \circ p_0$ agree on an open set in U they are equal everywhere.

We show next that φ is square-integrable. Let $S_0(r)$ be the subset of S_0 consisting of points whose distance from $F = S_0 - G_0$ is less than or equal to r. Define $S_t(r) \subset S_t$ in a similar way. If we exclude sets of measure 0, $S_0(r)$ is contained in $S_t(r)$ for small $t > 0$ in the sense that the corresponding $G_0(r)$ is contained in $G_t(r)$ and $S_t - G_t = S_0 - G_0 = F$. Therefore

$$\int_{S_0(r)} \varphi^2 dA \leqslant \lim_{t \to 0} \int_{S_t(r)} \varphi_t^2 \, dA \leqslant 1.$$

Since this is true for arbitrarily large r, $\varphi \in L^2(S_0)$.

To conclude the proof we show that φ is not identically zero. Roughly speaking, the only way for the limiting function φ to vanish is to have the functions φ_t concentrating in the degenerating handle. This is shown to be impossible unless $\lambda = 1/4$. Thus assume that $\varphi \equiv 0$. Then the functions φ_t and their gradients converge to zero uniformly on compact subsets of U. Again consider the configuration in Figure 2. Choose a point P on BE. This choice is independent of t. Let L_1 be the locus through P of points equidistant from the geodesic segment DD″, and let L_2 be the equidistant locus unit distance closer to DD″. Consider also the symmetric pair L_1', L_2' of curves equidistant from D‴D′ (see Figure 4). When t approaches 0 these equidistant loci converge to horocycles centered at E (cf. [Be], p. 163). Let M_t be the intersection of the bands between the two pairs of equidistant loci with $\tilde{G}_t$.

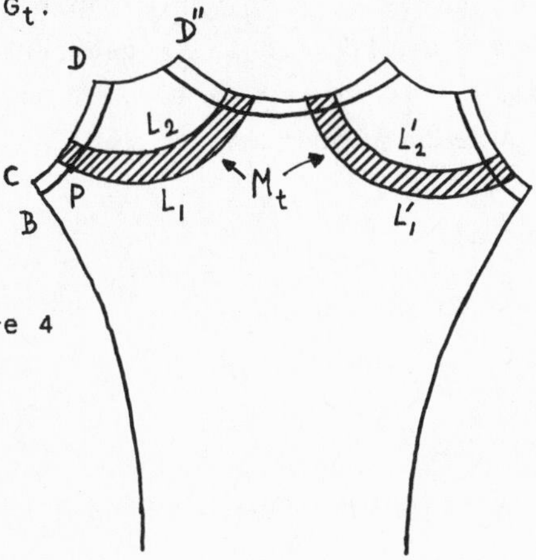

Figure 4

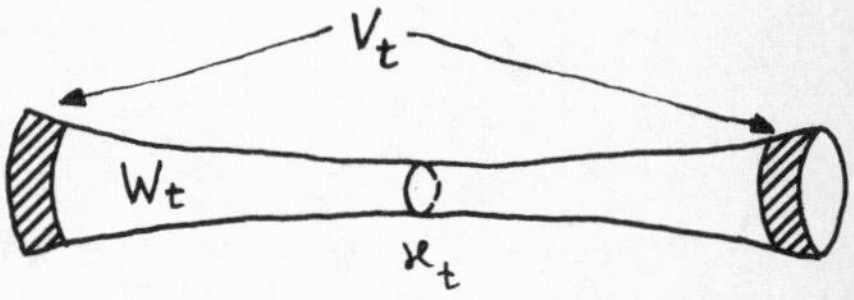

The union of sets M_t for $t \leq t_0$ is compact. It follows that φ_t and $\nabla\varphi_t$ are uniformly small on $V_t = p_t(M_t)$ and in fact on the complement of the tubular neighborhood W_t of the geodesic κ_t between the images of L_1 and L_1'. Since the area of S_t is constant, it follows that

$$\int_{W_t} \varphi_t^2 dA \approx 1$$

$$\int_{V_t} (\varphi_t^2 + |\nabla\varphi_t|^2) dA \approx 0,$$

for small t. By the argument of Lemma 3.3 of [DPRS] the energy of φ_t is bounded from below by a constant arbitrarily close to 1/4. It follows that $\lambda = \limsup \lambda_t = \limsup \int |\nabla\varphi_t|^2 dA = 1/4$. This contradicts our assumption that $\lambda < 1/4$ and concludes the proof.

<u>Remarks</u>. (a) If the surface S_0 has an even number of cusps we can group them in pairs and thicken every pair into a handle as described in Section 1 using the same parameter t for every pair of cusps. This

yields a family of surfaces S_t. Theorem 2.1 remains true with an almost identical proof.

(b) The following example suggests that the possibility that the eigenfunctions φ_t converge to zero cannot be ruled out. Let S_0 be the sphere with three punctures. Consider the family of surfaces S_t, where S_t is a three-holed sphere and each boundary component is a geodesic of length t. Let λ_t be the first positive eigenvalue for the Neumann problem on S_t, and let φ_t be a normalized eigenfunction belonging to λ_t. It is easy to see that $\lim_{t\to 0} \lambda_t = 1/4$. On the other hand, the spectrum of S_0 is contained in $[1/4,\infty)$ and 1/4 is <u>not</u> an eigenvalue. The argument in the proof of Theorem 2.1 implies that a subsequence of $\{\varphi_t\}$ converges to a limiting function. This limit has to be zero, for otherwise the limiting function would be an eigenfunction for S_0 with eigenvalue 1/4.

References

[B1] P. Buser, Cubic graphs and the first eigenvalue of a Riemann surface, Math. Z. 162 (1978), 87-99.

[B2] P. Buser, On the bipartition of graphs, Discrete Applied Mathematics, 9 (1984), 105-109.

[BJS] L. Bers, F. John, M. Schechter, <u>Partial Differential Equations</u>, AMS, Providence, RI 1979.

[Be] A.F. Beardon, <u>The Geometry of Discrete Groups</u>, Springer-Verlag, New York 1983.

[C] S. Y. Cheng, Eigenvalue comparison theorems and its geometric applications, Math. Z. 143 (1975), 289-297.

[E] D.B.A. Epstein, Curves on 2-manifolds and isotopies, Acta Math. 115 (1966), 83-107,

[JL] H. Jacquet, R. Langlands, <u>Automorphic Forms on GL(2)</u>, Lect. Notes in Math. 114, Springer-Verlag, Berlin 1970.

[DR] J. Dodziuk, B. Randol, Lower bounds for λ_1 on a finite volume hyperbolic manifold, J. Differential Geometry, 24 (1986), 133-139.

[DPRS] J. Dodziuk, T. Pignataro, B. Randol, D. Sullivan, Estimating small eigenvalues of Riemann surfaces, Contemporary Mathematics 64, (1987), 93-121.

[S] A. Selberg, On the estimation of Fourier coefficients of modular forms, in Proceedings of Symposia in Pure Mathematics, vol.8, AMS, Providence, RI 1965.

[Sh] G. Shimura, Introduction to the Arithmetic Theory of Automorphic Functions, Princeton University Press, Princeton NJ, 1971

[SWY] R. Schoen, S. Wolpert, S.-T. Yau, Geometric bounds on low eigenvalues of a compact surface, in Geometry of Laplace Operator, AMS (1980), 279-285.

Acknowledgement. Jozef Dodziuk is grateful to I. Kra and B. Maskit for helpful discussions.

*The research of Jozef Dodziuk was supported in part by the NSF Grant DMS-8500939 and by a grant from The City University of New York PSC-CUNY Research Award Program.

Cayley Graphs and Planar Isospectral Domains

Peter Buser
Département de Mathématiques
Ecole Polytechnique Fédérale de Lausanne
CH-1015 Lausanne, Switzerland

1. Introduction. In his paper [5] Sunada showed how certain finite groups lead to isospectral manifolds in a fairly simple way. These manifolds are cutting and pasting copies of each other and the isospectrality is due to the combinatorial structure of the finite group. Sunada's Theorem is as follows. Let G be a finite group which acts freely on a compact Riemannian manifold by isometries. If H_1, $H_2 \subseteq G$ are subgroups satisfying a certain condition then the two quotients $M_1 = H_1 \backslash M$ and $M_2 = H_2 \backslash M$ are isospectral (Thm.1 in [5]). The required property is that for any conjugacy class [g], $g \in G$, we have

$$\#([g] \cap H_1) = \#([g] \cap H_2)$$

where H denotes cardinality. A possible way to obtain such actions of G is to start with a compact Riemannian manifold (M_0, g_0) for which there exists a surjective group homomorphism $\pi_1 : (M_0) \rightarrow G$ and then to construct the corresponding finite covering manifold (M,g). We shall henceforth say that M, M_1, M_2, M_0 "are from diagram (*)" whenever they are obtained by this construction.

(*)
$$\begin{array}{ccc} & M & \\ \swarrow & & \searrow \\ M_1 & & M_2 \\ \searrow & & \swarrow \\ & M_0 & \end{array}$$

The aim of this paper is to visualize the combinatorial structure of Sunada's examples by looking at the so called *Cayley graph* of G. It seems that at least in some cases the examples become more transparent from this point of view. The paper is organized as follows. Section 2 gives a short introduction to Cayley graphs. In section 3 we consider the examples $\mathbf{Z}^*_8 \ltimes \mathbf{Z}_8$ and SL(3,2). They both give rise to graphs which can be drawn in a figure. In section 4 we obtain as a first application.

Theorem 1. *There exists a pair of flat metrics on the 9-holed sphere which are isospectral for Dirichlet boundary conditions but not isometric.*

The Theorem remains true if we replace "Dirichlet" by "Neumann". The possibility of having isospectrality on planar domains came out by looking at the graph structure of Brooks' examples of genus 3 in [1]. The domains can, however, not be embedded isometrically in $\mathbb{R}^2$. In section 5 we reprove Sunada's Theorem in a particular case by means of an algorithm for the length spectrum. Here the interplay between isospectrality and combinatorics comes out most clearly. In section 6 we go back to diagram (*) and ask for conditions for the metrics on M_0 such that we get *non-isometric* coverings M_1, M_2. Sunada showed that this holds for a generic metric on M_0 if the dimension is two or if M is simply connected. We shall see that we do not need any topological restriction for this fact.

Theorem 2. *The set of Riemannian metrics on* M_0 *for which the isospectral coverings* M_1, M_2 *are non-isometric comprises a residual set in the space of all Riemannian metrics on* M_0.

The idea is to reconstruct the graph structure which lead to the construction of M_1, M_2 out of the intrinsic geometry of M_1 and M_2.

2. Cayley graphs. In this section we give a very short introduction to the Cayley graph of a finite group. For more information we refer the reader to the book [3] in which he also will find numerous interesting examples.

A graph $\mathfrak{G}$ consists of a set of *vertices* and a set of *edges* together with a rule which associates to each edge of the graph a pair of vertices. If vertices P,Q belong to edge e we shall say that e *connects* P and Q. Graphs can be visualized by drawing figures. We shall represent vertices by thick points and edges by curves which connect these points. Examples of graphs are given in figures 1,2,3,4,8. To describe the construction of a graph it is useful to look at each edge as the union of two half edges. If e connects P,Q we shall say that one part or half edge of e emanates from P and one part from Q. To describe a graph we then may give a set of vertices, each vertex having a number of emanating half edges and then group these half edges into pairs. This way of constructing graphs is the analog of pasting together manifolds out of building blocks. The construction is illustrated in a simple example in fig. 1.

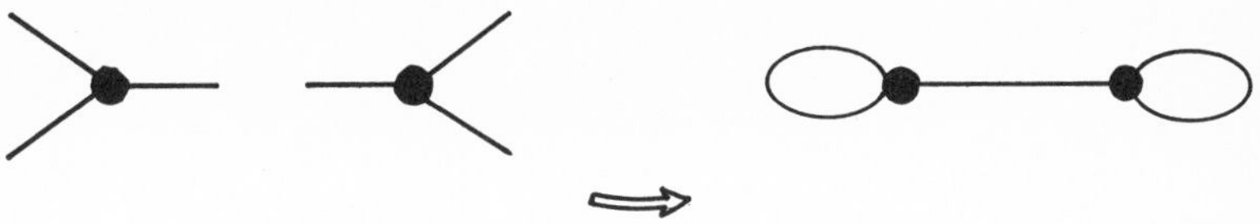

Fig. 1

Graphs may be given additional structures by adding attributes .In this paper the attributes will be *color* and *orientation* of the edges. An orientation of an edge e which connects P and Q may be given by saying that e goes *from* P *to* Q or *from* Q *to* P. In figures, the orientation will be indicated by an arrow. A graph is called *oriented* if some of its edges (or all) are oriented. Color is obtained by grouping edges into equivalence classes. Edges in the same class will be said "to have the same color or "to "be of the same type".

When are two graphs $\mathfrak{G}, \mathfrak{G}'$ "the same"? We shall say that $\phi : \mathfrak{G} \to \mathfrak{G}'$ is an *isomorphism* if ϕ is a one-to-one mapping of the vertex set of $\mathfrak{G}$ onto the vertex set of $\mathfrak{G}'$ such that whenever P and Q are connected by k edges then $\phi(P)$ and $\phi(Q)$ are connected by k edges and vice versa. If $\mathfrak{G}$ and $\mathfrak{G}'$ have color and orientation, an isomorphism ϕ is not requested to preserve this additional structure. But if it does, we shall say that ϕ is a *strong isomorphism*. Hence, for a strong isomorphism the number of edges of a given type from P to Q is always equal to the number of edges of the same type from $\phi(P)$ to $\phi(Q)$, and analog properties hold for the cases where only part of the edges are colored and/or part of the edges are oriented. Graphs $\mathfrak{G}, \mathfrak{G}'$ are *isomorphic* resp. *strongly isomorphic* if there exists an isomorphism resp. a strong isomorphism from $\mathfrak{G}$ to $\mathfrak{G}'$.

Now let G be a finite group with elements g,h,A,B etc. and let $A_1,\ldots,A_n$ be a list of pairwise different generators of G. The list need not be minimal. We define the *Cayley graph* $\mathfrak{G} = \mathfrak{G}[A_1,\ldots,A_n]$ of G with respect to the generators $A_1,\ldots,A_n$ as follows. Every $g \in G$ is a vertex with emanating half edges $a_1,\bar{a}_1, \ldots a_n,\bar{a}_n$. Half edge $\bar{a}_i$ of vertex g and half edge a_i of vertex g' form together an edge e if and only if $gA_i = g'$. The orientation of e is from g to gA_i, and e is said to be of type A_i. Depending on what we need we shall consider $\mathfrak{G}$ as colored and oriented graph or just as ordinary graph.

Example. Let $G = \mathbb{Z}_4 \times \mathbb{Z}_2$ with generators $A = (1,0)$, $B = (0,1)$. The Cayley graph $\mathfrak{G}[A,B]$ is shown in fig. 2. The horizontal edges are of type A, the vertical edges are of type B

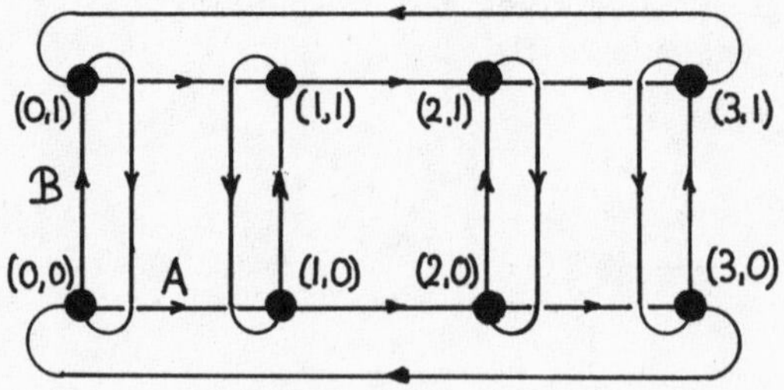

Fig. 2

Each $h \in G$ acts on the Cayley graph $\mathfrak{G}$ by left multiplication : If $g' = gA_i$ then $hg' = hgA_i$ and vice-versa. It follows that $G : \mathfrak{G} \to \mathfrak{G}$ acts by strong isomorphisms and, in fact, G is the full group of strong isomorphisms of $\mathfrak{G}$. As in the example of fig. 2, there may be additional non-strong isomorphisms.

For each subgroup $H \subseteq G$ we define the quotient graph $H\backslash\mathfrak{G}$ as follows. Each coset Hg is a vertex with emanating half edges $a_1, \bar{a}_1, \ldots a_n, \bar{a}_n$. Half edge $\bar{a}_i$ of vertex Hg and half edge a_i of vertex Hg' form together an edge of $H\backslash\mathfrak{G}$ if and only if $g' = hgA_i$ for same $h \in H$. The orientation is from Hg to HgA_i and the edge is said to be of type A_i.

Example. We consider again the graph of fig. 2. Let $H_1 = \{(0,0),(2,0)\}$ and $H_2 = \{(0,0),(0,1)\}$. The quotients $H_1\backslash\mathfrak{G}$ and $H_2\backslash\mathfrak{G}$ are given by fig. 3. Color and orientation are not indicated in this figure.

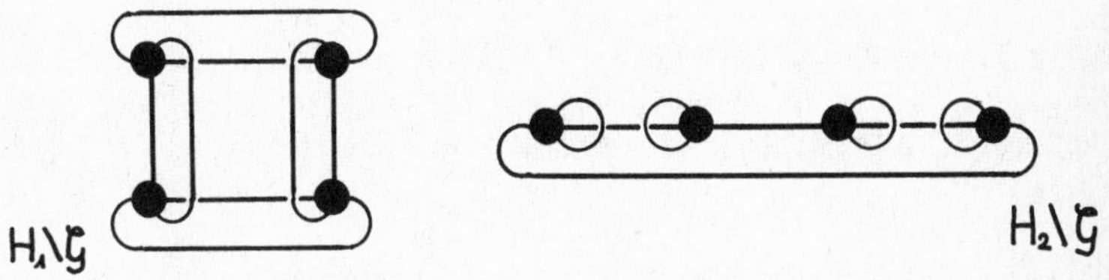

Fig. 3

3. The groups $\mathbb{Z}^*_8 \ltimes \mathbb{Z}_8$ and SL(3,2). We discuss the two examples of finite groups which yield particularly simple quotient graphs. The first example is the semi direct product

$$\mathbb{Z}^*_8 \ltimes \mathbb{Z}_8 = \{(x,y) \mid x = 1,3,5,7; \; y = 0,1,2,\ldots,7\}$$

with the product structure

$$(x,y) \bullet (x',y') = (xx', xy' + y).$$

The two subgroups are

$$H_1 = \{(1,0),(3,0),(5,0),(7,0)\}, \quad H_2 = \{(1,0),(3,4),(5,4),(7,0)\}.$$

It is easy to check that H_1, H_2 are not conjugate and that $\#([g] \cap H_1) = \#([g] \cap H_2)$ for all $g \in G = \mathbb{Z}^*_8 \ltimes \mathbb{Z}_8$. The following are generators,

$$A = (3,0), \quad B = (5,0), \quad C = (1,1).$$

We observe that $C^k = (1,k)$, $k = 0,\dots,7$ are pairwise non equivalent mod H_1 resp. mod H_2. This allows to represent the cosets in the form $H_\upsilon C^0,\dots,H_\upsilon C^7$, $(\upsilon=1,2)$. For each k there are unique exponents $\alpha_\upsilon(k)$, $\beta_\upsilon(k)$ such that

$$C^kA \in H_\upsilon\, C^{\alpha_\upsilon(k)},\ C^kB = H_\upsilon\, C^{\beta_\upsilon(k)},\ \upsilon=1,2,\ \ k = 0,\dots,7.$$

The exponents are given by the following table

Table 3

k	0	1	2	3	4	5	6	7
$\alpha_1(k)$	0	5	2	7	4	1	6	3
$\beta_1(k)$	0	3	6	1	4	7	2	5
$\alpha_2(k)$	4	1	6	3	0	5	2	7
$\beta_2(k)$	4	7	2	5	0	3	6	1

The corresponding quotients $H_1\backslash\mathfrak{G}$ and $H_2\backslash\mathfrak{G}$ for $\mathfrak{G} = \mathfrak{G}[A,B,C]$ are shown in fig. 8. The horizontal edges are of type C the dotted lines represent the edges of type B and the remaining edges are of type A. Observe that the two graphs are not isomorphic.

The second example is the group SL(3,2) of non singular 3×3-matrices with coefficients from $\mathbb{Z}_2$. Brooks used this group in [1] to construct isospectral surfaces of genus 3, 4 and 6 (and others). SL(3,2) has a number of interesting properties. It is e.g. the unique simple group with cardinality 168 and therefore isomorphic to PSL(2,7). For a list of other properties we refer to [6] pp 142-147. The following subgroups are non-conjugate and satisfy $\#([g]\cap H_1) = \#([g]\cap H_2)$ for all $g \in$ SL(3,2) :

$$H_1 = \text{subgroup of all} \begin{pmatrix} 1 & * & * \\ 0 & * & * \\ 0 & * & * \end{pmatrix}, \quad H_2 = \text{subgroup of all} \begin{pmatrix} 1 & 0 & 0 \\ * & * & * \\ * & * & * \end{pmatrix}$$

For a simple way of checking we refer to Brooks [l], (cf also Perlis [4]). The following are easily seen to be generators.

$$A = \begin{pmatrix} 0 & 1 & 1 \\ 0 & 1 & 0 \\ 1 & 0 & 0 \end{pmatrix} \qquad B = \begin{pmatrix} 1 & 0 & 0 \\ 0 & 0 & 1 \\ 0 & 1 & 1 \end{pmatrix}$$

The commutator $C = ABA^{-1}B^{-1}$ has order 7. Again we write the cosets in the form $H_\upsilon\, C^k$, $k = 0,\dots,6$ although C itself is not listed as generator. Note that $C^k \notin H_\upsilon$ for $\upsilon = 1,2$ unless $k \equiv 0 \pmod 7$, so that the cosets $H_\upsilon\, C^k$ are indeed pairwise different.

For each k there are unique exponents $\alpha_\upsilon(k)$, $\beta_\upsilon(k)$ such that $C^kA \in H_\upsilon\, C^{\alpha_\upsilon(k)}$, $C^kB \in H_\upsilon\, C^{\beta_\upsilon(k)}$, $\upsilon = 1,2;\ k = 0,\ldots,6$.

The exponents are given by the following table

Table 4

k	0	1	2	3	4	5	6
$\alpha_1(k)$	3	1	6	0	5	2	4
$\beta_1(k)$	0	2	5	6	3	1	4
$\alpha_2(k)$	1	6	5	0	4	2	3
$\beta_2(k)$	0	4	5	1	3	6	2

The graphs $H_1\backslash\mathfrak{G}$, $H_2\backslash\mathfrak{G}$ are given in fig. 4. For the convenience of the reader the vertices are labelled after the exponent k in $H_\upsilon C^k$. Dotted lines represent edges of type B, the remaining ones are of type A. Check that the figure coincides with the above table; e.g. an edge of type B goes from vertex 4 in $H_1\backslash\mathfrak{G}$ to vertex 3 because $\beta_1(4) = 3$, an edge of type A goes from vertex 6 to vertex 3 in $H_2\backslash\mathfrak{G}$ because $\alpha_2(6) = 3$ etc. Observe again that the two groups are not isomorphic.

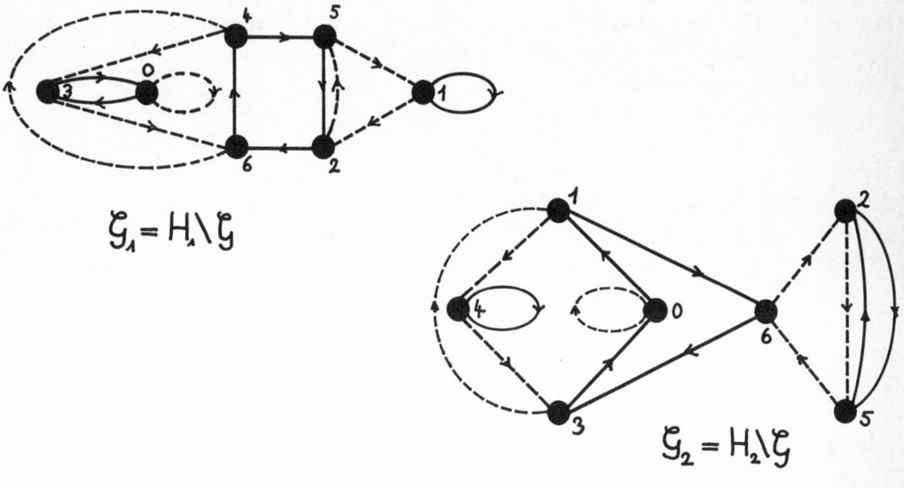

Fig. 4

4. Planar domains. In this section we build two planar isospectral domains out of the two graphs of fig. 4 in order to prove Theorem 1. The building block is a domain $D \subseteq \mathbb{R}^2$ with four rectangular ends as shown in Fig. 5. The sides $\bar{a}$, a, $\bar{b}$, b are named after the above generators A, B of SL(3,2) and have the same length. We may realize D by a paper model.

Gluing together sides a, ā and sides b, b̄ we obtain a smooth bordered surface S_o with smooth boundary. S_o carries a flat metric and is topologically a 3-holed sphere or 2-holed disc. We remark that gluing together a, b and ā, b̄ instead, we would get a one-holed torus.

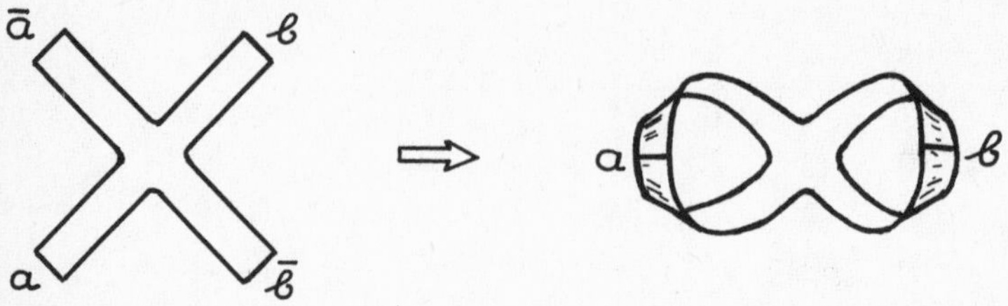

Fig. 5

We next construct the 168-sheeted covering surface S of S_o (for the discouraged reader: we shall never carry this out explicitly) on which $G = SL(3,2)$ acts by isometries. This may be obtained by replacing each vertex of the Cayley graph $\mathfrak{G} = \mathfrak{G}[A,B]$ with its emanating half edges a, ā, b, b̄ by a copy of D with sides ā, a, b̄, b. Then glue the copies together in just the same way as the vertices come together in $\mathfrak{G}$. The surface S obtained in this way is a geometric model of $\mathfrak{G}$ and G acts on S by isometries in the same way as it acts on $\mathfrak{G}$ by strong isomorphisms. Now let

$$S_1 = H_1\backslash S \quad S_2 = H_2\backslash S.$$

Since the action of G on S is compatible with the action on $\mathfrak{G}$, S_1 and S_2 can also be obtained by gluing together copies of D with respect to the graphs $H_1\backslash\mathfrak{G}$, $H_2\backslash\mathfrak{G}$ in fig. 4. By Sunada's Theorem, S_1 and S_2 are isospectral, e.g. for Dirichlet boundary conditions or for Neumann boundary conditions. (The theorem in [5] is not stated for bordered manifolds but the proof runs through without modifications). In order to prove that the topology of S_1 and S_2 is that of a 9-holed sphere or 8-holed disc we look again at fig. 4. First of all the two graphs are planar, i.e. they can be drawn in the plane without crossing edges. In addition (since planarity alone is not sufficient) the half edges are drawn in the same order at each vertex, namely, if we go around a vertex, say in the positive sense, we have always an incoming edge of type A followed by an outgoing edge of type B. This "superplanarity" of the graphs $\mathfrak{G}_1, \mathfrak{G}_2$ is sufficient to prove that S_1 and S_2 are planar surfaces. Finally, it is easy to read the graphs $\mathfrak{G}_1$, $\mathfrak{G}_2$ by looking at S_1, S_2. This proves that S_1 and S_2 are non-isometric.

5. An algorithm for the length spectrum. Instead of piecing together domains of the type used in the preceding section we may use a surface with closed boundary components as builing block, like for instance the one in figure 6 which is a 6-holed sphere with boundary components a, ā, b, b̄, c, c̄ named after the generators A,B,C, of the group $G = \mathbf{Z}*_8 \ltimes \mathbf{Z}_8$ of section 3.

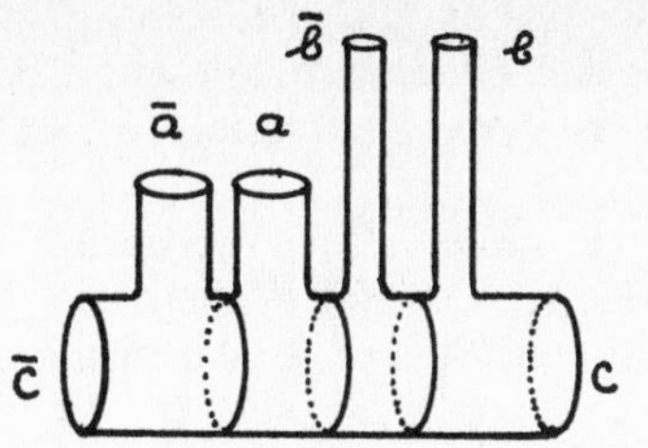

Fig. 6

In order to define the pasting correctly we assume that the boundary components a, ā, etc. are parametrized with constant speed in the form $t \mapsto a(t)$, $t \in [0,1]$, $a(0) = a(1)$ etc. The geometry of the building block B must be such that with the identifications

$$a(t) = \bar{a}(1-t), \quad b(t) = \bar{b}(1-t), \quad c(t) = \bar{c}(1-t), \quad t \in [0,1]$$

we get a smooth closed surface S_0 of genus 3. This may be obtained e.g. by starting with a surface of genus 3 and cutting it open along 3 simple closed non interesting geodesics from three different homotopy classes. In order to be able to reconstruct the building block we require that in addition *every closed geodesic* η *on* S_0 *which is different from the geodesics* a,b,c *is longer than* a,b,c. This too is not difficult to obtain. We point out that the 3-holed sphere construction of Riemann surfaces easily yields such domains which have constant negative curvature.

Now take 8 copies of B and glue them together with respect to the graph $\mathcal{G}_1$ resp. $\mathcal{G}_2$ of fig. 8 below. Each vertex is a copy of B. The horizontal half edges at each vertex correspond to the boundary components c, c̄ the half edges emanating on the upper side correspond to a, ā the half edges emanating from the lower side (dotted lines) correspond to b, b̄. Locally, the gluing is the same as for S_0, i.e. if e.g. boundary geodesic a of block B' is glued to ā of block B" then the identification is, as above, $a(t) = \bar{a}(1-t)$, $t \in [0,1]$ etc. Since S_0 is smooth, the resulting surfaces S_1, S_2 are smooth. The same reasoning as in section 4 proves that S_1, S_2 are non isometric: Since S_0 satisfies the above requirements for the lengths of closed geodescis, it suffices to cut open S_1 resp. S_2 along all simple closed geodesics of length $\leq \max\{\ell(a), \ell(b), \ell(c)\}$ in order to reconstruct $\mathcal{G}_1$, $\mathcal{G}_2$ out of the *intrinsic* geometry of S_1, S_2. Graphs $\mathcal{G}_1$, $\mathcal{G}_2$ are the quotients of $\mathcal{G}[A,B,C]$ with respect to the subgroups H_1, H_2 of $\mathbf{Z}*_8 \ltimes \mathbf{Z}_8$ of section 3. As in section 4 we can therefore conclude from Sunada's Theorem that S_1 and S_2 are isospectral, where "isospectral" may stand for the eigenvalues of the Laplacian or for the lengths of the primitive closed geodesics ([5] Theorem 2). The aim of this section is to read out the isospectrality for the length spectrum *directly from the graphs* in an algorithmic way. The algorithm has already been described in [2], but we hope that in its present form it is more accessible.

The idea is to copy each primitive closed geodesic from S_1 to S_2 and vice versa in order to established a one-to-one correspondence between the primitive closed geodesics of a given length on S_1 and the primitive closed geodesics of the same length on S_2. *Primitive* means that the geodesic is not the k-fold iterate ($k \geq 2$) of another closed geodesic with the same trace. The copy process is a follows.

We first observe that there is a bijection between the boundary geodesics of the building blocks in S_1 and those in S_2 which preserves lengths so that from now on we only consider geodesics which intersect the boundaries of the building blocks finitely many times.

Let δ on S_1 be such a geodesic. Parametrize it on the interval $[0,1]$ such that $p = \delta(0) = \delta(1)$ is an interior point of some building block. To avoid ambiguities choose p not to be a self intersection point of δ. (The copy will of course turn out to be independent of the choice of the parametrization). Geodesic δ is decomposed into a sequence of subarcs: $\delta = \delta_1\delta_2 ... d_N$ where δ_1 runs from p to the boundary of the block which contains p, each δ_i ($i = 2,...,N-1$) is contained in one of the building blocks with both end points on the boundary, and δ_N runs from the boundary of the initial building block back to p. There is an exception : if δ is contained in one of the blocks then δ consists only of δ_1.

Let B_k be the initial block, i.e the block containing segment δ_1. Use criterion (A) below to select the initial block B_k* in S_2 for the curve $\delta*$ which is to be defined. Then set

$$\delta_1^* = \varphi_{kk}*(\delta_1)$$

where $\varphi_{kk}*$: (int B_k in S_1) $\rightarrow$ (int B_k* in S_2) is the natural isometry from the interior of block B_k in S_1 onto the interior of block B_k* in S_2 (We cannot define $\varphi_{kk}*$ on the boundary of B_k because the gluings in S_1 and S_2 are different).

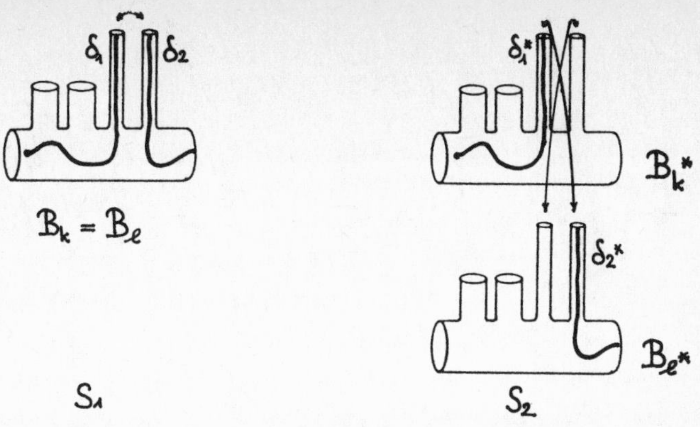

Fig. 7

If δ_2 is contained in B_ℓ we select $B_\ell{}^*$ such that the initial point of $\varphi_{\ell\ell}{}^*(\delta_2) := \delta_2{}^*$ coincides with the endpoint of $\delta_1{}^*$. This is illustrated in Fig. 7. Then, if δ_3 is on B_m we select $B_m{}^*$ such that the endpoint of $\delta_2{}^*$ is the initial point of $\delta_3{}^* := \varphi_{mm}{}^*(\delta_3)$, and so on. The geodesic $\delta^* = \delta_1{}^*\delta_2{}^*...\delta_N{}^*$ has the same length as δ, but it is not necessarily closed.

We now give an algorithm which tells how to select the initial block $B_k{}^*$ such that δ^* will be closed.

Denote by #a, #b, #c the number of times δ crosses one of the geodesics a,b,c respectively, which are on the boundary of the building blocks. Since looking at fig. 8 we see jumps, let us call each crossing of δ over a geodesic of type a an "a-jump" and define b-jumps, c-jumps similarly. Then #a, #b, #c are the number of a-jumps, b-jumps and c-jumps. We claim that the following initiation algorithm leads to *closed* copy curves.

$$\text{(A)} \qquad k^* = \begin{cases} k & \text{if } \#a \text{ even} \\ k+1 & \text{if } \#a \text{ odd and } \#b \text{ even} \\ k+2 & \text{if } \#a \text{ odd and } \#b \text{ odd} \end{cases}$$

(all integers mod 8).

Clearly, our claim is of purely combinatorial nature. For the proof it will therefore be sufficient to prove that the corresponding copying of closed circuits in graph $\mathfrak{G}_1$ always leads to closed circuits in $\mathfrak{G}_2$ and vice versa. All curves and copy curves may therefore be drawn on the graphs given in fig. 8. Before going to the proof one may enjoy testing the algorithm in a few examples. For better reference we state the claim as a lemma.

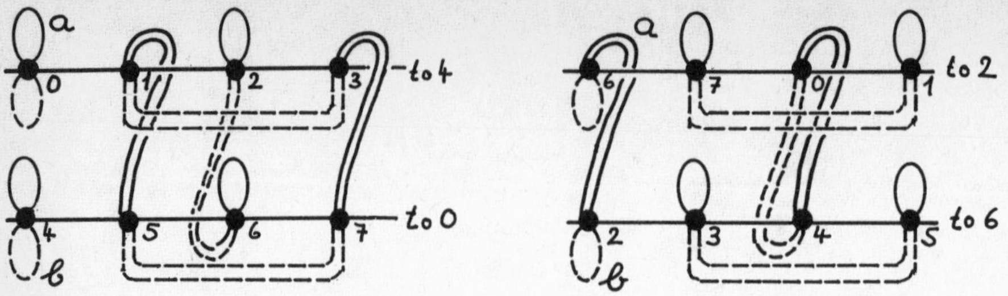

Fig. 8

Lemma 5. *With initiation* (A) *a path* δ *in* $\mathfrak{G}_1$ *is closed if and only if its copy path* δ^* *in* $\mathfrak{G}_2$ is closed ($\mathfrak{G}_1, \mathfrak{G}_2$ *are from* fig. 8).

Proof. Let two synchronized observers run along δ and δ^* with the same speed. They always jump simultaneously. Between the jumps their mouvements are identical, however with the first observer being on some block B_t in S_1 whereas the second observer is on some block B_s in S_2 with s different from t, in general. We call s-t the *shift* between the two observers. We have to prove that the shifts at the beginning and the end are the same. Since c-jumps do not change the shift we only have to consider a-jumps and b-jumps.

Case1: #a is even. The initial shift is zero. We claim that during the whole course the shift is always 0 or 4. In fact, as long as the shift is 0 or 4, b-jumps leave the shift invariant (check with fig. 8) and a-jumps change the shift from 0 to 4 resp from 4 to 0. Since #a is even the shift at the end is again zero.

Case 2. #a is odd and #b is even. The initial shift is 1. We claim that during the whole course the shift is always 1 or 3. In fact, as long as the shift is 1 or 3, a-jumps leave the shift invariant and b-jumps change it from 1 to 3 resp from 3 to 1. (Check with the dotted lines in fig. 8). Since #b is even the shift at the end is again 1.

Case 3. #a is odd and #b is odd. The initial shift is 2. We claim that the shift can only be 2 or -2 (always mod. 8). In fact, if the shift is 2 or -2 then a-jumps *and* b-jumps change the shift by 4. Since #a + #b is even we end up again with the initial shift. This proves the Lemma.

Lemma 5 says also that the inverse of the copy process maps closed geodesics (or paths) to closed ones. Hence we have a one-to-one correspondence between the primitive closed geodesics in S_1 and the primitive closed geodesics in S_2.

6. Non-isometry. We prove Theorem 2. As in Sunada [5] we use bumpy metrics on M_O. But here we want them such that we can recognize the colored graphs $\mathfrak{G}_1, \mathfrak{G}_2$ by looking at M_1, M_2. This will be sufficient since we have the following.

Proposition 6. *Let* $\mathfrak{G}$ *be the Cayley graph of a finite group* G *with respect to some generators. Let* H_1, H_2 *be subgroups of* G. *Then* H_1 *and* H_2 *are conjugate if and only if* $H_1\backslash\mathfrak{G}$ *and* $H_2\backslash\mathfrak{G}$ *are strongly isomorphic.*

Proof. Clearly, conjugate subgroups have strongly isomorphic quotients. Now assume $H_1\backslash\mathfrak{G}$ and $H_2\backslash\mathfrak{G}$ are strongly isomorphic. Then there exists a mapping ϕ from right cosets mod H_1 to right cosets mod H_2 such that whenever

$$\phi(H_1\, g) = H_2\, g'$$

and whenever A is a generator from the given set of generators, then we also have

$$\phi(H_1\, g\, A) = H_2\, g'\, A.$$

By induction on generators we find

$$\phi(H_1\, g\, f) = H_2\, g'\, f$$

for all $f \in G$. Now let $g_O \in G$ be the element for which $\phi(H_1) = H_2\, g_O$. For $h_1 \in H_1$ we then have

$$H_2\, g_O = \phi(H_1) = \phi(H_1 h_1) = H_2\, g_O\, h_1$$

i.e $g_O\, h_1^{-1} \in H_2$ for all $h_1 \in H_1$. By symmetry with respect to H_1, H_2 we have $g_O H_1 g_O^{-1} = H_2$. This proves the proposition.

Now let G, H_1,H_2 be again the triplet from Sunada's Theorem with the isospectral manifolds $M_1 = H_1\backslash M$, $M_2 = H_2\backslash M$ and quotient $M_O = G\backslash M$. In order to make the graphs $\mathfrak{G}_1 = H_1\backslash\mathfrak{G}$ and $\mathfrak{G}_2 = H_2\backslash\mathfrak{G}$ reconstructible from the intrinsic metric of M_1,M_2 it will suffice to have the following weakened form of a nowhere homogeneous metric on M_O.

Definition. A Riemannian metric g on a compact manifold M_O is said to have a *distinguished ball* if there exists a distance ball B in (M_O,g) whose radius is smaller than the injectivity radius of (M_O,g) such that the only isometry $\varphi : B \to \varphi(B) \subseteq M$ is $\varphi = id_B$.

The following is due to Sunada.

Proposition 7. *The set of all Riemannian metrics on* M_o *with a distinguished ball comprises a residual set in the space of all Riemannian metrics on* M_o.

Proof. This follows from [5] Proposition 1 because nowhere homogeneous metrics have distinguished balls.

Theorem 8. *If* (M_o,g) *has a distinguished ball then* M_1 *and* M_2 *are not isometric.*

Proof. We shall first "paint" $\mathfrak{G}_1$, $\mathfrak{G}_2$ upon M_1, M_2 in a way which needs knowledge of the above construction. Then we shall give an instruction which tells how to reproduce these drawings in a manner which uses only the *intrinsic* geometry of M_1, M_2.

Let B_{p_o} be a distinguished ball centered at some point $p_o \in M_o$, and let $p \in M$ be an inverse image of p_o with respect to the covering $M \to M_o$. Since G acts freely on M we may let the points g(p), $g \in G$ represent the vertices of the Cayley graph $\mathfrak{G} = \mathfrak{G}[A_1,...,A_n]$ of G where $A_1,...,A_n$ are the given generators of G. Draw a curve α_i from p to $A_i(p)$ for each of the generators $A_1,...,A_n$. Draw them such that color and orientation can thus be read out of the form of the curves. The images $g(\alpha_1)$,..., $g(\alpha_n)$ are curves from g(p) to $gA_i(p)$, i=1,...,n. Thus $\mathfrak{G}$ is represented by a drawing on M. Under the covering maps $M \to M_1$, $M \to M_2$ $\mathfrak{G}$ projects to drawings of $\mathfrak{G}_1$, $\mathfrak{G}_2$ on M_1, M_2. Finally let $\mathfrak{G}_o$ be the image of $\mathfrak{G}$ in M_o under the covering $M \to M_o$. Then $\mathfrak{G}_i$ is the lift of $\mathfrak{G}_o$ with respect to the covering $M_i \to M_o$, i=1,2.

It is easy to reconstruct the coverings $M_i \to M_o$: Consider e.g. M_1. Then B_{p_o} has lifts $B_{q_1},...,B_{q_r}$ In M_1. Since B_{p_o} is a distinguished ball these are the only balls in M_1 which are isometric to B_{p_o}, and for each of these balls there exists only one isometry to B_{p_o}, namely the covering map $M_1 \to M_o$. The following instructions are therefore executable if only the intrinsic metric of M_i is known (i=1,2).

- Find a ball $B \subseteq M_i$ which is isometric to B_{p_o} and select an isometry $\varphi : B \to B_{p_o}$.
- Extend φ to a covering $\varphi : M_i \to M_o$.
- Lift $\mathfrak{G}_o$ from M_o to M_i with respect to φ.

On M_1 the instruction yields the colored and oriented graph $\mathfrak{G}_1$. On M_2 the same instruction yields $\mathfrak{G}_2$. By Lemma 5 the results are different. Hence M_1 and M_2 are not isometric.

REFERENCES

[1] Brooks, R. and Tse, R., Isospectral surfaces of small genus, Nagoya Math. J., **107** (1987), 13-24.

[2] Buser, P., Isospectral Riemann surfaces, Ann. Inst. Fourier, Grenoble, **36** (1986), 167-192.

[3] Grossmann, I. and Magnus, W., Groups and their graphs, The Mathematical Association of America, **14**, 1964.

[4] Perlis, R., On the equation $\zeta_k(s) = \zeta_{k'}(s)$, Journal of Number Theory, **9** (1977), 342-360.

[5] Sunada, T., Riemannian coverings and isospectral manifolds, Annals of Mathematics, **121** (1985), 169-186.

[6] Weinstein, M., Examples of Groups, Polygonal Publishing House, 1977.

On the almost negatively curved 3-sphere

Peter Buser
Département de Mathématiques
EPF-Lausanne
CH-1015 Lausanne, Switzerland

Detlef Gromoll
Department of Mathematics
SUNY
Stony Brook, N.Y. 11794 (USA)

1. Introduction. By Hadamard's Theorem every simply connected Riemannian manifold with non positive curvature is diffeormorphic to $\mathbb{R}^n$. In view of the many pinching theorems of qualitative Riemannian geometry one might expect a similar theorem to hold if small amounts of positive curvature are permitted, e.g. if the upper bound of the sectional curvature is positive but small as compared to the maximal rank radius of the exponential map. Without additional assumptions this is not possible. In fact, Gromov points out in [3] that almost negatively curved metrics exist on the 3-sphere in the following sense.

Theorem. *For all* $\varepsilon > 0$ *there exists a Riemannian metric on* S^3 *with diameter* d *and upper sectional curvature bound* K *satisfying* $Kd^2 \leq \varepsilon$.

It follows among others that for a given point $p \in S^3$ the exponential map $\exp_p$: $T_pS^3 \to S^3$ has maximal rank within a ball in T_pS^3 whose radius is much larger than the diameter of S^3. We may therefore lift the interior of the cut locus of p from S^3 to T_pS^3 via $\exp_p^{-1}$ and obtain a tesselation of the ball with fundamental domains very much the same way as Hadamard manifolds are tesselated with fundamental domains of compact quotients.

Gromov's example has been generalized to all compact 3-manifolds by Bavard. [1] He uses open books which yield, in addition, control of the volume. In [2] Gao and Yau used a cutting and pasting technique similar to the one we are going to explain below, to prove that S^3 admits metrics with strictly negative Ricci curvature. Both papers are quite technical. The aim of this note is to give a simplified version of Gromov's original construction. Although we obtain no new results we hope that the note makes this interesting example more accessible.

2. Surgery in dimension 2. Let us first explain the idea in dimension 2 although we know in advance that it cannot work. The idea is to use surgery which produces many short cuts with the effect that the diameter shrinks. It is easy to do this without affecting the upper curvature bound.

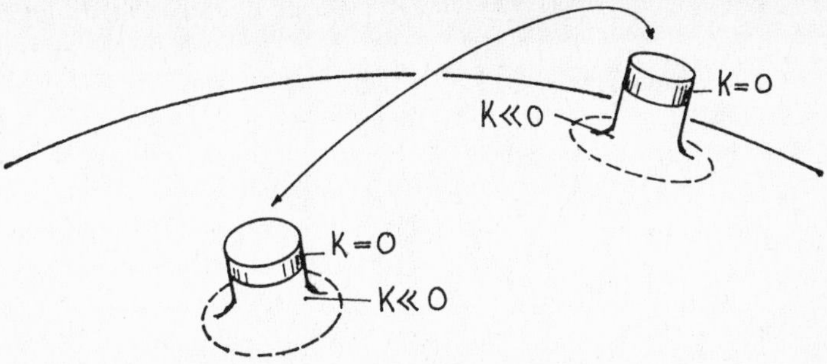

Fig. 1

From S^2 cut out two arbitrarily small discs which are at some distance apart. In the 2-holed sphere which remains we change the metric near each hole such that a neighbourhood of the hole becomes a flat cylinder. It takes highly negative curvature to adapt this metric smoothly to the standard metric outside some small annular region of the hole, but it is possible and we can do it such that the curvature remains bounded above by 1. Now identify the two boundaries (they are assumed to have the same length). This reduces the diameter. Unfortunately it produces a handle.

Of course we know that in dimension two we cannot reduce the diameter keeping the upper curvature bound and not change the topology: If K is an upper curvature bound and d the diameter then we have

Every Riemannian metric on S^2 *satisfies* $Kd^2 \geq \pi^2$

In fact, if $Kd^2 < \pi^2$ on some manifold M, then for $p \in M$, $\exp_p : T_pM \to M$ is a local diffeomorphism in a ball of radius greater than the diameter of M. We find therefore a fundamental domain for $\exp_p$, i.e. a simply connected compact domain whose boundary ∂ is mapped onto the cut locus of M and whose connected interior is mapped onto the interior of the cut locus. Since $\exp_p$ is a local diffeomorphism, $\exp_p : \partial \to M$ is an *immersion*. But if $M = S^2$ there is a topological obstruction: ∂ is an S^1, and any smoothly immersed image of S^1 in S^2 disconnects its complement. Hence we cannot have $Kd^2 < \pi^2$ on S^2.

In contrast to this we have no such obstruction in dimension 3.

3. Surgery in dimension 3. Let $\gamma : S^1 \to S^3$ and $\gamma_1 : S^1 \to S^2 \times S^1$ be closed geodesics of length

$$L = \ell(\gamma) = \ell(\gamma_1)$$

with respect to some given Riemannian metrics, and let γ_1 be homotopic to the S^1-factor of $S^2 \times S^1$. Assume that for small $r > 0$ the tubular neighbourhoods

$$\gamma^{3r} = \{p \in S^3 \mid \operatorname{dist}(p,\gamma) < 3r\}$$

and

$$\gamma_1{}^{3r} = \{p \in S^2 \times S^1 \mid \operatorname{dist}(p,\gamma_1) < 3r\}$$

are isometric to the flat tube

$$B^{3r} \times \mathbb{R} / [t \mapsto t + L]$$

where B^{3r} is the open disc of radius 3r in $\mathbb{R}^2$ and the metric is the product metric. In Fermi coordinates $p = p(\rho,\theta,t)$, the metric tensors are

$$ds^2 = d\rho^2 + \rho^2 d\theta^2 + dt^2.$$

($0 < \rho < 3r$ is the radial distance, θ is the angle coordinate and t is from $\mathbb{R} / [t \mapsto t + L]$)

The point is that *the interior* of $\gamma_1{}^r$ is *diffeomorphic* to the exterior. (This is true because γ_1 is homotopic to the S^1-factor of $S^2 \times S^1$ and the interior of a disc in S^2 is diffeomorphic to the exterior). Thus if we remove γ^r from S^3 and replace it by the exterior of $\gamma_1{}^r$ the topology does not change . We shall see that using the exterior of $\gamma_1{}^r$ will permit short cuts. But let us first see how to control smoothness and curvature. For $r \le \rho \le 3r$ we replace the factor ρ^2 of the metric tensor by a factor $\phi^2(\rho)$ where $\phi(\rho)$ is a smooth convex function satisfying $\phi(\rho) = 2.5\,r$ for $2r \le \rho \le 2.4r$, and $\phi(\rho) = \rho$ for $2.6r \le \rho \le 3r$. In $\gamma^{3r}-\gamma^r$ we have non positive curvature. $\gamma^{2r}-\gamma^r$ and $\gamma_1{}^{2r}-\gamma_1^r$ are isometric to the Riemannian product $[0,r] \times T$ where T is a flat torus.

Hence we can glue together $S^3 - \gamma^r$ and $S^2 \times S^1 - \gamma_1{}^r$ along their boundaries $T\gamma$ resp. $T\gamma_1$ (which are both isometric to T) using an arbitrary identifying isometry $T\gamma \to T\gamma_1$. The metric remains smooth in this pasting. By abuse of notation we shall write $T\gamma = T\gamma_1$ on the new manifold.

4. Short cuts. Short cuts may occur as follows. Let $p,q \in S^3 - \gamma^r$ be near $T\gamma$. Connect p, q with nearby points $p', q' \in T\gamma = T\gamma_1$. Assume that in $S^2 \times S^1 - \gamma_1^r$ points p' and q' have a short connecting curve. The three curves together then yield a connection from p to q which may be much shorter than the original distance of p and q in S^3.

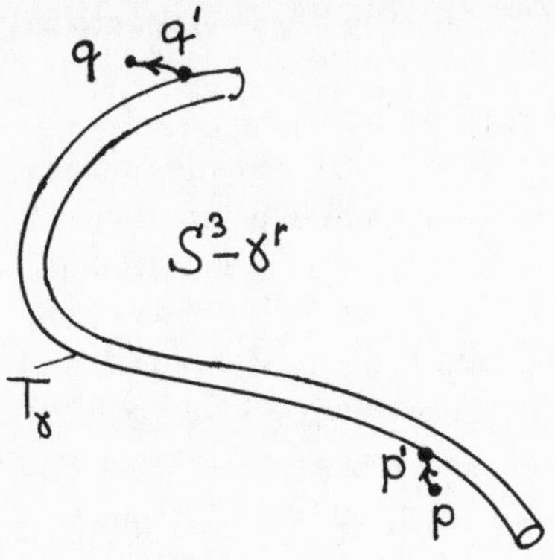

Fig. 2a

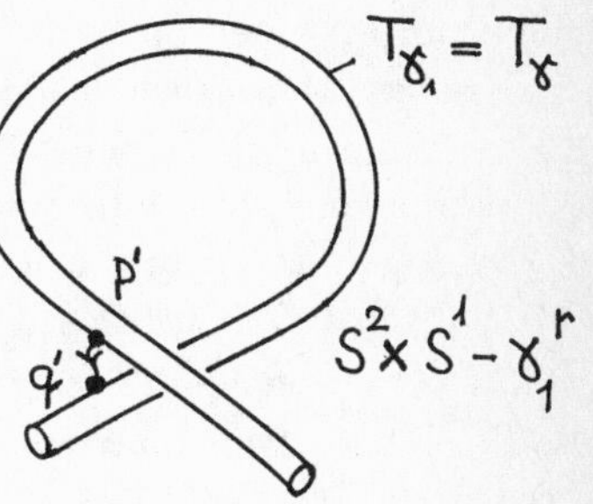

Fig. 2b

Now let us produce such short cuts. Let $\varepsilon > 0$ be arbitrarily small. Choose S^2 to be a rotational ellipsoid with curvature almost equal to 1 so that S^2 contains a closed geodesic γ_2 satisfying $\gamma_2^{\varepsilon} = S^2$. i.e every point of S^2 lies within distance ε from γ_2. Geodesic γ_2 consists of a number of segments, say m, of length approximately equal to 2π whose endpoints are within a distance $\leq \varepsilon$ from the north pole of S^2. Parametrize γ_2 in the form $t \mapsto \gamma_2(t)$, $0 \leq t \leq 1$ with $\gamma_2(0) = \gamma_2(1)$. Then take $S^1 = \mathbb{R}/[y \mapsto y + \varepsilon]$ and introduce the Riemannian product structure on the product $S^2 \times S^1$ of the ellipsoid and the small circle. The curve

$$t \mapsto \gamma_1(t) := (\gamma_2(t), \varepsilon t), \quad 0 \leq t \leq 1$$

is a simple closed geodesic on $S^2 \times S^1$ which is homotopic to the S^1 factor and satisfies $\gamma_1^{\varepsilon} = S^2 \times S^1$, i.e every point of $S^2 \times S^1$ lies within a distance ε from γ_1. Moreover, γ_1 has zero holonomy. By the lemma of section 5 below we can, for arbitrarily small $r > 0$, modify the Riemannian metric in γ_1^{4r} without affecting the radial distances (i.e. the distances from γ_1) such that the upper curvature bound does not exceed some universal constant K and such that γ_1^{3r} is isometric to the flat tube $B^{3r} \times \mathbb{R}/[t \mapsto t + L]$, L being the length of γ_1.

Now take S^3 with constant curvature such that the great circles have length L (we start with a big diameter to produce a small diameter). Let γ be a great circle and modify the Riemann metric in γ^{4r} as we did for γ_1^{4r}. Finally, perform the surgery of section 3.

Due to the choice of γ_2 on the ellipsoid we find points $p'_1,\dots,p'_m \in T\gamma = T\gamma_1$ whose distances *with respect to the metric of* S^3 are $\mathrm{dist}(p'_i,p'_{i+1}) = L/m$, $(i=1,\dots,m-1)$ but such that in $S^2 \times S^1 - \gamma_1{}^r$ they all lie within distance $\leq \varepsilon$ from the north pole of S^2. This is the situation of fig. 2b and we have produced short cuts, so far for the points $p'_1,\dots,p'_m \in T\gamma$. Choosing the identifying isometry $T\gamma \to T\gamma_1$ properly (cf section 3), we may prescribe $p'_1 \in T\gamma$ arbitrarily.

To reduce the diameter of S^3 fix some "north pole" $N \in S^3$ and distribute a finite set $\mathcal{R}$ of points on S^3 such that balls of radius ε around these points cover S^3. For any $p \in \mathcal{R}$ there exists a point $P \in S^3$ whose distance from p and from N is an integer multiple of L/m. We obtain a short cut from p to N as follows. Take great circles γ, η such that $\gamma \cap \eta = \emptyset$ with γ ε-close to p and P but $p, P \notin \gamma$ and with η ε-close to P,N and $P,N \notin \eta$. This is easy to do in S^3. Then take positive $r < \varepsilon$ so small that $\gamma^{4r} \cap \eta^{4r} = \emptyset$ and $p,P,N \notin \gamma^{4r} \cup \eta^{4r}$. With the above surgery we find P'_1, $P'_k \in T\gamma$ and P''_1, $P''_\ell \in T\eta$ whose respective distances from p, P, resp. P,N are $\leq \varepsilon$ with short cuts from P'_1 to P'_k and from P''_1 to P''_ℓ The two surgeries together yield a short cut from p via P to N (fig.3).

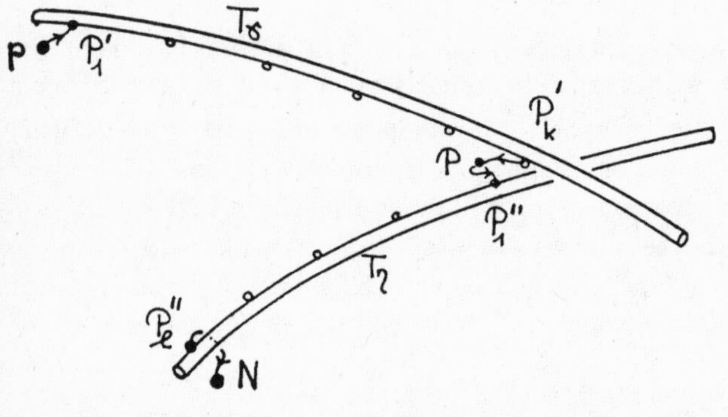

Fig. 3

Finally, we produce these short cuts for all p in the finite set $\mathcal{R}$ simultaneously, using pairwise disjoint great circles and sufficiently small r. On the new manifold, every point which comes from S^3 has distance $\leq 8\varepsilon$ from N, and every point which comes from one of the glued-in $S^2 \times S^1 - \gamma_1{}^r$ has distance $\leq \varepsilon$ from some point of S^3. Hence, the diameter is now $\leq 18\varepsilon$, the upper curvature bound is the universal constant K of the lemma below and the topology is still that of S^3.

5. The flattening of tubes. It remains to prove the following technical lemma.

Lemma. *Let* γ *be a simple closed geodesic on a Riemannian manifold* (M^n,g) *and let* $r>0$ *be arbitrarily small. There exists a new metric* $\tilde{g}$ *on* M *with the following properties :*

(i) γ *is again a geodesic and length and holonomy have not changed.*

(ii) *The distances to* γ *are the same under this new metric.*

(iii) $g = \tilde{g}$ *in* $M - \gamma^{4r}$.

(iv) γ^{3r} *is isometric to the Riemannian product* $B^{3r} \times \gamma$.

(v) *In* $\gamma^{4r} - \gamma^{3r}$ *the upper bound of the absolute curvature has increased at most by a factor* K *where* K *depends only on the dimension.*

Proof. Let $t \to V_1(t),\dots,V_n(t)$, with $V_n(t) = \dot{\gamma}(t)$ be parallel orthonormal vector fields along the unit speed geodesic γ and define

$$\phi(x_1,\dots,x_n) := \exp_{\gamma(x_n)}(x_1V_1(x_n) + \dots + x_{n-1}V_{n-1}(x_n))$$

for $x^2_1+\dots+x^2_{n-1} \le 4r$, $0 \le x_n = L = \ell(\gamma)$. If r is sufficiently small we may use ϕ^{-1} as a coordinate map for γ^{4r}. The components of g then are

$$g_{ij} = g(d\phi(\frac{\partial}{\partial x_i}), d\phi(\frac{\partial}{\partial x_j})), \quad i,j = 1,\dots,n.$$

Now consider a unit vector X at $(0,\dots,0,x_n) \in \mathbf{R}^n$, perpendicular to $\frac{\partial}{\partial x_n}$. The curve

$$\rho \to c(\rho) = \phi(\rho X)$$

is a geodesic with $\dot{c}(0) \perp \dot{\gamma}(x_n)$. The vectorfields

$$Y_i := d\phi_{\rho X}(\rho\frac{\partial}{\partial x_i}), \qquad i = 1,\dots,n-1; \qquad Y_n := d\phi_{\rho X}\frac{\partial}{\partial x_n}$$

are Jacobi fields along c with the initial conditions

$$Y_i(0) = 0, \qquad Y'_i(0) = \frac{\partial}{\partial x_i}, \qquad i = 1,\dots,n-1$$

$$Y_n(0) = \frac{\partial}{\partial x_n}, \qquad Y'_n(0) = 0$$

(' is the covariant derivative). From the Jacobi equation $Y_k'' + R(Y_k,\dot{c})\dot{c} = 0$, $k=1,\ldots,n$, where R is the curvature tensor, we find the Taylor series expansion

$$Y_i = \rho A_i - \frac{1}{6}\rho^3 B_i + \ldots$$

$$Y_n = A_n - \frac{1}{2}\rho^2 B_n + \ldots$$

where A_k, B_k are parallel vectorfields along c, satisfying

$$A_k(0) = V_k, \qquad B_k(0) = R(V_k,X)X, \qquad k = 1,\ldots,n.$$

Looking at $g(Y_i,Y_j)$ we find

$$\frac{\partial}{\partial \rho} g_{ij}(0) = 0, \qquad \left| \frac{\partial^2 g_{ij}}{\partial \rho^2}(0) \right| \le 2\,\|R\|$$

independently of X. Since $g_{ij}(0,\ldots,0,x_n) = \delta_{ij}$ we find

$$g_{ij} = \delta_{ij} + h_{ij}$$

where

$$|h_{ij}| \le \kappa\rho^2 + o(\rho^2), \qquad \left| \frac{\partial}{\partial x_k} h_{ij} \right| \le 4n\kappa\rho + 0(\rho^2)$$

$$\left| \frac{\partial^2}{\partial x_k \partial x_\ell} h_{ij} \right| \le 4\kappa + 0(\rho),$$

$i,j,k,\ell = 1,\ldots,n$, where κ is a *positive* upper bound of the norm $\|R\|$ of the curvature tensor along γ. To define $\mathfrak{g}$ we use a monotone smooth function φ: $[0,4] \to [0,1]$ such that $\varphi(t) = 0$ for $0 \le t \le 3$, $\varphi(t) = 1$ for $3.5 \le t \le 4$, $\varphi' \le 10$, $|\varphi''| \le 100$ and define

$$\mathfrak{g}_{ij} = \delta_{ij} + h_{ij}\,\varphi(\rho/r).$$

Here we have to assume that r is sufficiently small such that the above o- and 0-terms are negligible.Points (i) - (iv) of the Lemma are clear. For the curvature bound in (v) we first get from a direct computation

$$|g_{ij} - \delta_{ij}| \leq \text{const } \kappa\rho^2, \qquad |\frac{\partial}{\partial x_k} g_{ij}| \leq \text{const } \kappa\rho, \qquad |\frac{\partial^2}{\partial x_k \partial x_\ell} g_{ij}| \leq \text{const } \kappa$$

$$|g^{ij} - \delta_{ij}| \leq \text{const } \kappa\rho^2, \qquad |\frac{\partial}{\partial x_k} g^{ij}| \leq \text{const } \kappa\rho ,$$

where κ is the upper bound of $\|R\|$ from above, and "const" are various dimension constants.

Thus, we have $|\tilde{R}^m_{ijk}| <$ const κ for the curvature symbols of g in γ^{4r}. This proves the Lemma.

REFERENCES

[1] Bavard, Chr., Courbure presque négative en dimension 3, Compositio Mathematica **63** (1987) 223-236.

[2] Gao, L.Z., and Yau, S.T., The existence of negatively Ricci curved metrics on three manifolds, Invent. math. **85** (1986) 637-652.

[3] Gromov, M., Almost flat manifolds, J. Diff. Geom. **13** (1978) 231-241.

VANISHING THEOREMS FOR TENSOR POWERS OF A POSITIVE VECTOR BUNDLE

Jean-Pierre DEMAILLY
Université de Grenoble I,
Institut Fourier, BP 74,
Laboratoire associé au C.N.R.S. n° 188,
F-38402 Saint-Martin d'Hères

Abstract. — Let E be a holomorphic vector bundle of rank r over a compact complex manifold X of dimension n. It is shown that the Dolbeault cohomology groups $H^{p,q}(X, E^{\otimes k} \otimes (\det E)^l)$ vanish if E is positive in the sense of Griffiths and $p+q \geq n+1$, $l \geq r + C(n,p,q)$. The proof rests on the well-known fact that every tensor power $E^{\otimes k}$ splits into irreducible representations of $\mathrm{Gl}(E)$, each component being canonically isomorphic to the direct image on X of a positive homogeneous line bundle over a flag manifold of E. The vanishing property is then obtained by a suitable generalization of Le Potier's isomorphism theorem, combined with a new curvature estimate for the bundle of X–relative differential forms on the flag manifold of E.

0. Statement of results.

The aim of this work is to prove a rather general vanishing theorem for cohomology groups of tensor powers of a positive vector bundle.

Let X be a complex compact n–dimensional manifold and E a hermitian vector bundle of rank r over X. We denote by $C^\infty_{p,q}(X,E)$ the space of smooth E–valued differential forms of type (p,q) on X and by

$$D_E = D'_E + D''_E \;:\; C^\infty_{p,q}(X,E) \longrightarrow C^\infty_{p+1,q}(X,E) \oplus C^\infty_{p,q+1}(X,E)$$

(or simply $D = D' + D''$) the Chern connection of E. Let $(x_1, \ldots, x_n)$ be holomorphic coordinates on X and $(e_1, \ldots, e_n)$ a local C^∞ orthonormal frame of E. The Chern curvature tensor $c(E)$ is defined by $D^2 = c(E) \wedge \bullet$ and may be written

$$c(E) = \sum_{i,j,\lambda,\mu} c_{ij\lambda\mu}\, dx_i \wedge d\overline{x}_j \otimes e^\star_\lambda \otimes e_\mu \;\;,\;\; 1 \leq i,j \leq n \;,\; 1 \leq \lambda,\mu \leq r \;.$$

The curvature tensor $ic(E)$ is in fact a (1,1)–form with values in the bundle $\mathrm{Herm}(E,E)$ of hermitian endomorphisms of E, i.e. $c_{ij\lambda\mu} = \overline{c}_{ji\mu\lambda}$; thus $ic(E)$ can be identified with a hermitian form on $TX \otimes E$.

Let us recall that the bundle E is said to be *positive*, resp. *semi-positive* (in the sense of Griffiths [8]) if E can be endowed with a hermitian metric such that at every point $x \in X$ one has

$$ic(E)_x(\varsigma \otimes v, \varsigma \otimes v) = \sum_{i,j,\lambda,\mu} c_{ij\lambda\mu}(x)\, \varsigma_i \overline{\varsigma}_j v_\lambda \overline{v}_\mu > 0 \quad , \quad \text{resp.} \quad \geq 0$$

for all non zero vectors $\varsigma = \sum \varsigma_i \, \partial/\partial z_i \in T_x X$, $v = \sum v_\lambda e_\lambda \in E_x$. Every vector bundle E generated by sections is ≥ 0 .

Another important concept is that of ampleness, for which we refer to Hartshorne [9]; a vector bundle E is ample if and only if the line bundle $O_E(1)$ over $P(E^\star)$ is ample (or equivalently > 0) . It is also well-known that $E > 0$ implies E ample, but the converse is unknown.

In the case of a positive or ample vector bundle E of rank $r > 1$, only very few general and optimal vanishing results are available for the Dolbeault cohomology groups $H^{p,q}$ of tensor powers of E . For example, the famous Le Potier vanishing theorem [13] :

$$E \text{ ample} \implies H^{p,q}(X, E) = 0 \text{ for } p + q \geq n + r$$

does not extend to symmetric powers $S^k E$, even when $p = n$ and $q = n - 2$ (cf. [11]) . Nevertheless, the following result shows that the vanishing property is true for tensor powers involving a sufficiently large power of $\det E$.

THEOREM. — *Let L be a holomorphic line bundle over X . Assume that $E > 0$ and $L \geq 0$, or $E \geq 0$ and $L > 0$. For all integers p, q such that $p + q \geq n$, set*

$$A(n, p, q) = \frac{n(n+1)(p+1)(q+1)}{4(p+q-n+1)} \qquad \text{if } p < n$$

and $A(n, p, q) = 0$ if $p = n$. Let $h \in \{1, \ldots, r-1\}$ and let $\Gamma^a E$ be the irreducible tensor power representation of $\mathrm{Gl}(E)$ *of highest weight $a \in \mathbb{Z}^r$, with*

$$a_1 \geq a_2 \geq \ldots \geq a_h > a_{h+1} = \ldots = a_r = 0 \ .$$

If $p + q \geq n + 1$ then

$$H^{p,q}(X, \Gamma^a E \otimes (\det E)^l \otimes L) = 0 \quad \text{for } l \geq h + A(n, p, q) \ .$$

The proof of this theorem is based on analysis and differential geometry, but an analogous result can be obtained in a purely algebraic way (cf. [5]); in that case the positivity hypothesis can be replaced by ampleness, the semi-positivity hypothesis by the fact that the bundle is generated by its global sections; then, the condition required on l is $l \geq n - p + r - 1$. Both results overlap in most cases, but the above analytic result can be better if $r - h$ is very large.

Observing that S^kE is the irreducible representation of highest weight $(k,0,\ldots,0)$ and that $E^{\otimes k}$ splits into irreducible representations of the type $\Gamma^a E \otimes (\det E)^l$ with $h \le \min\{k, r-1\}$ (*cf.* formula (2.17)), we obtain :

COROLLARY. — *Under the positivity hypotheses of the theorem, then for all p,q such that $p+q \ge n+1$ one has*

(0.1) $H^{p,q}(X, S^kE \otimes (\det E)^l \otimes L) = 0$ *if* $l \ge 1 + A(n,p,q)$;

(0.2) $H^{p,q}(X, E^{\otimes k} \otimes (\det E)^l \otimes L) = 0$ *if* $l \ge \min\{k, r-1\} + A(n,p,q)$.

The special case $p = n$ of (0.1) is due to P. Griffiths [8] . For $p = n$ and arbitrary r , $k_0 \ge 2$, Peternell-Le Potier and Schneider [11] have constructed an example of a vector bundle $E > 0$ of rank r over a manifold X of dimension $n = 2r$ such that

(0.3) $$H^{n,n-2}(X, S^kE) \neq 0 \ , \quad 2 \le k \le k_0 \ .$$

This result shows that the lower bound $l \ge 1$ in (0.1) cannot be improved. More generally, the following example (for which we refer to [5]) shows that our condition $l \ge h$ in the theorem is optimal. This example gives a negative answer to a question of Sommese [15].

EXAMPLE. — *Let $X = G_r(V)$ be the Grassmannian of subspaces of codimension r of a vector space V of dimension d , and E the tautological quotient vector bundle of rank r over X (then $E \ge 0$ and $L = \det E$ is ample) . Let $h \in \{1,\ldots,r-1\}$ and $a \in \mathbb{Z}^r$, $\beta \in \mathbb{Z}^d$ be such that*

$$a_1 \ge \ldots \ge a_h \ge d-r \ , \quad a_{h+1} = \ldots = a_r = 0 \ ,$$
$$\beta = (a_1 - d + r, \ldots, a_h - d + r, 0, \ldots, 0) \ .$$

Set $n = \dim X = r(d-r)$, $q = (r-h)(d-r)$. Then

(0.4) $$H^{n,q}(X, \Gamma^a E \otimes (\det E)^h) = \Gamma^\beta V \otimes (\det V)^h \neq 0 \ .$$

Our approach is based on three well-known facts. First, every tensor power of E splits into irreducible representations of the linear group $\mathrm{Gl}(E)$. It is thus sufficient to consider "irreducible" tensor powers of E . Secondly, every irreducible tensor power of E appears in a natural way as the direct image on X of a *positive line bundle* over a suitable flag manifold of E . This follows from Bott's theory of homogeneous vector bundles [3]. The third fact is the isomorphism theorem of Le Potier [13], which relates the cohomology groups of E over X to those of the line bundle $\mathcal{O}_E(1)$ over $P(E^\star)$. We generalize here this isomorphism to the case of arbitrary flag bundles associated to E .

When $p = n$ the above-mentioned algebraic facts suffice to prove the theorem. However, when $p < n$, the generalized Borel-Le Potier spectral sequence does not degenerate at the E_1 level (*cf.* [12]). A possible way in order to overcome this difficulty is to establish a curvature estimate for the bundle of X–relative differential forms on the flag manifold of E , using the standard Kodaira-Akizuki-Nakano inequality [1] . Our estimate (*cf.* §4) measures in some sense how far is the spectral sequence from being E_1–degenerate. The following related problem is interesting, but its complete solution certainly requires a better understanding of the Borel-Le Potier spectral sequence for flag bundles.

PROBLEM. — *Given a dominant weight* $a \in \mathbb{Z}^r$ *with* $a_r = 0$, *determine the smallest admissible constant* $A(n,p,q)$ *in the theorem.*

It is shown in [5] that if the Borel-Le Potier spectral sequence degenerates in E_2 , then it is always sufficient to take $l \geq r-1+\min\{n-p, n-q\}$. In view of the above main theorem, one may suspect that the correct answer could be $l \geq h+\min\{n-p, n-q\}$.

The above results have been annouced in the note [4]. The author wishes to thank warmly Prof. Michel Brion, Friedrich Knopp, Thomas Peternell and Michael Schneider for valuable remarks which led to substantial improvements of this work.

1. Kodaira-Akizuki-Nakano vanishing theorem.

We recall here the basic Akizuki-Nakano inequality [1] which will be used several times in the sequel. Assume that X carries a Kähler metric ω , and let L be a hermitian line bundle over X . At each point $x \in X$, one can write

$$ic(L) = i \sum_{1 \leq j \leq n} \gamma_j \, dz_j \wedge d\overline{z}_j$$

where $\gamma_1 \leq \ldots \leq \gamma_n$ are the curvature eigenvalues of L , and where $(dz_1, \ldots, dz_n)$ is an ω–orthonormal basis of $T^\star X$. For every $v \in C^\infty_{p,q}(X, L)$ we have

$$(1.1) \qquad ||D''_L v||^2 + ||D''^\star_L v||^2 \geq \langle \Theta_L v \, , v \rangle \ ,$$

where Θ_L is the hermitian endomorphism defined by

$$(1.2) \qquad \langle \Theta_L v \, , v \rangle = \sum_{|I|=p, |J|=q} \big(\gamma_I + \gamma_J - \sum_{1 \leq j \leq n} \gamma_j\big) |v_{I,J}|^2 \ ,$$

with $\gamma_I = \sum_{m \in I} \gamma_m$. When L is > 0 , one can choose $\omega = ic(L)$ as the Kähler metric on X; in that case $\gamma_1 = \ldots = \gamma_n = 1$ and therefore

$$(1.3) \qquad ||D''_L v||^2 + ||D''^\star_L v||^2 \geq (p + q - n) ||v||^2 \ .$$

COROLLARY (Kodaira-Akizuki-Nakano). — *One has*

$$(1.4) \qquad L > 0 \implies H^{p,q}(X, L) = 0 \quad \textit{for} \quad p + q \geq n + 1 \ ,$$

$$(1.5) \qquad L < 0 \implies H^{p,q}(X, L) = 0 \quad \textit{for} \quad p + q \leq n - 1 \ .$$

2. Homogeneous line bundles over flag manifolds and irreducible representations of the linear group.

The aim of this section is to settle notations and to recall a few basic results on homogeneous line bundles over flag manifolds. The classical foundation works on this subject are Borel-Weil [2] and R. Bott [3] , which contain all the required material (*cf.* also Demazure [6] for a very simple proof of Bott's formula). We will give however an independent self-contained exposition in order to prepare the tools needed in the differential geometric approach of §4 .

Let B_r (resp. B^r) be the Borel subgroup of $\mathrm{Gl}_r = \mathrm{Gl}(\mathbb{C}^r)$ of lower (resp. upper) triangular matrices, $U_r \subset B_r$, $U^r \subset B^r$ the subgroups of unipotent matrices, and $T^r = B_r \cap B^r$ the complex torus $(\mathbb{C}^\star)^r$ of diagonal matrices. Let V be a complex vector space of dimension r . We denote by $M(V)$ the manifold of complete flags

$$V = V_0 \supset V_1 \supset \ldots \supset V_r = \{0\} \ , \quad \mathrm{codim}_{\mathbb{C}} V_\lambda = \lambda \ .$$

To every linear isomorphism $\varsigma \in \mathrm{Isom}(\mathbb{C}^r, V) : (u_1, \ldots, u_r) \longmapsto \sum_{1\leq\lambda\leq r} u_\lambda \varsigma_\lambda$, one can associate the flag $[\varsigma] \in M(V)$ defined by $V_\lambda = \mathrm{Vect}(\varsigma_{\lambda+1}, \ldots, \varsigma_r)$, $1 \leq \lambda \leq r$. This leads to the identification

$$M(V) = \mathrm{Isom}(\mathbb{C}^r, V)/B_r$$

where B_r acts on the right side. We denote simply by V_λ the tautological vector bundle of rank $r - \lambda$ on $M(V)$, and we consider the canonical quotient line bundles

$$\text{(2.1)} \qquad \begin{cases} Q_\lambda = V_{\lambda-1}/V_\lambda \ , & 1 \leq \lambda \leq r \ , \\ Q^a = Q_1^{a_1} \otimes \ldots \otimes Q_r^{a_r} \ , & a = (a_1, \ldots, a_r) \in \mathbb{Z}^r \ . \end{cases}$$

The linear group $\mathrm{Gl}(V)$ acts on $M(V)$ on the left, and there exist natural equivariant left actions of $\mathrm{Gl}(V)$ on all bundles V_λ, Q_λ, Q^a.

We compute now the tangent and cotangent vector bundles of $M(V)$. The action of $\mathrm{Gl}(V)$ on $M(V)$ yields

$$\text{(2.2)} \qquad TM(V) = \mathrm{Hom}(V, V)/W$$

where W is the subbundle of endomorphisms $g \in \mathrm{Hom}(V, V)$ such that $g(V_\lambda) \subset V_\lambda$, $1 \leq \lambda \leq r$. Using the self-duality of $\mathrm{Hom}(V, V)$ given by the Killing form $(g_1, g_2) \mapsto \mathrm{tr}(g_1 g_2)$, we find

$$\text{(2.3)} \qquad \begin{cases} T^\star M(V) = \big(\mathrm{Hom}(V, V)/W\big)^\star = W^\perp \\ W^\perp = \{g \in \mathrm{Hom}(V, V) \ ; \ g(V_{\lambda-1}) \subset V_\lambda \ , \ 1 \leq \lambda \leq r\} \ . \end{cases}$$

There exists a filtration of $T^\star M(V)$ by subbundles of the type

$$\{g \in \mathrm{Hom}(V, V) \ ; \ g(V_\lambda) \subset V_{\mu(\lambda)} \ , \ \lambda < \mu(\lambda) \ , \ 1 \leq \lambda \leq r\}$$

in such a way that the corresponding graded bundle is the direct sum of the line bundles $\mathrm{Hom}(Q_\lambda, Q_\mu) = Q_\lambda^{-1} \otimes Q_\mu$, $\lambda < \mu$; their tensor product is thus isomorphic to the canonical line bundle $K_{M(V)} = \det(T^\star M(V))$:

$$\text{(2.4)} \qquad K_{M(V)} = Q_1^{1-r} \otimes \ldots \otimes Q_\lambda^{2\lambda-r-1} \otimes \ldots \otimes Q_r^{r-1} = Q^c$$

where $c = (1 - r, \ldots, r - 1)$; c will be called the canonical weight of $M(V)$.

• *Case of incomplete flag manifolds.*

More generally, given any sequence of integers $s = (s_0, \ldots, s_m)$ such that $0 = s_0 < s_1 < \ldots < s_m = r$, we may consider the manifold $M_s(V)$ of incomplete flags

$$V = V_{s_0} \supset V_{s_1} \supset \ldots \supset V_{s_m} = \{0\} \ , \quad \mathrm{codim}_{\mathbb{C}} V_{s_j} = s_j \ .$$

On $M_s(V)$ we still have tautological vector bundles $V_{s,j}$ of rank $r - s_j$ and line bundles

$$\text{(2.5)} \qquad Q_{s,j} = \det(V_{s,j-1}/V_{s,j}) \ , \quad 1 \leq j \leq m \ .$$

For any r-tuple $a \in \mathbb{Z}^r$ such that $a_{s_{j-1}+1} = \ldots = a_{s_j}$, $1 \le j \le m$, we set

$$Q_s^a = Q_{s,1}^{a_{s_1}} \otimes \ldots \otimes Q_{s,m}^{a_{s_m}} \quad .$$

If $\eta : M(V) \to M_s(V)$ is the natural projection, then

$$(2.6) \qquad \eta^\star V_{s,j} = V_{s_j} \quad , \quad \eta^\star Q_{s,j} = Q_{s_{j-1}+1} \otimes \ldots \otimes Q_{s_j} \quad , \quad \eta^\star Q_s^a = Q^a \quad .$$

On the other hand, one has the identification

$$M_s(V) = \mathrm{Isom}(\mathbb{C}^r, V)/B_s$$

where B_s is the parabolic subgroup of matrices $(z_{\lambda\mu})$ with $z_{\lambda\mu} = 0$ for all λ, μ such that there exists an integer $j = 1, \ldots, m-1$ with $\lambda \le s_j$ and $\mu > s_j$. We define U_s as the unipotent subgroup of lower triangular matrices $(z_{\lambda\mu})$ with $z_{\lambda\mu} = 0$ for all λ, μ such that $s_{j-1} < \lambda \ne \mu \le s_j$ for some j , and we set $B^s = {}^tB_s$, $U^s = {}^tU_s$. In the same way as above, we get

$$(2.7) \qquad TM_s(V) = \mathrm{Hom}(V,V)/W_s \quad , \quad W_s = \{g \ ; \ g(V_{s_{j-1}}) \subset V_{s_j}\} \quad ,$$

$$(2.8) \qquad T^\star M_s(V) = W_s^\perp \quad ,$$

$$(2.9) \qquad K_{M_s(V)} = Q_{s,1}^{s_1 - r} \otimes \ldots \otimes Q_{s,j}^{s_{j-1}+s_j-r} \otimes \ldots \otimes Q_{s,m}^{s_{m-1}} = Q_s^{c(s)}$$

where $c(s) = (s_1 - r, \ldots, s_1 - r, \ldots, s_{j-1} + s_j - r, \ldots, s_{j-1} + s_j - r, \ldots)$ is the canonical weight of $M_s(V)$.

• *Curvature form of the line bundle Q^a .*

Assume now that V is a *hermitian* vector space. Then all our bundles carry a natural hermitian metric. We are going to compute the curvature of Q^a at any point $[e] \in M(V)$. Choose an orthonormal basis $(e_1, \ldots, e_r)$ of V which corresponds to the given point $[e]$. It is clear that $eB_re^{-1} \subset \mathrm{Gl}(V)$ is the isotropy subgroup of $[e]$, whereas $eU^re^{-1}.[e] = [eU^r]$ is an affine open subset of $M(V)$, corresponding to bases $(\varsigma_1, \ldots, \varsigma_r)$ of V such that

$$\varsigma_\mu = e_\mu + \sum_{\lambda<\mu} z_{\lambda\mu} e_\lambda \quad , \quad 1 \le \mu \le r \quad , \quad z_{\lambda\mu} \in \mathbb{C} \quad .$$

Then $(z_{\lambda\mu})_{1 \le \lambda < \mu \le r}$ is a coordinate system on $[eU^r]$ and the map

$$M(V) \ni [\varsigma] \longmapsto \tilde{\xi}_\mu := \varsigma_\mu \mod V_\mu$$

is a local section of $Q_\mu = V_{\mu-1}/V_\mu$. Hence

$$c(Q_\mu) = -d'd'' \log |\tilde{\xi}_\mu|^2 \quad .$$

Let us identify $\tilde{\xi}_\mu$ with the orthogonal projection of ς_μ on $V_{\mu-1} \cap V_\mu^\perp$. Then Gram-Schmidt's orthogonalization process yields

$$\tilde{\xi}_\mu = \varsigma_\mu - \sum_{\nu>\mu} \frac{\langle \varsigma_\mu, \tilde{\xi}_\nu \rangle}{|\tilde{\xi}_\nu|^2} \tilde{\xi}_\nu \quad , \quad |\tilde{\xi}_\mu|^2 = |\varsigma_\mu|^2 - \sum_{\nu>\mu} \frac{|\langle \varsigma_\mu, \tilde{\xi}_\nu \rangle|^2}{|\tilde{\xi}_\nu|^2} \quad .$$

Since $\langle \varsigma_\mu, \varsigma_\nu \rangle = \bar{z}_{\mu\nu} + \sum_{\lambda<\mu} z_{\lambda\mu}\bar{z}_{\lambda\nu}$, it follows by backward induction on ν that $\langle \varsigma_\mu, \tilde{\varsigma}_\nu \rangle = \bar{z}_{\mu\nu} + O(|z|^2)$ for $\nu > \mu$, hence

$$|\tilde{\varsigma}_\mu|^2 = 1 + \sum_{\lambda<\mu} |z_{\lambda\mu}|^2 - \sum_{\nu>\mu} |z_{\mu\nu}|^2 + O(|z|^3) .$$

We obtain therefore

$$(2.10) \qquad c(Q_\mu)_{[e]} = -\sum_{\lambda<\mu} dz_{\lambda\mu} \wedge d\bar{z}_{\lambda\mu} + \sum_{\nu>\mu} dz_{\mu\nu} \wedge d\bar{z}_{\mu\nu} ,$$

$$(2.11) \qquad c(Q^a)_{[e]} = \sum_\mu a_\mu c(Q_\mu) = \sum_{\lambda<\mu} (a_\lambda - a_\mu) dz_{\lambda\mu} \wedge d\bar{z}_{\lambda\mu} .$$

COROLLARY 2.12. — *Q^a is ≥ 0 if and only if $a_1 \geq a_2 \geq \ldots \geq a_r$. If there exists an index j such that $a_j < a_{j+1}$, then $H^0(M(V), Q^a) = 0$.*

Proof. — Only the second statement and the "only if" part of the first remain to be proved. Let us observe that the projection $\eta : M(V) \longrightarrow M_s(V)$, $s = (0, \ldots, j-1, j+1, \ldots, r)$, is a bundle with fibers $P(V_{j-1}/V_{j+1}) \simeq \mathbb{P}^1$. The restriction of Q_λ to each fiber is trivial if $\lambda \neq j, j+1$ whereas $Q_{j \restriction \mathbb{P}^1} \simeq O(1)$ and $Q_{j+1 \restriction \mathbb{P}^1} \simeq O(-1)$. Therefore $Q^a_{\restriction \mathbb{P}^1} \simeq O(a_j - a_{j+1})$ cannot have any non-zero section or any semi-positive metric if $a_j < a_{j+1}$. ■

When $a_1 \geq \ldots \geq a_r$, the bundle Q^a is not necessarily > 0 on $M(V)$; in fact one can write Q^a as the induced bundle $\eta^\star Q^a_s$ where $s_1 < \ldots < s_{m-1}$ is the sequence of integers $\lambda = 1, \ldots, r-1$ such that $a_{\lambda+1} > a_\lambda$. The affine open subset $[eU^s] \subset M_s(V)$ is a neighborhood of $[e]$, and $M_s(V)$ has local coordinates $(z_{\lambda\mu})$ where λ, μ are such that $\lambda \leq s_{j-1} < s_j \leq \mu$ for some j , i.e. $a_\lambda > a_\mu$. The curvature of Q^a_s is given formally by the same expression as (2.11) :

$$(2.13) \qquad c(Q^a_s)_{[e]} = \sum_{a_\lambda > a_\mu} (a_\lambda - a_\mu) dz_{\lambda\mu} \wedge d\bar{z}_{\lambda\mu} .$$

We see therefore that $Q^a_s > 0$ on $M_s(V)$.

• *Cohomology groups of Q^a .*

It remains now to compute $H^0(M_s(V), Q^a_s) \simeq H^0(M(V), Q^a)$ when $a_1 \geq \ldots \geq a_r$. Without loss of generality we may assume that $a_r \geq 0$, because $Q_1 \otimes \ldots \otimes Q_r = \det V$ is a trivial bundle.

PROPOSITION 2.14. — *For all integers $a_1 \geq a_2 \geq \ldots \geq a_r \geq 0$, there is a canonical isomorphism*

$$H^0(M(V), Q^a) = \Gamma^a V ,$$

where $\Gamma^a V \subset S^{a_1}V \otimes \ldots \otimes S^{a_r}V$ *is the set of polynomials* $f(\varsigma_1^\star, \ldots, \varsigma_r^\star)$ *on* $(V^\star)^r$ *which are homogeneous of degree* a_λ *with respect to* $\varsigma_\lambda^\star$ *and invariant under the left action of* U_r *on* $(V^\star)^r = \mathrm{Hom}(V, \mathbb{C}^r)$:

$$f(\varsigma_1^\star, \ldots, \varsigma_{\lambda-1}^\star, \varsigma_\lambda^\star + \varsigma_\nu^\star, \ldots, \varsigma_r^\star) = f(\varsigma_1^\star, \ldots, \varsigma_r^\star) \quad , \quad \forall \nu < \lambda \ .$$

Proof. — To any section $\sigma \in H^0(M(V), Q^a)$ we associate the holomorphic function f on $\mathrm{Isom}(V, \mathbb{C}^r) \subset (V^\star)^r$ defined by

$$f(\varsigma_1^\star, \ldots, \varsigma_r^\star) = (\varsigma_1^\star)^{a_1} \otimes \ldots \otimes (\varsigma_r^\star)^{a_r} . \sigma([\varsigma_1, \ldots, \varsigma_r])$$

where $(\varsigma_1, \ldots, \varsigma_r)$ is the dual basis of $(\varsigma_1^\star, \ldots, \varsigma_r^\star)$, and where the linear form induced by $\varsigma_\lambda^\star$ on $Q_\lambda = V_{\lambda-1}/V_\lambda \simeq \mathbb{C}\varsigma_\lambda$ is still denoted $\varsigma_\lambda^\star$. Let us observe that f is homogeneous of degree a_λ in $\varsigma_\lambda^\star$ and locally bounded in a neighborhood of every r–tuple of $(V^\star)^r \setminus \mathrm{Isom}(V, \mathbb{C}^r)$ (because $M(V)$ is compact and $a_\lambda \geq 0$) . Therefore f can be extended to a polynomial on all $(V^\star)^r$. The invariance of f under U_r is clear. Conversely, such a polynomial f obviously defines a unique section σ on $M(V)$. ∎

From the definition of $\Gamma^a V$, we see that

$$S^k V = \Gamma^{(k,0,\ldots,0)} V \ , \tag{2.15}$$

$$\Lambda^k V = \Gamma^{(1,\ldots,1,0,\ldots,0)} V \ . \tag{2.16}$$

For arbitrary $a \in \mathbb{Z}^r$, proposition 2.14 remains true if we set

$$\Gamma^a V = \Gamma^{(a_1 - a_r, \ldots, a_{r-1} - a_r, 0)} V \otimes (\det V)^{a_r} \quad \text{when } a \text{ is non-increasing} \ ,$$

$$\Gamma^a V = 0 \qquad \text{otherwise} \ .$$

The elements $a \in \mathbb{Z}^r$ will be ordered according to the partial ordering :

$$a \succcurlyeq b \quad \text{iff} \quad \sum_{1 \leq \lambda \leq \mu} a_\lambda \geq \sum_{1 \leq \lambda \leq \mu} b_\lambda \ , \quad 1 \leq \mu \leq r \ .$$

Bott's theorem [3] shows that $\Gamma^a V$ is an irreducible representation of $\mathrm{Gl}(V)$ of highest weight a; all irreducible representations of $\mathrm{Gl}(V)$ are in fact of this type (*cf.* Kraft [10]). In particular, since the weights of the action of a maximal torus $T^r \subset \mathrm{Gl}(V)$ on $V^{\otimes k}$ verify $a_1 + \ldots + a_r = k$ and $a_\lambda \geq 0$, we have a canonical $\mathrm{Gl}(V)$–isomorphism

$$V^{\otimes k} = \bigoplus_{\substack{a_1 + \ldots + a_r = k \\ a_1 \geq \ldots \geq a_r \geq 0}} \mu(a, k) \, \Gamma^a V \tag{2.17}$$

where $\mu(a, k) > 0$ is the multiplicity of the isotypical factor $\Gamma^a V$ in $V^{\otimes k}$.

Bott's formula (*cf.* also Demazure [6] for a very simple proof) gives in fact the expression of all cohomology groups $H^q(M(V), Q^a)$, but we will need them only in the case of *dominant* weights $a_1 \geq \ldots \geq a_r$.

PROPOSITION 2.18. — *Set* $N = \dim M(V)$, $N(s) = \dim M_s(V)$. *If* $a_{s_j} - a_{s_{j+1}} \geq 1$, *then*

(a) $H^{N(s),q}(M_s(V), Q_s^a) = 0$ *for all* $q \geq 1$,

(b) $H^{N(s),0}(M_s(V), Q_s^a) = \Gamma^{a+c(s)}V$.

Proof. — Under the assumption of (a), Q_s^a is > 0 by (2.13). The result follows therefore from the Kodaira-Akizuki-Nakano theorem. Now (b) is a consequence of proposition 2.14 since

$$H^{N(s),q}(M_s(V), Q_s^a) = H^q(M_s(V), K_{M_s(V)} \otimes Q_s^a) = H^q(M_s(V), Q_s^{a+c(s)}) \ . \blacksquare$$

3. An isomorphism theorem

Our aim here is to generalize Griffiths and Le Potier's isomorphism theorems ([8], [13]) in the case of arbitrary flag bundles, following the simple method of Schneider [14] .

Let E be a holomorphic vector bundle of rank r on a compact complex n–dimensional manifold X . For every sequence $0 = s_0 < s_1 < \ldots < s_m = r$, we associate to E its flag bundle $Y = M_s(E) \longrightarrow X$. If $a \in \mathbb{Z}^r$ is such that $a_{s_{j-1}+1} = \ldots = a_{s_j}$, $1 \leq j \leq m$, we may define a line bundle $Q_s^a \longrightarrow Y$ just as we did in §2 . Let us set

$$\Omega_X^p = \Lambda^p T^\star X \quad , \quad \Omega_Y^p = \Lambda^p T^\star Y \ .$$

One has an exact sequence

$$0 \longrightarrow \pi^\star \Omega_X^1 \longrightarrow \Omega_Y^1 \longrightarrow \Omega_{Y/X}^1 \longrightarrow 0 \tag{3.1}$$

where $\Omega_{Y/X}^1$ is by definition the bundle of relative differential 1–forms along the fibers of the projection $\pi : Y = M_s(E) \longrightarrow X$. One may then define a decreasing filtration of Ω_Y^t as follows :

$$F^{p,t} = F^p(\Omega_Y^t) = \pi^\star(\Omega_X^p) \wedge \Omega_Y^{t-p} \ . \tag{3.2}$$

The corresponding graded bundle is given by

$$G^{p,t} = F^{p,t}/F^{p+1,t} = \pi^\star(\Omega_X^p) \otimes \Omega_{Y/X}^{t-p} \ . \tag{3.3}$$

Over any open subset of X where E is a trivial bundle $X \times V$ with $\dim_{\mathbb{C}} V = r$, the exact sequence (3.1) splits as well as the filtration (3.2). Using proposition 2.18, we obtain the following lemma.

LEMMA. — *For every weight* $a \in \mathbb{Z}^r$ *such that* $a_{s_j} - a_{s_{j+1}} \geq 1$, $1 \leq j \leq m-1$, *the sheaf of sections of* $\Omega_{Y/X}^{N(s)} \otimes Q_s^a$ *has direct images*

$$\begin{cases} R^q\pi_\star(\Omega_{Y/X}^{N(s)} \otimes Q_s^a) = 0 \quad \text{for } q \geq 1 \ , \\ \pi_\star(\Omega_{Y/X}^{N(s)} \otimes Q_s^a) = \Gamma^{a+c(s)}E \ . \end{cases} \tag{3.4}$$

Let L be an arbitrary line bundle on X. Under the hypothesis $a_{s_j} - a_{s_{j+1}} \geq 1$, formulas (3.3) and (3.4) yield

$$R^q\pi_\star(G^{p,p+N(s)} \otimes Q^a_s \otimes \pi^\star L) = 0 \quad \text{for} \quad q \geq 1 \ ,$$
$$\pi_\star(G^{p,p+N(s)} \otimes Q^a_s \otimes \pi^\star L) = \Omega^p_X \otimes \Gamma^{a+c(s)}E \otimes L \ .$$

The Leray spectral sequence implies therefore :

THEOREM 3.5. — *If* $a_{s_j} - a_{s_{j+1}} \geq 1$, *then for all* $q \geq 0$

$$H^q(Y, G^{p,p+N(s)} \otimes Q^a_s \otimes \pi^\star L) \simeq H^{p,q}(X, \Gamma^{a+c(s)}E \otimes L) \ .$$

When $p = n$, $G^{n,n+N(s)}$ is the only non-vanishing quotient in the filtration of the canonical line bundle $\Omega^{n+N(s)}_Y$. We thus obtain the following generalization of Griffiths' isomorphism theorem [8] :

$$H^{n+N(s),q}(M_s(E), Q^a_s \otimes \pi^\star L) \simeq H^{n,q}(X, \Gamma^{a+c(s)}E \otimes L) \ . \tag{3.6}$$

In order to carry over results for line bundles to vector bundles, one needs the following lemma.

LEMMA 3.7. — *Assume that* $a_{s_1} > a_{s_2} > \ldots > a_{s_m} \geq 0$. *Then*

(a) $E \geq 0$ (*resp.* > 0) $\implies$ $Q^a_s \geq 0$ (*resp.* > 0) ;

(b) $E \geq 0$ *and* $L > 0$ $\implies$ $Q^a_s \otimes \pi^\star L > 0$;

(c) E *ample* $\implies$ Q^a_s *ample* .

Proof. — Part (a) will be proved in §4 (*cf.* formula (4.9)) and (b) follows from the fact that $c(Q^a_s) > 0$ along the fibers of π .

(c) By definition of an ample vector bundle (Hartshorne [9]) , S^kE is very ample for $k \geq k_0$ large enough. Hence $\Gamma^{ka}E$, which is a direct summand in $S^{ka_1}E \otimes \ldots \otimes S^{ka_r}E$, is also very ample for $k \geq k_0$. Now, formula (2.14) shows that $Q^a_s > 0$ along the fibers of π , hence Q^{ka}_s is very ample along each fiber for $k \geq k_1$. Since $\pi_\star(Q^{ka}_s) = \Gamma^{ka}E$, we conclude that Q^{ka}_s is very ample for $k \geq \max(k_0, k_1)$. ∎

We are now ready to attack the proof of the main theorem. We study first the special case $p = n$.

THEOREM 3.8. — *Let* $a \in \mathbb{Z}^r$ *be such that*

$$a_1 \geq a_2 \geq \ldots \geq a_h > a_{h+1} = \ldots = a_r = 0 \ , \quad 1 \leq h \leq r-1 \ .$$

Assume that E *is ample and* $L \geq 0$, *or* $E \geq 0$ *and* L *ample. Then*

$$H^{n,q}(X, \Gamma^a E \otimes (\det E)^l \otimes L) = 0 \quad \text{for} \quad q \geq 1 \ , \quad l \geq h \ .$$

Proof of theorem 3.8 and of the main theorem. —

Let $s_1 > \ldots > s_{m-1}$ be the sequence of integers $\lambda = 1, \ldots, r-1$ such that $a_{\lambda+1} > a_\lambda$. Then theorem 3.5 implies

$$H^{p,q}(X, \Gamma^a E \otimes (\det E)^l \otimes L) \simeq H^q(Y, G^{p,p+N(s)} \otimes Q_s^b \otimes \pi^\star L) \tag{3.9}$$

where $b = a - c(s) + (l, \ldots, l)$. The canonical weight $c(s)$ is non-decreasing and $c(s)_r = s_{m-1} = h$, hence

$$b_{s_1} > \ldots > b_{s_m} = l - h \ .$$

Lemma 3.7 shows that $Q_s^b \otimes \pi^\star L > 0$ if $l \geq h$. Now, it is clear that $F^{p,p+N(s)} = \Omega_Y^{p+N(s)}$. One gets thus an exact sequence

$$0 \longrightarrow F^{p+1,p+N(s)} \longrightarrow \Omega_Y^{p+N(s)} \longrightarrow G^{p,p+N(s)} \longrightarrow 0 \ . \tag{3.10}$$

The Kodaira-Akizuki-Nakano vanishing theorem (1.4) applied to $Q_s^b \otimes \pi^\star L$ with $\dim Y = n + N(s)$ yields

$$H^q(Y, \Omega_Y^{p+N(s)} \otimes Q_s^b \otimes \pi^\star L) = 0 \quad \text{for} \quad p + q \geq n + 1 \ .$$

The cohomology groups in (3.9) will therefore vanish if and only if

$$H^{q+1}(Y, F^{p+1,p+N(s)} \otimes Q_s^b \otimes \pi^\star L) = 0 \ . \tag{3.11}$$

This is obvious if $p = n$, for $F^{n+1,n+N(s)} = 0$. In the general case $p < n$, we will establish in §4 that (3.11) holds for $p + q \geq n$ and $b_r = l - h \geq A(n,p,q)$. This will be done by means of a suitable curvature estimate for the bundle $F^{p+1,p+N(s)}$. ∎

Remark 3.12. — If $p + q = n$, we still obtain some result, namely that the canonical map

$$H^{p+N(s),q}(Y, Q_s^b \otimes \pi^\star L) \longrightarrow H^{p,q}(X, \Gamma^a E \otimes (\det E)^l \otimes L)$$

is onto when $l \geq h + A(n,p,q)$.

Remark 3.13. — If the exact sequence (3.10) splits, then (3.11) is an immediate consequence of the Kodaira-Nakano-Akizuki theorem. However, Peternell-Le Potier and Schneider [11] , [12] have shown that in general the filtration $F^\bullet(\Omega_Y^t)$ does not split, and this is the reason why we have to introduce additional considerations in order to prove (3.11).

4. A curvature estimate for the subbundle $F^{p+1,p+N(s)}$.

We assume here that $E, L \longrightarrow X$ are hermitian vector bundles of respective ranks $r, 1$ and that $E \geq 0$ and $L > 0$, or $E > 0$ and $L \geq 0$. Let $a \in \mathbb{Z}^r$ be such that $a_1 \geq \ldots \geq a_r \geq 0$ and let $s_1 < \ldots < s_{m-1}$ be the sequence of integers $\lambda = 1, \ldots, r-1$ such that $a_{\lambda+1} > a_\lambda$. We set for simplicity

$$Y = M_s(E) \ , \ \Omega = \Omega_Y^{p+N(s)} \ , \ F = F^{p+1,p+N(s)} \ , \ G = G^{p,p+N(s)} = \Omega/F \ .$$

Our aim is to prove that the analogue of (3.11) :

$$H^{q+1}(Y, F \otimes Q_s^a \otimes \pi^\star L) = 0 \tag{4.1}$$

holds when $p+q \geq n$ and $a_r \geq A(n,p,q)$. Let us consider the exact sequences of vector bundles over Y :

$$\begin{cases} 0 \longrightarrow F \longrightarrow \Omega \longrightarrow G \longrightarrow 0 \ , \\ 0 \longrightarrow F(a) \longrightarrow \Omega(a) \longrightarrow G(a) \longrightarrow 0 \ , \end{cases} \tag{4.2}$$

where the second sequence arises from the first one after taking tensor products with the line bundle $Q_s^a \otimes \pi^\star L$. Then Y can be equipped with the Kähler metric $\omega = ic(Q_s^a \otimes \pi^\star L)$; the positivity of ω is a consequence of (4.9) below. To every smooth form v of type $(p+N(s), q+1)$ with values in $Q_s^a \otimes \pi^\star L$, let us apply the Akizuki-Nakano inequality (1.3) , where $\dim Y = n + N(s)$:

$$||D''_{\Omega(a)} v||^2 + ||D''^\star_{\Omega(a)} v||^2 \geq (p+q-n+1)) \ ||v||^2 \ . \tag{4.3}$$

With respect to the orthogonal C^∞–splitting $\Omega \simeq F \oplus G$, the Chern connections of Ω , F , G are related by the well-known formula (cf. [8]) :

$$D_\Omega = \begin{pmatrix} D_F & -\beta^\star \wedge \bullet \\ \beta \wedge \bullet & D_G \end{pmatrix} \ , \ \beta \in C^\infty(\Lambda^{1,0} T^\star Y \otimes \mathrm{Hom}(F,G)) \ ;$$

$\beta^\star$ is a D''–closed (0,1)–form with values in $\mathrm{Hom}(G,F)$, and its cohomology class is the obstuction to the existence of a global splitting of (4.2). We obtain therefore

$$D''_{\Omega(a)} = \begin{pmatrix} D''_{F(a)} & -\beta^\star \wedge \bullet \\ 0 & D''_{G(a)} \end{pmatrix} \ , \ D''^\star_{\Omega(a)} = \begin{pmatrix} D''^\star_{F(a)} & 0 \\ -\beta \ \lrcorner \ \bullet & D''^\star_{G(a)} \end{pmatrix} \ ,$$

where $\lrcorner$ denotes the interior product of differential forms combined with the evaluation map $\mathrm{Hom}(F,G) \times F \to G$ (note that $\beta \lrcorner \bullet = (\beta^\star \wedge \bullet)^\star$). For every $(0, q+1)$–form f with values in $F(a)$ we get

$$D''_{F(a)} f = D''_{\Omega(a)} f \ , \ ||D''^\star_{F(a)} f||^2 = ||D''^\star_{\Omega(a)} f||^2 - ||\beta \lrcorner f||^2 \ . \tag{4.4}$$

From (4.3) and (4.4), we see that the vanishing property (4.1) will hold if

$$|\beta \lrcorner f|^2 < (p+q-n+1) \, |f|^2 \tag{4.5}$$

at every point of Y . We are going to compute β explicitly in suitable coordinate systems on Y . Let $y^0 \in Y$ be an arbitrary point and $(x_1, \ldots, x_n)$ local coordinates on X centered at the point $x^0 = \pi(y^0)$.

LEMMA. — *There exists a local holomorphic frame* $(e_1, \ldots, e_r)$ *of* E *such that* y^0 *coincides with the flag* $[e_1(x^0), \ldots, e_r(x^0)]$ *and*

$$\langle e_\lambda(x), e_\mu(x)\rangle = \delta_{\lambda\mu} - \sum_{i,j} c_{ij\lambda\mu}\, x_i\overline{x}_j + O(|x|^3) \ , \tag{4.6}$$

where $(c_{ij\lambda\mu})$ *is the curvature tensor of* E .

Proof. — Choose a holomorphic frame $(\varepsilon_1, \ldots, \varepsilon_r)$ of E such that $(\varepsilon_1(x^0), \ldots, \varepsilon_r(x^0))$ is orthonormal and $[\varepsilon_1(x^0), \ldots, \varepsilon_r(x^0)] = y^0$. Then the inner product $\langle \varepsilon_\lambda(x), \varepsilon_\mu(x)\rangle$ has a Taylor expansion of the type

$$\begin{aligned}\langle \varepsilon_\lambda(x), \varepsilon_\mu(x)\rangle = \delta_{\lambda\mu} &+ \sum_i (\gamma_{i\lambda\mu}x_i + \overline{\gamma}_{i\mu\lambda}\overline{x}_i)\\ &+ \sum_{i,j}(\gamma_{ij\lambda\mu}x_i\overline{x}_j + \gamma'_{ij\lambda\mu}x_ix_j + \overline{\gamma}'_{ij\mu\lambda}\overline{x}_i\overline{x}_j) + O(|x|^3) \ .\end{aligned}$$

This expansion can be reduced to (4.6) (with suitable coefficients $c_{ij\lambda\mu}$) if one sets

$$e_\lambda(x) = \varepsilon_\lambda(x) - \sum_{i,\mu}\gamma_{i\lambda\mu}x_i\, e_\mu - \sum_{i,j,\mu}\gamma'_{ij\lambda\mu}x_ix_j\, e_\mu \ .$$

Now (4.6) implies

$$\begin{aligned} De_\lambda &= -\sum_{i,j,\mu} c_{ij\lambda\mu}\overline{x}_j dx_i \otimes e_\mu + O(|x|^2) \ ,\\ D^2 e_\lambda &= \sum_{i,j,\mu} c_{ij\lambda\mu} dx_i \wedge d\overline{x}_j \otimes e_\mu + O(|x|) \ ,\end{aligned}$$

showing that the $c_{ij\lambda\mu}$'s are precisely the curvature coefficients at x^0 . ∎

Let us denote by $z = (z_{\lambda\mu})$ the affine coordinates on the fiber $M_s(E_x) \subset Y$ associated to the basis $(e_1(x), \ldots, e_r(x))$. Then $(x_1, \ldots, x_n, z_{\lambda\mu})$ define local coordinates on Y in a neighborhood of y^0 . Assume first that $Y = M_s(E) = M(E)$ is the manifold of complete flags of E . Then we have tautological subbundles $V_\lambda = \mathrm{Vect}(e_{\lambda+1}, \ldots, e_r) \subset \pi^\star E$ and the map

$$Y \ni (x,z) \longmapsto \zeta_\mu = e_\mu(x) + \sum_{\lambda<\mu} z_{\lambda\mu} e_\lambda(x) \tag{4.7}$$

is a local section of $V_{\mu-1}$. Let us denote by $\widetilde{\zeta}_\mu$ the image of ζ_μ in $Q_\mu = V_{\mu-1}/V_\mu$, represented by the orthogonal projection of ζ_μ on $V_{\mu-1} \cap (V_\mu)^\perp$. As in §2, one finds

$$\begin{cases} \widetilde{\zeta}_\lambda = \zeta_\lambda - \displaystyle\sum_{\mu>\lambda} \frac{\langle \zeta_\lambda, \widetilde{\zeta}_\mu\rangle}{|\widetilde{\zeta}_\mu|^2}\widetilde{\zeta}_\mu \ , \quad |\widetilde{\zeta}_\lambda|^2 = |\zeta_\lambda|^2 - \displaystyle\sum_{\mu>\lambda} \frac{|\langle \zeta_\lambda, \widetilde{\zeta}_\mu\rangle|^2}{|\widetilde{\zeta}_\mu|^2} \ ,\\ \langle \zeta_\lambda, \widetilde{\zeta}_\mu\rangle = \overline{z}_{\lambda\mu} - \displaystyle\sum_{i,j} c_{ij\lambda\mu}x_i\overline{x}_j \mod(z^2, x^3, x^2 z) \ \text{for} \ \lambda < \mu \ .\end{cases} \tag{4.8}$$

We need a Taylor expansion of $c(Q_\mu) = -d'd'' \log|\widetilde{\varsigma}_\mu|^2$ up to order 1, hence of $|\widetilde{\varsigma}_\mu|^2$ up to order 3. Moreover, pure terms x^3 , z^3 will not play any role because the Kähler property of $c(Q_\mu)$ enables one to get rid of the terms $O(|x|dx \wedge d\overline{x})$, $O(|z|dz \wedge d\overline{z})$. Therefore, we are interested only in terms of degree ≤ 2 and in mixed terms xz^2 , x^2z . Thanks to formulas (4.7) and (4.8), we get the following equalities modulo the ideal (x^3, z^3, x^2z^2) :

$$|\widetilde{\varsigma}_\lambda|^2 \sim |\varsigma_\lambda|^2 - \sum_{\mu>\lambda} |\langle\varsigma_\lambda, \widetilde{\varsigma}_\mu\rangle|^2 \ ,$$

$$\begin{aligned}|\varsigma_\lambda|^2 \sim 1 &- \sum_{i,j} c_{ij\lambda\lambda} x_i \overline{x}_j + \sum_{\mu<\lambda} |z_{\mu\lambda}|^2 \\ &- \sum_{i,j,\mu<\lambda} c_{ij\lambda\mu} x_i \overline{x}_j \overline{z}_{\mu\lambda} - \sum_{i,j,\mu<\lambda} \overline{c}_{ij\lambda\mu} \overline{x}_i x_j z_{\mu\lambda} \ ,\end{aligned}$$

$$|\langle\varsigma_\lambda, \widetilde{\varsigma}_\mu\rangle|^2 \sim |z_{\lambda\mu}|^2 - \sum_{i,j} c_{ij\lambda\mu} x_i \overline{x}_j z_{\lambda\mu} - \sum_{i,j} \overline{c}_{ij\lambda\mu} \overline{x}_i x_j \overline{z}_{\lambda\mu} \ .$$

We have now

$$c(Q^a) = \sum_\lambda a_\lambda c(Q_\lambda) = d'd''\Big(-\sum_\lambda a_\lambda \log|\widetilde{\varsigma}_\lambda|^2\Big) \ ,$$

$$\begin{aligned}-\sum_\lambda a_\lambda \log|\widetilde{\varsigma}_\lambda|^2 \sim &\sum_{i,j,\lambda} a_\lambda c_{ij\lambda\lambda} x_i \overline{x}_j + \sum_{\lambda<\mu} (a_\lambda - a_\mu)|z_{\lambda\mu}|^2 \\ &- \sum_{i,j,\lambda<\mu} (a_\lambda - a_\mu) c_{ij\lambda\mu} x_i \overline{x}_j z_{\lambda\mu} \\ &- \sum_{i,j,\lambda<\mu} (a_\lambda - a_\mu) \overline{c}_{ij\lambda\mu} \overline{x}_i x_j \overline{z}_{\lambda\mu} \ .\end{aligned}$$

We find therefore

$$\begin{aligned}c(Q^a) = &\sum_{i,j} \Big(\sum_\lambda a_\lambda c_{ij\lambda\lambda} + O(|x|)\Big) dx_i \wedge d\overline{x}_j \\ &+ \sum_{\lambda<\mu} (a_\lambda - a_\mu) dz_{\lambda\mu} \wedge d\overline{z}_{\lambda\mu} + O(|z|dz \wedge d\overline{z}) \\ &- \sum_{i,j,\lambda<\mu} (a_\lambda - a_\mu) c_{ij\lambda\mu} \big(z_{\lambda\mu} dx_i \wedge d\overline{x}_j + x_i dz_{\lambda\mu} \wedge d\overline{x}_j\big) \\ &- \sum_{i,j,\lambda<\mu} (a_\lambda - a_\mu) \overline{c}_{ij\lambda\mu} \big(\overline{z}_{\lambda\mu} dx_j \wedge d\overline{x}_i + \overline{x}_i dx_j \wedge d\overline{z}_{\lambda\mu}\big) + O(|x|^2 + |z|^2).\end{aligned}$$

Since $Q^a = \pi_s^\star Q_s^a$, the same identity holds for Q_s^a . At the point y^0 we get

$$c(Q_s^a)_{y^0} = \sum_{i,j,\lambda} a_\lambda c_{ij\lambda\lambda} dx_i \wedge d\overline{x}_j + \sum_{a_\lambda > a_\mu} (a_\lambda - a_\mu) dz_{\lambda\mu} \wedge d\overline{z}_{\lambda\mu} \ . \tag{4.9}$$

Now, $\omega = i(\pi^\star c(L) + c(Q_s^a))$ is Kähler on $Y = M_s(E)$, thus in particular along the fiber $x = 0$ and along the local section $z = 0$. It follows that one can find coordinate changes $x \mapsto x'$, $z'_{\lambda\mu} = \sqrt{a_\lambda - a_\mu}\, z_{\lambda\mu}$ mod z^2 such that the terms $O(|x| dx \wedge d\overline{x})$ and $O(|z| dz \wedge d\overline{z})$ disappear in the expansion of ω , and such that

$$\omega_{ij}(y^0) = c(L)_{ij}(x^0) + \sum_\lambda a_\lambda c_{ij\lambda\lambda} = \delta_{ij} \quad . \tag{4.10}$$

We obtain therefore

$$\begin{aligned}\frac{1}{i}\omega = &\sum_j dx'_j \wedge d\overline{x}'_j + \sum_{\lambda<\mu} dz'_{\lambda\mu} \wedge d\overline{z}'_{\lambda\mu} \\ &- \sum_{i,j,\lambda<\mu} \sqrt{a_\lambda - a_\mu}\; c_{ij\lambda\mu}(z'_{\lambda\mu} dx'_i \wedge d\overline{x}'_j + x'_i dz'_{\lambda\mu} \wedge d\overline{x}'_j) \\ &- \sum_{i,j,\lambda<\mu} \sqrt{a_\lambda - a_\mu}\; \overline{c}_{ij\lambda\mu}(\overline{z}'_{\lambda\mu} dx'_j \wedge d\overline{x}'_i + \overline{x}'_i dx'_j \wedge d\overline{z}'_{\lambda\mu}) + O(|x'|^2 + |z'|^2).\end{aligned}$$

Omitting the primes in the coordinates x' , z' for simplicity, we see that the norms of the basis elements of TY with respect to ω are given modulo $O(|x|^2 + |z|^2)$ by

$$\begin{aligned}\langle \frac{\partial}{\partial x_i}, \frac{\partial}{\partial x_j} \rangle &\sim \delta_{ij} - \sum_{\lambda<\mu} \sqrt{a_\lambda - a_\mu}\; (c_{ij\lambda\mu} z_{\lambda\mu} + c_{ij\mu\lambda} \overline{z}_{\lambda\mu}) \quad , \\ \langle \frac{\partial}{\partial z_{\lambda\mu}}, \frac{\partial}{\partial z_{\lambda'\mu'}} \rangle &\sim \delta_{\lambda\lambda'} \delta_{\mu\mu'} \quad , \\ \langle \frac{\partial}{\partial z_{\lambda\mu}}, \frac{\partial}{\partial x_j} \rangle &\sim -\sum_i \sqrt{a_\lambda - a_\mu} c_{ij\lambda\mu} x_i \quad .\end{aligned}$$

By duality, we get

$$\begin{aligned}\langle dx_i, dx_j \rangle &\sim \delta_{ij} + \sum_{\lambda<\mu} \sqrt{a_\lambda - a_\mu}\; (\overline{c}_{ij\lambda\mu} \overline{z}_{\lambda\mu} + \overline{c}_{ij\mu\lambda} z_{\lambda\mu}) \quad , \\ \langle dz_{\lambda\mu}, dz_{\lambda'\mu'} \rangle &\sim \delta_{\lambda\lambda'} \delta_{\mu\mu'} \quad , \\ \langle dz_{\lambda\mu}, dx_j \rangle &\sim \sum_i \sqrt{a_\lambda - a_\mu}\; \overline{c}_{ij\lambda\mu} \overline{x}_i \quad .\end{aligned}$$

Taking the exterior derivative in the above estimates, we find that the Chern connection D on $\Omega^1_Y = T^\star Y$ is given in terms of the basis vectors dx_i , $dz_{\lambda\mu}$ by

$$\begin{aligned}D(dx_j) &= \sum_{i,\lambda<\mu} \sqrt{a_\lambda - a_\mu}\; c_{ij\lambda\mu}(dz_{\lambda\mu} \otimes dx_i + dx_i \otimes dz_{\lambda\mu}) + O(|x| + |z|) \quad , \\ D(dz_{\lambda\mu}) &= \ \ 0 \ \ + O(|x| + |z|) \quad .\end{aligned}$$

The subbundle $F = F^{p+1,N(s)}$ (resp. the quotient bundle $G = G^{p,p+N(s)}$) admits at y^0 the orthonormal basis

$dx_I \wedge dz_J$ with $|I|+|J| = p+N(s)$, $|I| \geq p+1$ (resp. $|I| \leq p$, $|J| = N(s)$) .

Let $v = \sum v_{I,J}\, dx_I \wedge dz_J$ be a C^∞ section of F . The (1,0)–form $\beta \wedge v$ is nothing else than the projection of Dv on $G = \Omega/F$. From this observation, one obtains the expression of β at y^0 :

$$(4.11)\quad \beta \wedge v = \sum_{i,j,\lambda<\mu} \sqrt{a_\lambda - a_\mu}\; c_{ij\lambda\mu}\, dx_i \otimes \big(dz_{\lambda\mu} \wedge (\frac{\partial}{\partial x_j} \lrcorner\, v)\big) \mod F \ ,$$

where $\xi \lrcorner\, v$ means contraction of the differential form v by the tangent vector ξ . In fact any differentiation of a factor dx_j in a term $D(v_{I,J}\, dx_I \wedge dz_J)$ decreases of one unity the partial degree $|I|$ when dx_j is differentiated into $c_{ij\lambda\mu}\, dx_i \otimes dz_{\lambda\mu}$. The corresponding part of the differential is thus in G if $|I| = p+1$. For every $(0,q+1)$–form $f = \sum f_{I,J,K,L}\, dx_I \wedge dz_J \wedge d\overline{x}_K \wedge d\overline{z}_L$ with values in $F(a)$, $|I|+|J| = p+N(s)$, $|I| \geq p+1$, $|K|+|L| = q+1$, we obtain consequently

$$(4.12)\ \beta \lrcorner\, f = \sum_{i,j,\lambda<\mu} \sqrt{a_\lambda - a_\mu}\; c_{ij\lambda\mu} \frac{\partial}{\partial \overline{x}_i} \lrcorner\, \big(dz_{\lambda\mu} \wedge (\frac{\partial}{\partial x_j} \lrcorner\, f)\big) \mod F(a) \ .$$

The only terms of f that contribute to the expression of $\beta \lrcorner\, f$ are those for which $|I| = p+1$ and $|J| = N(s)-1$. Let us write $g = \beta \lrcorner\, f$ under the form

$$g = \sum g_{I',J',K',L'}\, dx_{I'} \wedge dz_{J'} \wedge d\overline{x}_{K'} \wedge d\overline{z}_{L'} \ ,$$

where $|I'| = p$, $|J'| = N(s)$, $|K'|+|L'| = q$. Formula (4.12) implies

$$g_{I',J',K',L'} = \sum_{i,j,\lambda<\mu} \pm\sqrt{a_\lambda - a_\mu}\; c_{ij\lambda\mu}\, f_{jI',J'\setminus\{\lambda\mu\},iK',L'} \ ,$$

$$|g_{I',J',K',L'}|^2 \leq \Big(\sum_{i,j,\lambda<\mu} (a_\lambda - a_\mu)|c_{ij\lambda\mu}|^2\Big) \sum_{i,j,\lambda<\mu} |f_{jI',J'\setminus\{\lambda\mu\},iK',L'}|^2 \ ,$$

and $$\sum_{I',J',K',L'} \ \sum_{i,j,\lambda<\mu} |f_{jI',J'\setminus\{\lambda\mu\},iK',L'}|^2 \leq (p+1)(q+1) \sum_{I,J,K,L} |f_{I,J,K,L}|^2 \ .$$

We obtain therefore the inequality

$$(4.13)\qquad |\beta \lrcorner\, f|^2 \leq (p+1)(q+1)\Big(\sum_{i,j,\lambda<\mu} (a_\lambda - a_\mu)|c_{ij\lambda\mu}|^2\Big)\, |f|^2 \ .$$

The main point now is to find an estimate of the sum $\sum_{i,j,\lambda<\mu}(a_\lambda - a_\mu)|c_{ij\lambda\mu}|^2$ under condition (4.10).

LEMMA 4.14. — *Let $(h_{\lambda\mu})_{1\leq\lambda,\mu\leq r}$ be a semi-positive hermitian matrix and let $\alpha_1 \leq \ldots \leq \alpha_r$ be real numbers. Then*

$$\sum_{\lambda<\mu} (\alpha_\mu - \alpha_\lambda)|h_{\lambda\mu}|^2 \leq \frac{1}{4}(\alpha_r - \alpha_1)\Big(\sum_\lambda h_{\lambda\lambda}\Big)^2 \ .$$

Proof. — Use Cauchy-Schwarz inequality $|h_{\lambda\mu}|^2 \leq h_{\lambda\lambda}h_{\mu\mu}$ and take $t_\lambda = h_{\lambda\lambda}$ in the identity

$$\frac{1}{4}(\alpha_r - \alpha_1)\Big(\sum_\lambda t_\lambda\Big)^2 - \sum_{\lambda<\mu}(\alpha_\mu - \alpha_\lambda)\, t_\lambda t_\mu$$
$$= \frac{1}{4}\sum_{1\leq\lambda<r}(\alpha_{\lambda+1} - \alpha_\lambda)(t_1 + \ldots + t_\lambda - t_{\lambda+1} - \ldots - t_r)^2 \geq 0 \ . \ \blacksquare$$

LEMMA 4.15. — *Under condition (4.10) one has*

$$\sum_{i,j,\lambda<\mu}(a_\lambda - a_\mu)|c_{ij\lambda\mu}|^2 \leq \frac{1}{4}n(n+1)\Big(\frac{1}{a_r} - \frac{1}{a_1}\Big) \ .$$

Proof. — Let us apply lemma 4.14 to

$$h_{\lambda\mu} = \sqrt{a_\lambda a_\mu}\sum_{i,j} c_{ij\lambda\mu}t_i\bar{t}_j \ , \quad \alpha_\lambda = \frac{1}{a_\lambda} \ ,$$

where $t = (t_1, \ldots, t_n)$ are arbitrary complex numbers. The Griffiths semi-positivity assumption on $c(E)$ means that $(h_{\lambda\mu})$ is semi-positive for all t . We get

$$\sum_\lambda h_{\lambda\lambda} = \sum_{i,j,\lambda} a_\lambda c_{ij\lambda\lambda}t_i\bar{t}_j \leq |t|^2$$

by condition (4.10), thus

$$(4.16) \qquad \sum_{\lambda<\mu}(a_\lambda - a_\mu)\Big|\sum_{i,j} c_{ij\lambda\mu}t_i\bar{t}_j\Big|^2 \leq \frac{1}{4}\Big(\frac{1}{a_r} - \frac{1}{a_1}\Big)|t|^4 \ .$$

Apply now inequality (4.16) to $t = (e^{i\theta_1}, \ldots, e^{i\theta_n}) \in \mathbb{T}^n$ and integrate the result over $\mathbb{T}^n$. Parseval's identity for Fourier series yields

$$\sum_{\lambda<\mu}(a_\lambda - a_\mu)\Big(\sum_{i\neq j}|c_{ij\lambda\mu}|^2 + \Big|\sum_i c_{ii\lambda\mu}\Big|^2\Big) \leq \frac{1}{4}\Big(\frac{1}{a_r} - \frac{1}{a_1}\Big)n^2 \ .$$

Inequality (4.16) applied to each vector of the standard basis of $\mathbb{C}^n$ yields in the same way

$$\sum_{\lambda<\mu}(a_\lambda - a_\mu)|c_{ii\lambda\mu}|^2 \leq \frac{1}{4}\Big(\frac{1}{a_r} - \frac{1}{a_1}\Big)$$

for all i , and lemma 4.15 follows. ∎

Combining inequality (4.13) with lemma 4.15 we get

$$|\beta \lrcorner f|^2 < \frac{1}{4a_r}n(n+1)(p+1)(q+1)|f|^2 \ ,$$

and using criterion (4.5) we see that $H^{q+1}(Y, F\otimes Q_s^a \otimes \pi^\star L) = 0$ for

$$a_r \geq \frac{n(n+1)(p+1)(q+1)}{4(p+q-n+1)} \quad , \quad p+q \geq n \ .$$

The proof of the main theorem is therefore achieved.

5. On the Borel-Le Potier spectral sequence.

Denote as before $\pi : Y = M_s(E) \longrightarrow X$ the projection. To every integer t and every coherent analytic sheaf $\mathcal{S}$ on Y, one may associate the complex

$$D'' : \mathcal{K}^q = \Gamma\big(Y, C_Y^\infty(\Omega_Y^t \otimes \overline{\Omega}_Y^q) \otimes_{\mathcal{O}_Y} \mathcal{S}\big) \longrightarrow \mathcal{K}^{q+1}$$

of C^∞–differential forms of type (t,q) with values in $\mathcal{S}$. This Dolbeault complex is filtered by the decreasing sequence of subcomplexes

$$D'' : \mathcal{K}_p^q = \Gamma\big(Y, C_Y^\infty(F^p(\Omega_Y^t) \otimes \overline{\Omega}_Y^q) \otimes_{\mathcal{O}_Y} \mathcal{S}\big) \longrightarrow \mathcal{K}_p^{q+1} \ .$$

This gives rise to a spectral sequence which we shall name after Borel and Le Potier, whose E_0, E_1 terms are

$$(5.1) \qquad \begin{cases} E_0^{p,q-p} = \Gamma\big(Y, C_Y^\infty(G^{p,t} \otimes \overline{\Omega}_Y^q) \otimes_{\mathcal{O}_Y} \mathcal{S}\big) \\ E_1^{p,q-p} = H^q(Y, G^{p,t} \otimes \mathcal{S}) \ . \end{cases}$$

The limit term $E_\infty^{p,q-p}$ is the p–graded module corresponding to the filtration of $H^q(\mathcal{K}^\bullet) = H^q(Y, \Omega_Y^t \otimes \mathcal{S})$ by the canonical images of the groups $H^q(\mathcal{K}_p^\bullet)$. Assume that the spectral sequence degenerates in E_2, *i.e.* $d_r : E_r^{p,q-p} \to E_r^{p+r,q+1-(p+r)}$ is zero for all $r \geq 2$ (by Peternell, Le Potier and Schneider [12], the spectral sequence does not degenerate in general in E_1). Then $E_2^{p,q-p} = E_\infty^{p,q-p}$. This equality means that the q–th cohomology group of the E_1–complex

$$d_1 : H^q(Y, G^{p,t} \otimes \mathcal{S}) \longrightarrow H^{q+1}(Y, G^{p+1,t} \otimes \mathcal{S})$$

is the p-graded module corresponding to a filtration of $H^q(Y, \Omega_Y^t \otimes \mathcal{S})$. By Kodaira-Akizuki-Nakano, we get therefore :

PROPOSITION 5.1. — *Assume that E is ample and $L \geq 0$, or $E \geq 0$ and L ample, and that the E_2–degeneracy occurs for the ample invertible sheaf $\mathcal{S} = Q_s^a \otimes \pi^\star L$ on Y. Then the complex*

$$d_1 : H^q(Y, G^{p,t} \otimes Q_s^a \otimes \pi^\star L) \longrightarrow H^{q+1}(Y, G^{p+1,t} \otimes Q_s^a \otimes \pi^\star L)$$

is exact in degree $q \geq n + N(s) + 1 - t$.

This result would be a considerable help for the proof of vanishing theorems. For example, it is shown in [5] that the main vanishing theorem would be true with $l \geq r - 1 + \min\{n-p, n-q\}$.

Since d_1 is the coboundary operator associated to the exact sequence $0 \longrightarrow G^{p+1,t} \longrightarrow F^{p,t}/F^{p+2,t} \longrightarrow G^{p,t} \longrightarrow 0$, it is easy to see that $d_1 = -\beta^\star \wedge \bullet$ where $\beta^\star$ is the D''–closed (0,1)–form of §4, reinterpreted as a (0,1)–section of $\mathrm{Hom}(G^{p,t}, G^{p+1,t})$. Our hope is that the E_2–degeneracy can be proved in all cases by a suitable deepening of the analytic method of §4.

References

[1] Y. AKIZUKI and S. NAKANO. — *Note on Kodaira-Spencer's proof of Lefschetz theorems*, Proc. Jap. Acad., **30** (1954), 266–272.

[2] A. BOREL and A. WEIL. — *Représentations linéaires et espaces homogènes kählériens des groupes de Lie compacts*, Séminaire Bourbaki (exposé n° 100 par J.-P. Serre), (mai 1954), 8 pages.

[3] R. BOTT. — *Homogeneous vector bundles*, Ann. of Math., **66** (1957), 203–248.

[4] J.-P. DEMAILLY. — *Théorèmes d'annulation pour la cohomologie des puissances tensorielles d'un fibré positif*, C. R. Acad. Sci. Paris Sér. I Math., **305** (1987), à paraître.

[5] J.-P. DEMAILLY. — *Vanishing theorems for tensor powers of an ample vector bundle*, Prépublication Institut Fourier, Univ. de Grenoble I (August 1987), submitted to Invent. Math.

[6] B. DEMAZURE. — *A very simple proof of Bott's theorem*, Invent. Math., **33** (1976), 271–272.

[7] R. GODEMENT. — *Théorie des faisceaux*, Hermann, Paris, 1958.

[8] P.A. GRIFFITHS. — *Hermitian differential geometry, Chern classes and positive vector bundles*, Global Analysis, Papers in honor of K. Kodaira, Princeton Univ. Press, Princeton (1969), 185–251.

[9] R. HARTSHORNE. — *Ample vector bundles*, Publ. Math. I.H.E.S., **29** (1966), 63–94.

[10] H. KRAFT. — *Geometrische Methoden in der Invariantentheorie*, Aspekte der Mathematik, Band D1, Braunschweig, Vieweg & Sohn, 1985.

[11] Th. PETERNELL, J. LE POTIER and M. SCHNEIDER. — *Vanishing theorems, linear and quadratic normality*, Invent. Math., **87** (1987), 573–586.

[12] Th. PETERNELL, J. LE POTIER and M. SCHNEIDER. — *Direct images of sheaves of differentials and the Atiyah class*, preprint (1986).

[13] J. LE POTIER. — *Annulation de la cohomologie à valeurs dans un fibré vectoriel holomorphe de rang quelconque*, Math. Ann., **218** (1975), 35–53.

[14] M. SCHNEIDER. — *Ein einfacher Beweis des Verschwindungssatzes für positive holomorphe Vektorraumbündel*, Manuscripta Math., **11** (1974), 95–101.

[15] A.J. SOMMESE. — *Submanifolds of abelian varieties*, Math. Ann., **233** (1978), 229–256.

Decay of eigenfunctions on
Riemannian manifolds

Harold Donnelly
School of Mathematics
Institute for Advanced Study
Princeton, N.J. 08540

1. Introduction

Let M be a complete Riemannian manifold and Δ its Laplacian acting on functions. The unbounded positive self adjoint operator Δ may have both discrete and essential spectrum, all contained in $[0,\infty)$. The discrete spectrum consists of isolated eigenvalues having finite multiplicity. Its complement is called the essential spectrum. Thus, the essential spectrum consists of cluster points of the spectrum and eigenvalues of infinite multiplicity. Suppose that the smallest real number lying in the essential spectrum is $\alpha > 0$ and that there are discrete eigenvalues less than α . Therefore, one has $u \in L^2M$, $\Delta u = \lambda u$, where $\lambda < \alpha$.

There have been several recent results concerning decay of such eigenfunctions u . Upper bounds for the decay in L^2-norm were proved in [6], valid on any complete M . If the Ricci curvature of M is bounded below, these L^2 estimates were converted to L^∞ estimates. The article [5] contains related theorems for quotients of hyperbolic space by geometrically finite groups. If λ_0 is the infimum of the spectrum of Δ , then the eigenspace for λ_0 is one dimensional and contains positive eigenfunctions u_0 . When the Ricci curvature of M is bounded below, one has a pointwise lower bound for the ground state u_0 , [4], [8]. If $\lambda > \lambda_0$, it seems more difficult to obtain lower bounds for u . Since u must be orthogonal to u_0, a pointwise lower bound is rather problematic. Suppose M is simply connected and negatively curved with sectional curvature K approaching -1 at infinity, where specific decay conditions are satisfied for $K+1$ and its derivatives. Under these hypotheses, lower bounds for the spherical averages of u were proved in [8].

The present paper contains two parts. In the first section, we give a more detailed exposition of the earlier results summarized above. The remainder of the manuscript is more technical. We show that one can remove the hypotheses on derivatives of $K+1$ and still establish lower bounds as given in [8] for spherical averages of eigenfunctions. Crucial use is made of the Poisson kernel estimates [9] for elliptic operators in divergence form.

2. Review of earlier work

Suppose that M is any complete Riemannian manifold with basepoint p . The symbol B_r will denote a geodesic ball of radius r centered at p . Let

$\alpha = \inf \operatorname{ess} \operatorname{spec} \Delta$ be positive by hypothesis. Assume that $u \in L^2M$, $\Delta u = \lambda u$, $\lambda < \alpha$. One has the following general L^2 estimate:

Theorem 2.1 For any $\varepsilon > 0$,

$$\int_{M-B_r} |u|^2 \leqq C_1 e^{-2\sqrt{\alpha-\lambda}\, r + \varepsilon r}$$

The constant C_1 may depend upon p and ε.

To proceed further, we assume in addition that the Ricci curvature of M is bounded below by $-(n-1)c$, for some constant $c > 0$. If $x \in M$ is a variable point, then $\mathrm{Vol}(B_1(x))$ denotes the volume of a geodesic ball having radius one and centered at x. As above, $r(x)$ is the distance of x from our basepoint p. We may write

Corollary 2.2 For any $\varepsilon > 0$,

$$|u(x)| \leqq C_2 \mathrm{Vol}^{-1/2}(B_1(x)) e^{(-\sqrt{\alpha-\lambda} + \varepsilon) r(x)}$$

Consequently, if $\mathrm{Vol}(B_1(x))$ is uniformly bounded below for $x \in M$, then $u(x)$ must decay exponentially. This holds for example on the universal covers of compact manifolds. In general, $u(x)$ need not decay exponentially in the L^∞ sense. Just consider the constant function 1 on a manifold of finite volume, such as a Riemann surface with its hyperbolic metric.

The proof of Theorem 2.1 in [6] employs an extension of Agmon's method for the Schrödinger operator [1] in R^n. It seems remarkable that this technique remains valid for any complete Riemannian manifold. To convert this L^2 upper bound to an L^∞ upper bound, Corollary 2.2, one must control the Sobolev constant. According to the usual argument [3], this requires a lower bound for the Ricci curvature.

Suppose that $\lambda_0 = \inf \operatorname{spec} \Delta$. By hypothesis, we assume $\lambda_0 < \alpha$. There exists a positive ground state eigenfunction u_0, $\Delta u_0 = \lambda_0 u_0$. Moreover, any other eigenfunction with eigenvalue λ_0 is a constant multiple of u_0. If $-(n-1)c$ is a lower bound for the Ricci curvature of M, then $\alpha \leq (n-1)^2 c/4$. Set $\beta_0 = (n-1)\sqrt{c}/2 + [(n-1)^2 c/4 - \lambda_0]^{1/2}$. The ground state admits a lower bound:

Theorem 2.3 $u_0(x) \geq C_3 e^{-\beta_0 r(x) - \varepsilon r(x)} u_0(p)$ for any $\varepsilon > 0$, with constant C_3 depending upon ε and p.

The estimate of Theorem 2.3 is reasonable in certain examples. For instance, let M be obtained by a compactly supported perturbation, containing a sufficiently large Euclidean ball, of the metric on a simply connected complete space of constant negative curvature. To prove Theorem 2.3, one employs a lower bound for the heat kernel and uniqueness of positive solutions to the heat equation [8]. The Harnack inequality of [4] yields a similar result with a less satisfactory value for the exponent β_0.

It seems more difficult to derive lower bounds for the eigenfunctions u with eigenvalue $\lambda > \lambda_0$. Clearly, u is orthogonal to u_0 and this forces u to vanish on a subset of M. Apparently, a pointwise lower bound for u is hard to formulate. Under restrictive conditions on M, we proved [8] a lower bound for certain spherical averages of u. Suppose M is simply connected and negatively curved with sectional curvatures K approaching -1 at infinity. The essential spectrum of Δ then coincides with the interval $[(n-1)^2/4,\infty)$. Choose spherical coordinates $[r,\omega]$ centered at p and define

$$A(r) = \left(\int u^2(r,\omega)d\omega\right)^{1/2}$$

Since λ lies below the essential spectrum, $A(r) > 0$ for r sufficiently large.

Let ∇ denote the covariant derivative of S^{n-1} with respect to its standard metric. The sectional curvature K along radial geodesics may be considered as a function on S^{n-1}. Suppose that K satisfies the following decay conditions:

$$\begin{aligned}
&\text{i)}\ \int_0^\infty r|K+1|dr < d_1\\
&\text{ii)}\ \lim_{r\to\infty} r|K+1| = 0\text{, uniformly on } S^{n-1}\\
&\text{iii)}\ \int_0^\infty |\nabla K|e^{2r}dr < d_2\\
&\text{iv)}\ \int_0^\infty |\nabla^2 K|e^{2r}dr < d_3\\
&\text{v)}\ \lim_{r\to\infty} r\left|\frac{\partial K}{\partial r}\right| = 0\text{, uniformly on } S^{n-1}
\end{aligned} \tag{2.4}$$

for constants d_1, d_2, d_3. Under these hypotheses, one knows [8]:

Theorem 2.5 Given $\varepsilon > 0$,

$$A(r) > C_4 \exp\left(-\left[\frac{(n-1)^2}{4} - \lambda\right]^{1/2} r - \left[\frac{n-1}{2} + \varepsilon\right] r\right)$$

with C_4 depending upon ε and p. One may construct simple examples with arbitrarily many eigenfunctions having eigenvalue $\lambda < (n-1)^2/4$. Just take $K = -1/2$ on a large compact set and apply the minimax principle.

3. Lower bounds for the spherical averages

Suppose that M^n is a complete simply connected negatively curved Riemannian manifold whose sectional curvatures converge to -1 at infinity. Let u be a non-zero eigenfunction of Δ with eigenvalue $\lambda < (n-1)^2/4$ lying below the essential spectrum. Our goal is to remove the hypotheses (2.4) iii), iv), v) and still establish the conclusion of Theorem 2.5. The rest of the paper will be devoted to this purpose. For the sake of clarity, we present complete details of the proof, although this necessitates some repetition of the material in [8].

Fix a basepoint $p \in M$. The exponential map $\exp: T_pM \to M$ is a diffeomorphism by the Hadamard Cartan theorem. For each $\omega \in S^{n-1}$, let $\gamma(\omega,t)$ be the unit speed geodesic starting at p with direction ω. If $\alpha \in S^{n-1}$ is perpendicular to ω, we suppose that $K(\alpha,\omega,t)$ is the sectional curvature of the plane obtained by parallel translation of the (α,ω) plane along $\gamma(\omega,t)$. Assume the following conditions:

$$\text{i)} \int_0^\infty t|K+1|\,dt < d_1 \qquad\qquad (3.1)$$

$$\text{ii)} \lim_{t\to\infty} t|K+1| = 0 \text{ , uniformly in } \omega$$

for a constant d_1, independent of ω.

One employs the method of Jacobi fields to deduce the asymptotic behavior of the metric in geodesic spherical coordinates. A vector field y, along γ, is called a Jacobi field [11] if

$$y'' + R(y,v)v = 0$$

Here y'' is the second covariant derivative along γ, R is the curvature tensor, and v is the unit tangent vector to γ. We are interested in Jacobi fields which are normal to γ. Suppose $\Gamma(\omega,t)$ is a straight line in T_pM which maps to $\gamma(\omega,t)$ under $\exp: T_pM \longrightarrow M$. Assume that $E_1, E_2, \ldots, E_{n-1}$ is a Euclidean parallel frame field for the normal space to Γ in T_pM. Set $y_i(t) = \exp_*(tE_i)$. It follows from the variational characterization of Jacobi fields that $y_i(t)$ is the unique Jacobi field along γ satisfying $y_i(0) = 0$, $y_i'(0) = E_i$.

We say that a vector field is constant if it is parallel along γ. If the decay conditions (3.1) are satisfied, then one has

<u>Proposition 3.2</u> There exists a constant non-zero vector z_i so that for large t:

$$\text{i)} \; y_i(t) = e^t(z_i+o(t^{-1})) \text{ , } y_i'(t) = e^t(z_i+o(t^{-1}))$$

$$\text{ii)} \; y_i''(t) = e^t(z_i+o(t^{-1}))$$

<u>Proof</u>: The estimates for y_i and y_i' follow from (3.1), i), using the method of asymptotic integrations [12]. The vector z_i is non-zero since M has negative curvature everywhere. The asymptotic formula for y_i'' is then deduced from the Jacobi equation and (3.1), ii).

Let g_{ij} be the components of the metric tensor in spherical normal coordinates. Here i and j vary from two to n. By proper choice of frame E_i, we may assume that $g_{ij} = \langle y_i, y_j \rangle$. One has

<u>Proposition 3.3</u> There exists a positive definite tensor h_{ij} so that

$$\text{i)} \; g_{ij} = e^{2t}(h_{ij}+o(t^{-1})) \text{ , } g_{ij}' = 2e^{2t}(h_{ij}+o(t^{-1}))$$

$$\text{ii)} \; g_{ij}'' = 4e^{2t}(h_{ij}+o(t^{-1}))$$

<u>Proof</u>: Let $h_{ij} = \langle z_i, z_j \rangle$. Since M has negative curvature everywhere, the z_i are linearly independent, and h_{ij} is non-degenerate. Clearly $g_{ij} = \langle y_i, y_j \rangle$ implies that $g_{ij}' = \langle y_i', y_j \rangle + \langle y_i, y_j' \rangle$. So i) follows from Proposition 3.2, i). Similarly (ii) follows by differentiation and substitution from Proposition 3.2, ii).

Suppose g^{ij} is the inverse matrix of h_{ij}. One has

Proposition 3.4 Assume that h^{ij} is the inverse matrix of h_{ij}. Then

i) $g^{ij} = e^{-2t}(h^{ij}+o(t^{-1}))$, $(g^{ij})' = -2e^{-2t}(h^{ij}+o(t^{-1}))$

ii) $(g^{ij})'' = 4e^{-2t}(h^{ij}+o(t^{-1}))$

Proof: The estimate $g^{ij} = e^{-2t}(h^{ij}+o(t^{-1}))$ follows from Proposition 3.3 and the standard formula for the inverse matrix in terms of cofactors. The other assertions follow from Proposition 3.3 and successive differentiation of the relation $\Sigma\, g^{ij}g_{jk} = \delta_{ik}$.

Finally, let $\theta = \sqrt{\det g_{ij}}$ be the volume element in geodesic spherical coordinates. Define $\theta_\infty = \sqrt{\det h_{ij}}$. We may write

Proposition 3.5 θ_∞ is not zero. Moreover,

i) $\theta = e^{(n-1)t}(\theta_\infty+o(t^{-1}))$, $\theta' = (n-1)e^{(n-1)t}(\theta_\infty+o(t^{-1}))$

ii) $\theta'' = (n-1)^2e^{(n-1)t}(\theta_\infty+o(t^{-1}))$

Proof: θ_∞ is non-zero since h_{ij} is positive definite. The asymptotic formula for θ follows from Proposition 3.3, i). Note that $\det(g_{ij})$ is a polynomial of order $n-1$ in the g_{ij}. The derivative estimates follow by repeated differentiation of the definition $\theta = \sqrt{\det g_{ij}}$ and the corresponding parts of Proposition 3.3.

Let $\Delta u = \lambda u$, $u \in L^2M$, $\lambda < (n-1)^2/4$. If (r,ω) are spherical coordinates at p, then define

$$A(r) = \left(\int u^2(r,\omega)d\omega\right)^{1/2}$$

Since λ lies below the essential spectrum, $[(n-1)^2/4,\infty)$, of Δ, one has $A(r) > 0$ for sufficiently large r. We shall establish the following improvement of Theorem 2.5:

Theorem 3.6 Assume that the decay conditions (3.1),i),ii) are satisfied. Given $\varepsilon > 0$, one has

$$A(r) > c_1\exp\left(-\left[\frac{(n-1)^2}{4} - \lambda\right]^{1/2} r - \left[\frac{n-1}{2} + \varepsilon\right]r\right)$$

for some constant $c_1 > 0$ and sufficiently large $r > r_0$.

To prepare for the proof of Theorem 3.6 we compute the Laplacian in geodesic spherical coordinates on M. Suppose that $(x_2,x_3,\dots,x_n)$ is a local coordinate system for the standard unit sphere. The Riemannian measure of S^{n-1} may be written as $d\mathrm{vol}(\omega) = J\,dx_2dx_3\dots dx_n$. Consequently, the Riemannian measure of M is given by $\theta(r,\omega)dr\,d\mathrm{vol}(\omega) = \theta J\,dr\,dx_2\dots dx_n$. In spherical coordinates, one has

$$\Delta f = -\theta^{-1}\frac{\partial}{\partial r}\left(\theta\frac{\partial f}{\partial r}\right) - \theta^{-1}J^{-1}\sum_{i,j=2}^{n}\frac{\partial}{\partial x_i}\left(\theta J\,g^{ij}\frac{\partial f}{\partial x_j}\right)$$

The sphere S^{n-1} may be covered by a finite number of coordinate systems $(x_2,x_3,\dots,x_n)$.

Define $D = e^{cr}\Delta e^{-cr}$, $c = (n-1)/2$, and $w = e^{cr}f$. Then one computes

$$Dw = \frac{-\partial^2 w}{\partial r^2} + S\frac{\partial w}{\partial r} - J^{-1}\theta^{-1}\sum_{i,j=2}^{n}\frac{\partial}{\partial x_i}\left(\theta J\, g^{ij}\frac{\partial w}{\partial x_j}\right) + Vw \tag{3.7}$$

where

$$S = c - e^{cr}\theta^{-1}\frac{\partial}{\partial r}(\theta e^{-cr})$$

$$V = ce^{cr}\theta^{-1}\frac{\partial}{\partial r}(\theta e^{-cr})$$

Assume that the decay conditions (3.1) are satisfied. We may write

Lemma 3.8 As $r \to \infty$,

i) $V = \frac{(n-1)^2}{4} + o(r^{-1})$, $V' = o(r^{-1})$

ii) $S = o(r^{-1})$, $S' = o(r^{-1})$

Proof: This follows by substituting the estimates of Proposition 3.5 into the defining formulas for V , V' and S , S' .

It will be useful to write (3.7) in a more invariant form. For each $r > 0$, there is a global inner product on $C^\infty(\Lambda^1 S^{n-1})$ induced by the metric of M . If β, ζ are one forms supported in the domain of the coordinates $(x_2, x_3, \ldots, x_n)$ then

$$<\beta,\zeta> = \int \Sigma\, g^{ij}\beta_i\zeta_j\theta J\, dx_2\ldots dx_n$$

Let $d_T : C^\infty(\Lambda^0 S^{n-1}) \longrightarrow C^\infty(\Lambda^1 S^{n-1})$ be the exterior derivative along S^{n-1} and d_T^* its adjoint with respect to the above inner product. One has

$$Dw = \frac{-\partial^2 w}{\partial r^2} + S\frac{\partial w}{\partial r} + d_T^* d_T w + Vw \tag{3.9}$$

4. Simplification

The next two sections are devoted to the proof of Theorem 3.6. If the spherical averages $A(u)$ of our eigenfunction decay rapidly, we presently show that the averages $A(|du|)$ must also decay on concentric spheres. This reduces the original problem to obtaining a lower bound for $A(u) + A(|du|)$.

Let M be a simply connected negatively curved manifold with $K \to -1$ at infinity, according to the decay conditions (3.1). Fix a basepoint $p \in M$ and use the symbol $B(r)$ to denote the ball of radius r centered at p . Suppose that $\Delta u = \lambda u$ for some $u \in L^2 M$ and $\lambda < (n-1)^2/4$. Since λ lies below the essential spectrum of Δ , the spherical averages $A(u)$ are positive for sufficiently large $r > r_0$. Similarly, we may assume that the spectrum of Δ on $M-B(r_0)$, with Dirichlet boundary conditions, lies in $[\lambda+\varepsilon_1, \infty)$, for some $\varepsilon_1 > 0$.

Suppose r_1 is greater than r_0 . Denote $X = \partial B(r_1)$, which is S^{n-1} with the induced metric as a submanifold of M . We identify $Y = X \times [0,1]$ isometrically with the spherical shell $B(r_1+1)-\mathrm{Int}\, B(r_1)$, using the exponential map. Let f be the restriction of u to $\partial B(r_1)$. Consider the Dirichlet problem:

$$\Delta h = 0 \qquad \text{on } Y$$

$$h(x,0) = f(x)\ ,\ h(x,1) = 0 \quad \text{on } \partial Y$$

This problem is uniquely solvable and one has the following key estimate:

Proposition 4.1 $\|h\|_{L^2 Y} \leq C_1 \|f\|_{L^2 X}$, where the constant C_1 is independent of r_1 .

Proof: We rescale the metric by defining $\bar{g}_{ij} = e^{-2r_1} g_{ij}$ on the fixed spherical shell $S^{n-1} \times [0,1]$. Consider the Dirichlet problem for the divergence form elliptic operator $B = \frac{\partial}{\partial x_i}\left(B^{ij}\frac{\partial}{\partial x_j}\right)$ with $B^{ij} = \sqrt{\bar{g}}\,\bar{g}^{ij}$, using geodesic spherical coordinates. Given $\varepsilon_2 > 0$, Propositions 3.4 and 3.5 yield $|B^{ij}-B_0^{ij}| \leq \varepsilon_2$, $|\partial B^{ij}/\partial r| \leq C_2$. Here B_0^{ij} is a fixed non-degenerate operator and C_2 is a constant, both independent of r_1 . The result now follows from the work of [9, pp. 124, 138-140]. Note that our rescaling has a commensurable effect on the measures of X and Y .

We use the previous result to establish

Lemma 4.2 There exists a smooth function w with compact support in $X \times [0,1)$ satisfying:

i) $w(x,0) = f(x)$

ii) $\|w\| + \|\Delta w\| \leq C_6 \|f\|$, in L^2-norm

The constant C_6 is independent of r_1 .

Proof: Suppose that $m(r)$ is a smooth cutoff function with $m(r) = 1$ on a neighborhood of $\partial B(r_1)$ and $m(r) = 0$ on $M\text{-}B(r_1+1/2)$. Let $0 \leq p(r) \leq 1$ be identically one on the support of dm and assume that $p(r)$ itself has compact support in $X \times (0,1)$. By the Hessian comparison theorem, we may assume that $|\nabla m|$, $|\nabla p|$, and $|\nabla^2 m|$ are uniformly bounded, independent of r_1 .

Define $w = mh$. By Proposition 4.1, $\|w\| \leq C_7 \|f\|$. Moreover, $\Delta w = h\Delta m + 2\langle dm, dh\rangle$. So $\|\Delta w\| \leq C_8 \|h\| + C_9 \|p\,dh\|$. Since p is compactly supported in $X \times (0,1)$, integration by parts gives:

$$\int p^2 |dh|^2 \leq 2 \int p|h|\ |dp|\ |dh|$$
$$\leq 2 \left(|h|^2 |dp|^2\right)^{1/2} \left(\int p^2 |dh|^2\right)^{1/2}$$

So $\|p\,dh\| \leq C_{10} \|f\|$. Thus $\|w\| + \|\Delta w\| \leq C_7 \|f\| + C_8 \|h\| + C_{10} \|f\| \leq C_6 \|f\|$, using Proposition 4.1 again.

We extend w to $M\text{-}B(r_1)$ be letting $w(x) = 0$ for $x \in M\text{-}B(r_1+1)$. Define the function $\phi = u-w$. Clearly $\phi \in L^2(M\text{-}B(r_1))$ and ϕ vanishes on the boundary of $M\text{-}B(r_1)$. If $z = \lambda w - \Delta w$, then one computes $(\Delta-\lambda)\phi = z$.

Let $(\Delta-\lambda)^{-1}$ be the resolvent for the Laplacian with Dirichlet boundary conditions on $M\text{-}B(r_1)$. Since $\phi(x) = 0$, for $x \in \partial B(r_1)$, ϕ lies in the domain of Δ. Therefore $\phi = (\Delta-\lambda)^{-1} z$. By Lemma 4.2, $\|\phi\| \leq C_{11} \|f\|$.

The definitions give $u = \phi + w$ and $\Delta u = \lambda u$. So $\|u\| + \|\Delta u\| \leq (1+\lambda)\|u\| \leq (1+\lambda)(\|\phi\| + \|w\|)$. Using Lemma 4.2 and the above estimate for $\|\phi\|$ yields

$$\|u\| + \|\Delta u\| \leq C_{12}\|f\| \tag{4.3}$$

Here $\|u\|$ denotes the norm of $u \in L^2(M-B(r_1))$.

Our primary goal in this section is to establish:

Proposition 4.4 For some $1 < a < 2$, we have

$$\int_{\partial B(r_1+a)} |u|^2 + |du|^2 \leq C_{13} \int_{\partial B(r_1)} |u|^2$$

The constant C_{13} is independent of r_1.

Proof: Suppose that $0 \leq q(r) \leq 1$ is a smooth cutoff function and that $q(r)$ is identically one on $B(r_1+2)-B(r_1+1)$ and has compact support in $M-B(r_1)$. Then

$$\int q^2|du|^2 \leq \int q^2|u|\,|\Delta u| + 2\int q|u|\,|dq|\,|du|$$

$$\leq \lambda\|u\|^2 + C_{14}\|q\,du\|\,\|u\|$$

So $\|q\,du\| \leq C_{15}\|u\|$

Recall that $f = u$ on $\partial B(r_1)$. Using (4.3), we deduce

$$\int_{B(r_1+2)-B(r_1+1)} |u|^2 + |du|^2 \leq C_{13} \int_{\partial B(r_1)} |u|^2$$

Proposition 4.4 now follows from Fubini's theorem.

5. Estimate of the spherical averages

In this section, the proof of Theorem 3.6 will be completed. We follow the scheme of [2]. Suppose that $\Delta u = \lambda u$, $u \in L^2M$, and u is not identically zero. If $c = (n-1)/2$, then set $w = e^{cr}u$. The function w is square integrable with respect to the measure $dr\, d\mathrm{vol}(\omega)$. Moreover, by (3.9) one has

$$-w'' + Sw' + d_T^*d_Tw + (V-\lambda)w=0$$

Here w'' denotes the second derivative in r.

Suppose that r_0 is a sufficiently large positive number and $r_1 > r_0$. Let $g(r)$ be a smooth cutoff function which satisfies $g(r) = 0$ for $r \leq r_0$ and $g(r) = 1$ for $r \geq r_1$. Define $y = gw$. Then we may write

$$-y'' + Sy' + d_T^*d_Ty + (V-\lambda)y = \sigma_1$$

Here σ_1 has support in the spherical shell $B(r_1)-B(r_0)$.

Define $M = [\frac{(n-1)^2}{4} - \lambda + \delta]^{1/2}$, where $\delta > 0$ is sufficiently small. If k is a positive integer, then let $h = r^{-k}e^{-Mr}$. The symbols σ_i will denote various functions with support in the spherical shell $B(r_1)-B(r_0)$. The functions σ_i may depend upon u and M, but not upon k.

Set $z = h^{-1}y$. A calculation verifies that

$$z'' + Az - d_T^* d_T z - Sz' - Bz' = r^k \sigma_2 \tag{5.1}$$

where

$$A = \frac{h''}{h} + \lambda - V - S\frac{h'}{h} = (kr^{-1}+M)^2 + kr^{-2} + \lambda - V + S(M+kr^{-1})$$

$$B = \frac{-2h'}{h} = 2(M+kr^{-1})$$

We now take the square of (5.1)

$$(z'' + Az - d_T^* d_T z - Sz')^2 - 2(z'' + Az - d_T^* d_T z)Bz'$$
$$+ (B^2+2BS)(z')^2 = r^{2k}\sigma_2^2$$

Clearly, Lemma 3.8 implies $B^2 + 2BS > 0$, for r_1 sufficiently large. Retaining only the cross terms leads to an inequality

$$-(z'' + Az - d_T^* d_T z) \cdot Bz' \leq r^{2k}\sigma_3 \tag{5.2}$$

We suppose that Theorem 3.6 fails and argue by contradiction. From Proposition 4.4, there is an $\eta > 0$ and a sequence $s_m \to \infty$ with

$$\int_{\partial B(s_m)} (|z|^2 + |dz|^2)\,dvol(\omega) \leq C_1 \exp(-\eta s_m)$$

Here $|dz|$ is the pointwise norm in the metric of M . The sequence s_m may depend upon k . Let $Q_j(s_m)$ denote related quantities which are also forced to decay exponentially.

Multiply (5.2) by $r^3\theta e^{-p(r)}$. Here $p(r) = 2cr(1+o(1))$, as $r \to \infty$, is to be chosen. Integrate to obtain

$$\int_0^{s_m} \int_{\partial B(r)} -r^3 B(z'' + Az - d_T^* d_T z)z'\theta e^{-p(r)}\,dvol(\omega)dr$$
$$\leq \iint r^{2k}\sigma_4 \tag{5.3}$$

The left hand side is the sum of three terms $\mathcal{M}_1$, $\mathcal{M}_2$, and $\mathcal{M}_3$.

The first term is

$$\mathcal{M}_1 = \int_0^{s_m} \int_{\partial B(r)} -r^3 Bz''z'\theta e^{-p(r)}\,dvol(\omega)dr$$

Define $\gamma = \partial/\partial r \log(\theta e^{-p(r)})$. By Proposition 3.5, we may choose p , so that $0 < \gamma = o(r^{-1})$. Integrating by parts gives

$$\mathcal{M}_1 = \frac{1}{2}\int_0^{s_m} \int_{\partial B(r)} [(r^3B)' + \gamma r^3 B](z')^2\theta e^{-p(r)}\,dvol(\omega)dr + Q_1(s_m)$$

Observe that $\frac{1}{2}[(r^3B)' + \gamma r^3 B] = 3Mr^2 + 2kr + [2(M+kr^{-1})]o(r^2) > 0$, for r_1 sufficiently large and independent of k . So one has

$$\mathcal{M}_1 \geq Q_1(s_m) \tag{5.4}$$

Similarly, one deals with

$$\mathcal{M}_2 = \int_0^{s_m} \int_{\partial B(r)} -ABr^3 zz'\theta e^{-p(r)} dvol(\omega)dr$$

Partial integration yields

$$\mathcal{M}_2 = \frac{1}{2} \int_0^{s_m} \int_{\partial B(r)} [(ABr^3)' + \gamma ABr^3] z^2 \theta e^{-p(r)} dvol(\omega)dr + Q_2(s_m)$$

One calculates

$$\frac{1}{2}(ABr^3)' = 3M(M^2 + \lambda - V - \frac{1}{3} V'r)r^2 +$$
$$(6M^2 + 2(\lambda - V) - V'r)kr + 3Mk^2 + kM + (rS[M^2r^2 + 2kMr + k^2])'$$

Using Lemma 3.8 and the definition of M, we deduce the inequality $\frac{1}{2}(ABr^3)' + \frac{1}{2}\gamma ABr^3 > \varepsilon r^2$. Here $\varepsilon > 0$ is independent of k. The condition $\gamma > 0$ guarantees that the k^3 terms in γABr^3 are positive.

Therefore, we have

$$\mathcal{M}_2 \geq \int_0^{s_m} \int_{\partial B(r)} \varepsilon r^2 z^2 \theta e^{-p(r)} dvol(\omega)dr + Q_2(s_m) \tag{5.5}$$

The third term from (5.3) is

$$\mathcal{M}_3 = \int_0^{s_m} \int_{\partial B(r)} r^3 B(d_T^* d_T z) z' \theta e^{-p(r)} dvol(\omega)dr$$

Let Φ be the pointwise inner product for the induced metric on the cotangent space of $\partial B(r)$. One may write

$$\mathcal{M}_3 = \int_0^{s_m} \int_{\partial B(r)} r^3 B \Phi(d_T z, d_T z') \theta e^{-p(r)} dvol(\omega)dr$$

Here it is crucial that B and p are functions of r alone.

Set $\Psi = r^2\Phi$. Proposition 3.4 gives $\Psi' = (-2 + O(r^{-1}))\Psi$. In particular, $\Psi' \leq -3\Psi/2$, for large r. Thus

$$2\Psi(d_T z, d_t z') = \frac{\partial}{\partial r} \Psi(d_T z, d_T z) - \Psi'(d_T z, d_T z)$$
$$\geq \frac{\partial}{\partial r} \Psi(d_T z, d_T z) + \frac{3}{2} \Psi(d_T z, d_T z)$$

It follows that

$$\mathcal{M}_3 \geq \int_0^{s_m} \int_{\partial B(r)} \frac{1}{2} rB \frac{\partial}{\partial r} \Psi(d_T z, d_T z) \theta e^{-p(r)} dvol(\omega)dr$$
$$+ \int_0^{s_m} \int_{\partial B(r)} \frac{3}{4} rB \Psi(d_T z, d_T z) \theta e^{-p(r)} dvol(\omega)dr$$

Partial integration in the first summand yields

$$\mathcal{M}_3 \geq \int_0^{s_m} \int_{\partial B(r)} [\frac{3}{4} rB - \frac{1}{2}(rB)' - \frac{1}{2}\gamma rB] \Psi(d_T z, d_T z) \theta e^{-p(r)} dvol(\omega)dr + Q_3(s_m)$$

One computes $\frac{3}{4} rB = \frac{3}{2} Mr + \frac{3}{2} k$ and $\frac{1}{2}(rB)' = M$. Since $\gamma = o(\frac{1}{r})$, we have for r_2 sufficiently large

$$\mathcal{M}_3 \geq Q_3(s_m) \tag{5.6}$$

Substituting (5.4), (5.5), and (5.6) back into (5.3) gives

$$\int_0^{s_m} \int_{\partial B(r)} \varepsilon r^2 z^2 \theta e^{-p(r)} d\mathrm{vol}(\omega) dr \le \iint r^{2k} \sigma_4 + Q_4(s_m)$$

and thus

$$\int_0^{s_m} \int_{\partial B(r)} z^2 \theta e^{-p(r)} d\mathrm{vol}(\omega) dr \le \iint r^{2k} \sigma_5 + Q_5(s_m)$$

Letting $s_m \to \infty$, we obtain

$$\int_0^{\infty} \int_{\partial B(r)} z^2 \theta e^{-p(r)} d\mathrm{vol}(\omega) dr \le \iint r^{2k} \sigma_5$$

Recall that σ_5 is supported in the spherical shell $B(r_1) - B(r_0)$. Therefore

$$\int_0^{\infty} \int_{\partial B(r)} z^2 \theta e^{-p(r)} d\mathrm{vol}(\omega) dr \le r_1^{2k} \iint \sigma_5$$

Suppose that $r_2 > r_1$. Recalling that $z = r^k e^{Mr} y$ yields, for r_2 sufficiently large,

$$r_2^{2k} \int_{r_2}^{\infty} \int_{\partial B(r)} y^2 d\mathrm{vol}(\omega) dr \le r_1^{2k} \iint \sigma_6$$

We divide by r_2^{2k} and let $k \to \infty$. Since σ_6 is independent of k , it follows that

$$\int_{r_2}^{\infty} \int y^2 d\mathrm{vol}(\omega) dr = 0$$

So $y = ge^{cr}u$ is identically zero in $M - B(r_2)$. By unique continuation, the eigenfunction u is identically zero on M . This contradiction completes the proof of Theorem 3.6.

Bibliography

1. Agmon, S., Lectures on exponential decay of solutions of second order elliptic equations, bounds on eigenfunctions of n-body Schrödinger operators, Princeton University Press, Princeton, N.J., 1982.

2. Bardos, C. and Merigot, M., Asymptotic decay of the solution of a second order elliptic equation in an unbounded domain, applications to the spectral properties of a Hamiltonian, Proceedings of the Royal Society, Edinburgh, 76A (1977), 323-344.

3. Cheeger, J., Gromov, M., and Taylor, M., Finite propagation speed, kernel estimates for functions of the Laplace operator, and the geometry of complete Riemannian manifolds, J. Differential Geometry, 17 (1982), 15-53.

4. Cheng, S. Y. and Yau, S. T., Differential equations on Riemannian manifolds and their geometric applications, Communications on Pure and Applied Mathematics, 28 (1975), 333-354.

5. Davies, E. B., Simon, Barry, and Taylor, M., L^p spectral theory of Kleinian groups, Preprint.

6. Donnelly, H., Eigenforms of the Laplacian on complete Riemannian manifolds, Communications in Partial Differential Equations, 9 (1984), 1299-1321.

7. Donnelly, H., On the essential spectrum of a complete Riemannian manifold, Topology, 20 (1981), 1-14.

8. Donnelly, H., Lower bounds for eigenfunctions on Riemannian manifolds, Preprint.

9. Fabes, E., Jerison, D., and Kenig, C., Necessary and sufficient conditions for absolute continuity of elliptic-harmonic measure, Annals of Math., 119 (1984), 121-141.

10. Glazeman, I. M., Direct methods of qualitative spectral analysis of singular differential operators, Daniel Davey, N.Y., 1965.

11. Greene, R. and Wu, H., Function theory on manifolds which possess a pole, Springer-Verlag Lecture Notes in Mathematics, Vol. 699, Berlin, Heidelberg, N.Y., 1979.

12. Hartman, P., Ordinary Differential Equations, Wiley, N.Y., 1964.

13. Reed, M. and Simon, B., Methods of Modern Mathematical Physics IV, Analysis of Operators, Academic Press, N.Y., 1978.

Stability and Negativity for Tangent Sheaves of Minimal Kähler Spaces

Dedicated to Professor Shingo Murakami on his sixtieth birthday

Ichiro ENOKI

Department of Mathematics
College of General Education, Osaka University
Osaka, 560, Japan

1. Introduction

Let X be a minimal Kähler space and K_X its canonical bundle. Our aim is to show that tensor powers of the tangent sheaf of X has certain stability and negativity. In particular they are semistable in the sense of Mumford-Takemoto if K_X is either big or numerically trivial.

We state our result using a desingularization M of X. Let TM be the holomorphic tangent bundle of a complex manifold M. A complex representation ρ of $GL(n,\mathbb{C})$, $n = \dim M$, induces a holomorphic vector bundle $(TM)^{\rho}$ associated to TM ; let $d(\rho)$ be a non-negative integer such that

$$\det (TM)^{\rho} = (\det TM)^{\otimes d(\rho)} .$$

Let $\mathcal{T}_M = \mathcal{O}(TM)$ and $\mathcal{T}_M^{\rho} = \mathcal{O}((TM)^{\rho})$.

Theorem 1.1. *Let X be an n-dimensional minimal Kähler space and $\pi : M \longrightarrow X$ a desingularization of X. Let Φ_X be a Kähler form on X. Then for any $t > 0$ and a coherent subsheaf $\mathcal{S} \subset \mathcal{T}_M^{\rho}$ of positive rank, we have*

$$\frac{1}{r}\int_M c_1(\mathcal{S}) \wedge \Psi_t^{n-1} \le \frac{1}{\ell}\int_M c_1(\mathcal{T}_M^{\rho}) \wedge \Psi_t^{n-1}$$
$$+ \left(\frac{1}{r} - \frac{1}{\ell}\right)\int_M d(\rho) t\pi^*\Phi_X \wedge \Psi_t^{n-1}$$

where $r = \mathrm{rank}(\mathcal{S})$, $\ell = \mathrm{rank}(\mathcal{T}_M^{\rho})$ *and* $\Psi_t = t\pi^*\Phi_X + c_1(\pi^*K_X)$.

Let $\nu(X)$ be the numerical Kodaira dimension of X. Then, compairing the lowest order terms in t of the inequality above, we obtain

Corollary 1.2. *Let X be a minimal Kähler space.*

i) *If $\nu(X) = 0$, then the tangent sheaf of X is Φ_X-semistable for any Kähler form Φ_X on X.*

ii) *If $\nu(X) = \dim X$, then the tangent sheaf of X is K_X-semistable.*

For a smooth X, the K_X-semistability has been already shown by Tsuji[Ts].

Consider next the case: $0 < \nu(X) < \dim X$. Then the lowest order terms are of degree $n - \nu - 1$. If the equality holds in this level, then an inequality between terms of degree $n - \nu$ holds. Namely:

Corollary 1.3. *Let X be an n-dimensional minimal Kähler space with $0 < \nu = \nu(X) < n$. Let Φ_X be a Kähler form on X and $\pi : M \longrightarrow X$ a desingularization. Then for any coherent subsheaf $\mathcal{G} \subset \mathcal{T}_M$ of rank $r > 0$,*

$$\int_M c_1(\mathcal{G}) \wedge c_1(\pi^* K_X)^{\nu} \wedge \pi^* \Phi_X{}^{n-\nu-1} \leq 0.$$

If the equality holds, then we obtain further

$$\nu \int_M c_1(\mathcal{G}) \wedge c_1(\pi^* K_X)^{\nu-1} \wedge \pi^* \Phi_X{}^{n-\nu}$$
$$\leq (r + \nu - n) \int_M c_1(\mathcal{T}_M) \wedge c_1(\pi^* K_X)^{\nu-1} \wedge \pi^* \Phi_X{}^{n-\nu} .$$

The first inequality in Corollaly 1.3 above is the semistability condition relative to $K_X{}^{\nu} \Phi_X{}^{n-\nu-1}$, which can be regarded also as certain negativity. This kind of negativity holds under weaker condition as follows.

Theorem 1.4. *Let M be an n-dimensional compact Kähler manifold with Kähler form Φ. Assume that*

(1.5) *the canonical line bundle K_M of M decomposes into a nef $\mathbb{Q}$-line bundle L and an effective $\mathbb{Q}$-divisor D, $K_M = L + D$, as $\mathbb{Q}$-line bundle.*

Then

$$\int_M c_1(\mathcal{G}) \wedge \Phi^{n-1} \leq 0$$

for any coherent subsheaf $\mathcal{G} \subset \mathcal{T}_M{}^{\otimes m}$ of positive rank.

Thus $\mathcal{T}_M^{\otimes m}$ can be called Φ-*seminegative*. This property will be used in [En], where we will need to consider $\mathcal{T}_M^{\otimes m}$, $m \geq 1$, as well as $\mathcal{T}_M$. If restricted to $\mathcal{T}_M$ itself (i.e., to the case: $m = 1$) with M projective algebraic, Theorem 1.4 is a weak form of Miyaoka's *generic* seminegativity theorem [Mi1, p.564, Cor. 6.4][Mi2], though our proof is completely different from Miyaoka's.

We conclude this section by reviewing some basic definitions. Let X be a complex analytic space. We always assume that X is reduced and irreducible. X is called a *Kähler space* if there is a 0-cochain $\{\varphi_i\}$ of C^∞ strictly pluri-subharmonic functions on an open covering $\{U_i\}$ of X such that $\varphi_i - \varphi_j$ is pluri-harmonic on $U_i \cap U_j$; thus $\{\sqrt{-1}\,\partial\bar{\partial}\,\varphi_i\}$ defines a global C^∞ (1,1)-form Φ_X, which is called a Kähler form on X. If X is a Kähler space, then every proper modification of X is also a Kähler space (see [Ca]).

Let X be a compact Kähler space and $\pi : M \longrightarrow X$ a resolution of singularities. Let $\varphi \in H^2(M, \mathbb{R})$ be a Kähler class, namely a deRham cohomology class of a Kähler form on M. A holomorphic line bundle L over X is *nef* if for any $\varepsilon > 0$ deRham cohomology class $c_1(\pi^*L)_{\mathbb{R}} + \varepsilon\varphi$ contains a C^∞ d-closed (1,1)-form which is positive definite everywhere. This definition is independent of the choice of the resolution M and the Kähler class φ; it is equivalent to the usual one [Re] if X is projective algebraic.

If X is $\mathbb{Q}$-Gorenstein, the canonical bundle K_X of X is defined as $\mathbb{Q}$-line bunlde, i.e., an element of $H^1(X, \mathcal{O}_X^*)\otimes\mathbb{Q}$. A *minimal* Kähler space X is by definition a compact normal $\mathbb{Q}$-Gorenstien Kähler space such that every desingularization $\pi : M \longrightarrow X$ of X satisfies (1.5) with $L = \pi^*K_X$ and D supported on the exceptional set of π.

The *numerical Kodaira* dimension $\nu(X)$ of a minimal Kähler space X is defined by

$$\nu(X) = \max \{\, k \mid c_1(\pi^*K_X)^k \neq 0 \quad \text{in} \quad H^{2k}(M, \mathbb{R}) \,\} ,$$

where $\pi : M \longrightarrow X$ is a desingularization.

2. Curvature of Subbundles

Let M be a compact n-dimensional Kähler manifold with Kähler form Φ. For a (1,1)-from φ, define $\Lambda\varphi$ by

$$(\Lambda\varphi)\ \Phi^n = n\ \varphi \wedge \Phi^{n-1}.$$

The action of Λ is extended to bundle-valued forms.

Let (E, h) be a hermitian vector bundle over M and let R^E be the curvature of the hermitian connection of (E, h). Then R^E is an End(E)-valued (1,1)-form and we have

$$\Lambda\ \mathrm{tr} R^E = \mathrm{tr}\ \Lambda R^E .$$

Proposition 2.1. *Let* $0 \longrightarrow S \longrightarrow E \longrightarrow Q \longrightarrow 0$ *be an exact sequence of holomorphic vector bundles defined over an open subset of* M. *Then*

$$\Lambda\ \mathrm{tr} R^S \le \mathrm{tr}(\mathrm{pr}_S \circ \Lambda R^E{}_{|S}) ,$$

where R^S *is the curvature of* S *equipped with the induced metric and* $\mathrm{pr}_S : E \longrightarrow S$ *is the orthogonal projection. Moreover if the equlity holds, then the exact sequence above splits.*

Proof. This is a modification of Proposition (8.2) in [Ko, p.176]. In fact, the curvatures R^S and R^E are related by the Gauss-Codazzi equation (6.13) in [Ko, p.23]. In particular

$$\Lambda\ \mathrm{tr} R^S + \|A\|^2 = \mathrm{tr}(\mathrm{pr}_S \circ \Lambda R^E{}_{|S}),$$

where A is the second fundamental form of the subbundle $S \subset E$ and $\|A\|$ its norm. Thus Proposition 2.1 follows from this and Proposition (6.14) in [Ko, p.23].

Let $\mathcal{S} \subset \mathcal{O}(E)$ be a coherent subsheaf of rank$(\mathcal{S}) > 0$. Let $W(\mathcal{S})$ be the maximal subset of M such that $\mathcal{S}_{|M - W(\mathcal{S})}$ defines a holomorphic subbundle $S \subset E_{|M - W(\mathcal{S})}$, namely $\mathcal{S}_{|M - W(\mathcal{S})} = \mathcal{O}(S)$.

Proposition 2.2. *For any coherent subsheaf* $\mathcal{S} \subset \mathcal{O}(E)$ *of positive rank,*

$$2\pi \int_M c_1(\mathcal{S}) \wedge \Phi^{n-1} \le \int_{M - W(\mathcal{S})} \sqrt{-1}\ \mathrm{tr}\ R^S \wedge \Phi^{n-1} ,$$

where R^S *denote the curvature of the subbundle* $S \subset E_{|M - W(\mathcal{S})}$ *corresponding to* $\mathcal{S}$. *If the equality holds, then* $\mathrm{codim}\ W(\mathcal{S}) \ge 2$.

Proof. We refer to [Ko], from the end of p.172 to p.182.

3. Degenerate Monge-Ampere Equations

Let M be a compact n-dimentional Kähler manifold with Kähler form Φ. Assume (1.5) so that $K_M = L + D$ with L a nef $\mathbb{Q}$-line bundle and D an effective $\mathbb{Q}$-divisor. Write D as $D = kD_0$ with D_0 effective (usual $\mathbb{Z}$-coefficients) divisor and $k > 0$. Fix a hermitian metric $\| \|^2$ and a holomorphic section s of $[D_0]$ which defines the divisor D_0. For each $\varepsilon > 0$, define

$$\gamma(D, \varepsilon) := \sqrt{-1}\ \bar\partial\partial \log \|s\|^{2k} - \sqrt{-1}\ \bar\partial\partial \log(\|s\|^2 + \varepsilon)^k .$$

For real (1,1)-forms φ and ψ we mean by $\varphi \le \psi$ that $\varphi - \psi$ is negative semi-definite everywhere.

Lemma 3.1. $-\gamma(D, \varepsilon) \le \chi_\varepsilon \Phi$, *where* $\chi_\varepsilon > 0$ *is uniformly bounded and* $\chi_\varepsilon \to 0$ *in* L^1*-sense as* $\varepsilon \to 0$.

Proof. $\chi_\varepsilon = C k\varepsilon/(\|s\|^2 + \varepsilon)$ has this property for some constant $C > 0$.

Proposition 3.2. *Let* $K_M = L + D$ *as (1.5) and let* $\gamma(D, \varepsilon)$, χ_ε *be as above. Fix a Kähler form* Φ *on* M. *Then for each* ε, $t > 0$ *there are Kähler forms* $\Phi(\varepsilon, t)$, $\Psi(\varepsilon, t)$ *on* M *with the following properties:*

a) $\Phi(\varepsilon, t)$ *is cohomologous to* Φ *and*

$$\sqrt{-1}\ \mathrm{Ric}(\Phi(\varepsilon, t)) \le (\chi_\varepsilon + t)\ \Phi ;$$

b) $\Psi(\varepsilon, \delta) - t\Phi$ *represents* $2\pi\ c_1(L)_{\mathbb{R}}$ *and*

$$\sqrt{-1}\ \mathrm{Ric}(\Psi(\varepsilon, t)) = -\Psi(\varepsilon, t) - \gamma(D, \varepsilon) + t\Phi ;$$

c) $\Phi(\varepsilon, t)$ *and* $\Psi(\varepsilon, t)$ *remain bounded on* M *as* $\varepsilon \to 0$ *whenever* t *is fixed.*

Proof. Fix $t > 0$. Since L is nef, the deRham cohomology class $2\pi c_1(L)_{\mathbb{R}}$ contains a real d-closed (1,1)-form $\gamma(t)$ such that

$$\Psi(t) = \gamma(t) + t\Phi$$

is a Kähler form on M. Then we have a C^∞ function F_t on M with

$$\sqrt{-1}\ \mathrm{Ric}(\Phi) = -\gamma(t) + \sqrt{-1}\ \partial\bar\partial \log \|s\|^{2k} + \sqrt{-1}\ \partial\bar\partial\ F_t .$$

Using holomorphic local coordinates $z^1, \ldots, z^n$, we write

$$\Phi = \sqrt{-1}\,\Sigma\, g_{i\bar{j}}\, dz^i \wedge d\bar{z}^j, \qquad \Psi(t) = \sqrt{-1}\,\Sigma\, g(t)_{i\bar{j}}\, dz^i \wedge d\bar{z}^j .$$

By [Ya], for each $\varepsilon > 0$ there are C^∞ solutions $\varphi(\varepsilon, t)$, $\psi(\varepsilon, t)$ of the following equations:

$$\det(g_{i\bar{j}} + \partial_i \bar{\partial}_j\, \varphi(\varepsilon, t)) = C_\varepsilon\, (\|s\|^2 + \varepsilon)^k \exp(F_t) \det(g_{i\bar{j}}) ,$$

$$\det(g(t)_{i\bar{j}} + \partial_i \bar{\partial}_j\, \psi(\varepsilon, t))$$

$$= C_{\varepsilon,t}\, (\|s\|^2 + \varepsilon)^k \exp(F_t + \psi(\varepsilon, t)) \det(g(t)_{i\bar{j}}) ,$$

where ∂_i denotes $\partial/\partial z^i$ and constants C_ε, $C_{\varepsilon,t}$ are choosen suitably. Moreover C^2-norms of $\varphi(\varepsilon, t)$ and $\psi(\varepsilon, t)$ remain bounded as $\varepsilon \longrightarrow 0$ whenever t is fixed [Ya, §5, §7]. Thus

$$\Phi(\varepsilon, t) := \Phi + \sqrt{-1}\,\Sigma\, \partial_i \bar{\partial}_j\, \varphi(\varepsilon, t)\, dz^i \wedge d\bar{z}^j \qquad \text{and}$$

$$\Psi(\varepsilon, t) := \Psi(t) + \sqrt{-1}\,\Sigma\, \partial_i \bar{\partial}_j\, \psi(\varepsilon, t)\, dz^i \wedge d\bar{z}^j$$

have the required properties.

4. Stability and Negativity

Let M be a compact n-dimensional Kähler manifold with Kähler form Ψ. The metric g associated to Ψ defines an isomorphism $\bar{T}^*M \simeq TM$. That is, a (0,1)-form η corresponds to a (1,0)-vector $\eta^\wedge$ so that

$$\eta(\bar{u}) = g(\eta^\wedge, u) \qquad \text{for } u \in TM.$$

The action of " $\wedge$ " is extended to bundle-valued forms. In particular, regarding a (1,1)-form φ as T^*M -valued (0,1)-form, we have an endomorphism $\varphi^\wedge$ of TM with $\mathrm{tr}(\varphi^\wedge) = \Lambda\varphi$ (cf. §2). Since M is Kähler, the hermitian connectioin of TM coincides with the Levi-Civita connecction. Then Bianchi's first identity implies

$$\text{(4.1)} \qquad \Lambda R^{TM} = \mathrm{Ric}^\wedge ,$$

where Ric is the Ricci (1,1)-form of the metric.

Proof of Theorem 1.1. Let X be an n-dimensional minimal Kähler space and $\pi : M \longrightarrow X$ a desingularization. Then the canonical bundle K_M of M satisfies (1.5):

$$K_M = L + D, \qquad L = \pi^* K_X \ ,$$

with D an effective $\mathbb{Q}$-divisor supported on the exceptional set of π. Let Φ_M be a Kähler form on M and let $\Psi := \Psi(\varepsilon, t)$ be the Kähler form given in Proposition 3.2 so that Ψ is cohomologous to $t\Phi_M + c_1(\pi^* K_X)$. Let $\rho : GL(n, \mathbb{C}) \longrightarrow GL(\ell, \mathbb{C})$ be a complex representation and ρ_* the induced Lie algebra homomorphism. We equip TM with the metric associate to the Kähler form Ψ and let $E = (TM)^\rho$ as hermitian vector bundle. Then by the Bianchi identity (4.1),

$$(4.2) \qquad \Lambda R^E = \rho_*(\Lambda R^{TM})$$
$$= - \mathrm{id}_E - \rho_*(\gamma(D, \varepsilon)^\wedge) + t\rho_*(\Phi_M{}^\wedge) \ ,$$

where Λ and " $\wedge$ " are relative to Ψ.

Let $\mathcal{S} \subset \mathcal{O}(E)$ be a coherent subsheaf with $r = \mathrm{rank}(\mathcal{S}) > 0$. Then $\mathcal{S}$ defines a holomorphic subbundle $S \subset E$ outside the analytic subset $W(\mathcal{S})$ of M. By Propositions 2.1 and 2.2,

$$2\pi n \int_M c_1(\mathcal{S}) \wedge \Psi^{n-1} \le \int_{M-W(\mathcal{S})} \mathrm{tr}(\mathrm{pr}_S \circ \Lambda R^E|_S)\ \Psi^n \ .$$

Moreover by (4.2) and Lemma 3.1,

$$\mathrm{tr}(\mathrm{pr}_S \circ \Lambda R^E|_S)\ \Psi^n \le \{ - r + (\chi_\varepsilon + t)\mathrm{tr}(\mathrm{pr}_S \circ \rho_*(\Phi_M{}^\wedge)|_S)\}\ \Psi^n$$
$$\le \{ - r + (\chi_\varepsilon + t)\mathrm{tr}(\rho_*(\Phi_M{}^\wedge))\}\ \Psi^n$$
$$= - r\Psi^n + (\chi_\varepsilon + t)d(\rho)n\ \Phi_M \wedge \Psi^{n-1} \ ,$$

since all eigen values of $\rho_*(\Phi_M{}^\wedge)$ are positive. Let $\ell = \mathrm{rank}(E)$. Then, since $\Lambda \circ \mathrm{tr} = \mathrm{tr} \circ \Lambda$, it follows from (4.2) that:

$$- \ell\Psi^n = \sqrt{-1}\ n \mathrm{tr} R^E \wedge \Psi^{n-1} + d(\rho)n\gamma(D, \varepsilon) \wedge \Psi^{n-1} - d(\rho)nt\Phi_M \wedge \Psi^{n-1} \ .$$

Combinig these together, we obtain

$$\frac{2\pi}{r} \int_M c_1(\mathcal{S}) \wedge \Psi^{n-1} \le \frac{2\pi}{\ell} \int_M c_1(E) \wedge \Psi^{n-1} + \left(\frac{1}{r} - \frac{1}{\ell} \right) \int_M d(\rho)t\Phi_M \wedge \Psi^{n-1}$$
$$+ \frac{d(\rho)}{\ell} \int_M \gamma(D, \varepsilon) \wedge \Psi^{n-1} + \frac{d(\rho)}{r} \int_M \chi_\varepsilon \Phi_M \wedge \Psi^{n-1} \ .$$

By Lemma 3.1 and c) of Proposition 3.2, the last term tends to 0 as $\varepsilon \longrightarrow 0$; the other terms are independent of ε (they depend only on cohomology classes of Φ_M and Ψ). Replace Φ_M with $2\pi(\alpha\Phi_M + \pi^*\Phi_X)$ so that Ψ is now cohomlogous to

$$2\pi\{\ t(\alpha\Phi_M + \pi^*\Phi_X) + c_1(\pi^*K_X)\}\ , \qquad\qquad \alpha > 0.$$

Then, as $\alpha \longrightarrow 0$,

$$\int_M \gamma(D,\ \varepsilon) \wedge \Psi^{n-1} \longrightarrow (2\pi)^n \int_D \{t\pi^*\Phi_X + c_1(\pi^*K_X)\}^{n-1} = 0\ ,$$

because $\dim \pi(\mathrm{spt}D) \le n - 2$. Thus, letting $\alpha \longrightarrow 0$, we obtain the desired inequality.

Proof of Theorem 1.5. This is parallel to that of Theorem 1.1. Let $\Psi = \Phi(\varepsilon,t)$ be the Kähler form given in Proposition 3.2 (instead of $\Psi(\varepsilon,t)$ in the proof of Theorem 1.1) so that Ψ is now cohomologous to the given Kähler form Φ on M. Let $E = (TM)^{\otimes m}$ with the metric induced by Ψ. Then by the Bianchi identity we have

$$\Lambda R^E = (\Lambda R^{TM})^{\otimes m}$$

By Proposition 3.2 it follows

$$\mathrm{tr}(\mathrm{pr}_S \circ \Lambda R^E{}_{|S})\ \Psi^n \le (\chi_\varepsilon + t)mn\Phi \wedge \Psi^{n-1}$$

for a holomorphic subbundle $S \subset E$ defined on an open subset of M. Therefoere, by Propositions 2.1 and 2.2

$$2\pi\int_M c_1(\mathcal{S}) \wedge \Psi^{n-1} \le \int_M (\chi_\varepsilon + t)m\Phi \wedge \Psi^{n-1}$$

for each coherent subsheaf $\mathcal{S} \subset \mathcal{O}(E)$ of positive rank. Thus we obtain the desired inequality by letting ε and t tend to zero.

References

[Ca] Campana, F.: "Applicatiuon de l'espace des cycles à la classification biméromorphe des espaces analytiques Kählériens compacts", prepublication of Univ. de Nancy 1, May 1980.

[En] Enoki, I.: Kodaira dimension and higher cohomology of nef line bundles over compact Kähler manifolds, in preparation.

[Ko] Kobayashi, S.: "Differential Geomerty of Complex Vector Bundles" (Publication of the Math. Soc. of Japan Vol.15), Iwanami Shoten , Tokyo; Princeton Univ. Press, 1987.

[Mi1] Miyaoka, Y.: The Chern classes and Kodaira dimension of a minimal variety, in "Algebraic Geometry, Sendai, 1985",

Advanced Studies in Pure Mathematics 10 (T. Oda, ed.), Kinokuniya, Tokyo ; Noth-Holland, Amsterdam, 1987, 449-476.

[Mi2] —— : Deformations of a morphism along a foliation and applications to appear in Proc. of Symp. in Pure Math.

[Re] Reid, M.: Minimal models of canonical 3-folds, in "Algebraic Varieties and Analytic Varieties" Advanced Studies in Pure Mathematics 1 (S. Iitaka ed.), Kinokuniya, Tokyo ; Noth-Holland, Amsterdam, 1983, 131-180.

[Ts] Tsuji, H.: Stability of tangent bundle of minimal algebraic varieties, preprint, Harvard Univ., 1987, to appear in Topology.

[Ya] Yau, S.-T.: On the Ricci curvature of a compact Kähler manifold and the complex Monge-Ampere equation, I, Comm. Pure Appl. Math. 31 (1978), 339-411.

AN OBSTRUCTION CLASS AND A REPRESENTATION OF HOLOMORPHIC AUTOMORPHISMS

Dedicated to Professor Shingo Murakami on his sixtieth birthday

BY

AKITO FUTAKI AND TOSHIKI MABUCHI

0. INTRODUCTION

Let N be an n-dimensional compact connected Kähler manifold, Aut(N) the group of holomorphic automorphisms of N, and $Aut^0(N)$ its identity component. In [F1] and [M1], both authors independently defined a group character

$$\psi: \ Aut(N) \to \mathbb{R}_+$$

such that the corresponding real Lie algebra homomorphism

$$\psi_*: H^0(N, \mathcal{O}(TN)) \to \mathbb{R}$$

coincides with Futaki's obstruction for N to admit an Einstein-Kähler metric. Recall that the Albanese map α: N → Alb(N) of N naturally induces the Lie group homomorphism $\tilde{\alpha}$: $Aut^0(N) \to Aut^0(Alb(N)) (\simeq Alb(N))$. Then by a theorem of Fujiki, Ker $\tilde{\alpha}$ has a natural structure of a linear algebraic group (defined over $\mathbb{C}$). Moreover, the Chevalley decomposition allows us to express Ker $\tilde{\alpha}$ as a semidirect product $S \ltimes U$ of a reductive algebraic subgroup S of Ker $\tilde{\alpha}$ and the unipotent radical U of Ker $\tilde{\alpha}$.

Now the purpose of this note is to introduce a representation

$$\phi: Aut(N) \to GL_{\mathbb{C}}(V)$$

such that, if $c_1(N) > 0$ (which obviously implies Ker $\tilde{\alpha} = Aut^0(N)$), we have:

(0.1) ϕ <u>is an algebraic group homomorphism</u> (<u>defined over</u> $\mathbb{C}$), <u>and therefore</u> $\det\circ\phi$ <u>is trivial when restricted to</u> U;

(0.2) $\psi(s) = |\det \phi(s)|^\gamma$ <u>for all</u> $s \in S$ <u>with some positive rational constant</u> γ <u>depending only on</u> n;

(0.3) V <u>is a finite dimensional</u> $\mathbb{C}$-<u>vector space written in the form</u> $\oplus_\nu m_\nu H^0(N, \mathcal{O}(K_N^{-\nu}))$ <u>for some nonzero integers</u> m_ν <u>depending only on</u> n,

where for a positive integer m and a $\mathbb{C}$-vector space W, we denote by mW (resp. -mW) the $\mathbb{C}$-vector space obtained as a direct sum of m-copies of W (resp. W*). Note here that, if $\dim_{\mathbb{C}} N = 2$ and $c_1(N) > 0$, then V is nothing but $H^0(N, \mathcal{O}(K_N^{-1}))$ and the map ϕ is induced by the natural action of Aut(N) on $H^0(N, \mathcal{O}(K_N^{-1}))$.

Parts of this note come out of stimulating discussions at Katata. We wish to express our sincere gratitude to the Taniguchi Foundation for promoting the symposium at Katata. Special thanks are due also to Dr. I. Enoki for valuable suggestions and to Professors S. Murakami and T. Sunada for organizing the symposium so nicely.

1. NOTATION, CONVENTIONS AND PRELIMINARIES.

Throughout this note, we fix a holomorphic line bundle L over N and a complex Lie subgroup G of Aut(N) such that G acts holomorphically on L as bundle isomorphisms covering the G-action on N. (If N is algebraic, we further assume that G is an algebraic group acting algebraically on N.) Let $\mathcal{H}$ be the set of all C^∞ Hermitian fibre metrics of L over N. For each $h \in \mathcal{H}$, we denote by $c_1(L;h)$ the corresponding first Chern form $(\sqrt{-1}/2\pi)\bar{\partial}\partial(\log h)$. Note that G acts on $\mathcal{H}$ (from the left) by

$$G \times \mathcal{H} \to \mathcal{H}, \quad (g, h) \mapsto g \cdot h := (g^{-1})^* h,$$

where $(g^{-1})^*h$ is defined by $((g^{-1})^*h)(\ell_1, \ell_2) := h(g^{-1}\ell_1, g^{-1}\ell_2)$ for all $\ell_1, \ell_2 \in L$ in the same fibres of L over N.

Let $\mathrm{Lie}(G)\ (\subset H^0(N, \mathcal{O}(TN)))$ be the Lie algebra of G. To each $Y \in \mathrm{Lie}(G)$, we associate the real vector field $Y_{\mathbb{R}} := Y + \bar{Y}$ on N. Furthermore, put $\mathrm{Lie}(G)_{real} := \{Y_{\mathbb{R}} \mid Y \in \mathrm{Lie}(G)\}$. Then $Y \mapsto Y_{\mathbb{R}}$ defines an isomorphism of the complex Lie algebras $(\mathrm{Lie}(G), \sqrt{-1}) \simeq (\mathrm{Lie}(G)_{real}, J)$, where J is the complex structure of N. For every $t\ (= t_1 + \sqrt{-1}\, t_2) \in \mathbb{C}$ (where $t_1, t_2 \in \mathbb{R}$), we set

$$\exp(tY) := \exp\{\tfrac{1}{2}(t_1 Y_{\mathbb{R}} + t_2 J(Y_{\mathbb{R}}))\}.$$

Then $\{\exp(tY) \mid t \in \mathbb{C}\}$ is naturally regarded as a holomorphic one-parameter subgroup of G.

Now, to each $g \in G$, we associate

$$\psi_L(g) := \exp\{ \int_0^1 (\int_N \tfrac{1}{2} h_t^{-1} (\partial h_t/\partial t)\ c_1(L;h_t)^n) dt\},$$

where $\{h_t \mid 0 \leqq t \leqq 1\}$ is an arbitrary piecewise smooth path in $\mathcal{H}$ such that $h_0 = g \cdot h_1$. Then $\psi_L(g)$ is independent of the choice of the path $\{h_t\}$ and therefore well-defined. Note in particular that

$$\psi_L : G \to \mathbb{R}_+$$

is a Lie group homomorphism (see for instance [M2]). The corresponding Lie algebra homomorphism $(\psi_L)_* : \mathrm{Lie}(G) \to \mathbb{R}$ is given by

(1.1) $(\psi_L)_*(Y) = \frac{1}{2}\int_N h^{-1}(Y_{\mathbb{R}}h)\ c_1(L;h)^n, \qquad Y \in \mathrm{Lie}(G),$

where the right-hand side does not depend on the choice of h in H, and

$$Y_{\mathbb{R}}h := \frac{\partial}{\partial t}\Big|_{t=0}(\exp(tY_{\mathbb{R}}))^*h = \frac{\partial}{\partial t}\Big|_{t=0}(\exp(-tY_{\mathbb{R}}))\cdot h\ .$$

If $L = K_X^{-1}$ and $G = \mathrm{Aut}(N)$ where the G-action on L is the natural one, then ψ_L above coincides with ψ in the introduction, i.e.,

(1.2) $\psi_{K^{-1}} = \psi.$

2. G-REPRESENTATIONS ON THE SPACES OF COHOMOLOGY GROUPS.

Throughout this note, we only consider left G-actions. For instance, G acts on $\wedge^p T^*N \otimes \wedge^q \overline{T^*N}$ by

$$G \times \wedge^p T^*N \otimes \wedge^q \overline{T^*N} \to \wedge^p T^*N \otimes \wedge^q \overline{T^*N}, \qquad (g, \theta) \mapsto (g^{-1})^*\theta.$$

In view of the G-action on L, we see that G acts naturally on the vector bundle $\wedge^q \overline{T^*N} \otimes L^\nu$ ($\nu \in \mathbb{Z}$). This further induces the action

$$G \times H^q(N,\mathcal{O}(L^\nu)) \to H^q(N,\mathcal{O}(L^\nu)), \qquad (g,[\eta]) \mapsto (\rho_{q,\nu}(g))([\eta]) := [g\cdot\eta],$$

where for each L^ν-valued $\bar\partial$-closed (0,q)-form η, we denote by $[\eta]$ the cohomology class in $H^q(N,\mathcal{O}(L^\nu))$ represented by η. The map

$$\rho_{q,\nu}: G \to GL_{\mathbb{C}}(H^q(N,\mathcal{O}(L^\nu)))$$

is holomorphic for $q = 0$. In general, however, $\rho_{q,\nu}$ is only a C^∞ map. We now put

$$X(L^\nu) := \Big(\bigoplus_{q:\mathrm{even}} H^q(N,\mathcal{O}(L^\nu))\Big) \oplus \Big(\bigoplus_{q:\mathrm{odd}} (H^q(N,\mathcal{O}(L^\nu)))^*\Big),$$

$$\rho(L^\nu) := \Big(\bigoplus_{q:\mathrm{even}} \rho_{q,\nu}\Big) \oplus \Big(\bigoplus_{q:\mathrm{odd}} {}^t(\rho_{q,\nu})^{-1}\Big),$$

where ${}^t(\rho_{q,\nu})^{-1}: G \to GL_{\mathbb{C}}((H^q(N,\mathcal{O}(L^\nu)))^*)$ is the contragredient representation of $\rho_{q,\nu}$. Then

$$\rho(L^\nu): G \to GL_{\mathbb{C}}(X(L^\nu))$$

is a Lie group homomorphism (which is not necessarily holomorphic). By the Kodaira vanishing theorem, the following quasi-classical fact holds:

(2.1) THEOREM. <u>If</u> ν <u>is such that</u> $c_1(L^\nu \otimes K_N^{-1}) > 0$, <u>then</u> $X(L^\nu) = H^0(N,\mathcal{O}(L^\nu))$ <u>and therefore</u> $\rho(L^\nu): G \to GL_{\mathbb{C}}(X(L^\nu))$ <u>is an algebraic group homomorphism</u> (<u>defined over</u> $\mathbb{C}$).

Let W be a $\mathbb{C}$-vector space on which G acts by

$$\lambda : G \to GL_{\mathbb{C}}(W),$$

where λ is a Lie group homomorphism which is not necessarily holomorphic. Note that det: $GL_{\mathbb{C}}(W) \to \mathbb{C}^*$ is the character which sends each $B \in GL_{\mathbb{C}}(W)$ to $\det(B)$. As in the introduction, for every nonnegative integer m, let mW (resp. $-mW$) be the $\mathbb{C}$-vector space

$$\overset{m}{\oplus} W \quad (\text{resp. } \overset{m}{\oplus} W^*),$$

by which we mean the space $\{0\}$ if $m = 0$. Corresponding to this, we have the Lie group homomorphism

$$\overset{m}{\oplus} \lambda : G \to GL_{\mathbb{C}}(mW) \quad (\text{resp. } \overset{m}{\oplus} {}^t\lambda^{-1} : G \to GL_{\mathbb{C}}(-mW)),$$

which is simply denoted by $m\lambda$ (resp. $-m\lambda$). We now recall, for each $g \in G$, that the Lefschetz number $\Lambda_g(L^\nu)$ of g on $H^*(N, \mathcal{O}(L^\nu))$ is given by

$$\Lambda_g(L^\nu) = \Sigma_q \ (-1)^q \operatorname{Tr}(\rho_{q,\nu}(g)) = \operatorname{Tr}(\rho(L^\nu)(g)).$$

Let e_ν $(\nu \in \mathbb{Z})$ be integers which are nonzero only for a finite number of ν's. Then denoting by ξ the Lie group homomorphism

$$\underset{\nu}{\oplus} e_\nu \rho(L^\nu) : G \to GL_{\mathbb{C}}(\underset{\nu}{\oplus} e_\nu X(L^\nu)),$$

we have the character $\det \circ \xi : G \to \mathbb{C}^*$. The associated Lie algebra homomorphism $(\det \circ \xi)_* : \operatorname{Lie}(G) \to \mathbb{C}$ satisfies, for each $Y \in \operatorname{Lie}(G)$, the following:

$$(2.2) \qquad \begin{aligned} (\det \circ \xi)_*(Y) &= \operatorname{Tr}(\xi_*(Y)) = \Sigma_\nu \ e_\nu \operatorname{Tr}(\rho(L^\nu)_*(Y)) \\ &= \Sigma_\nu \ e_\nu \left.\frac{d}{dt}\right|_{t=0} \operatorname{Tr}(\rho(L^\nu)(\exp(tY))) \\ &= \Sigma_\nu \ e_\nu \left.\frac{d}{dt}\right|_{t=0} \Lambda_{\exp(tY)}(L^\nu). \end{aligned}$$

3. THE SERRE DUALITY.

For each $\theta \in H^q(N, \mathcal{O}(L^\nu))$, we denote by θ' the element of $(H^{n-q}(N, \mathcal{O}(K_N \otimes L^{-\nu})))^*$ defined by

$$\theta'(\eta) := \int_N \theta \wedge \eta, \qquad \eta \in H^{n-q}(N, \mathcal{O}(K_N \otimes L^{-\nu})),$$

where $\theta \wedge \eta$ is regarded as an element of $H^{n,n}(N)$ via the natural isomorphism $H^n(N, \mathcal{O}(K_N)) \simeq H^{n,n}(N)$. Then by the Serre duality theorem, the mapping $\theta \mapsto \theta'$ induces the G-equivariant $\mathbb{C}$-linear isomorphism

$$H^q(N, \mathcal{O}(L^\nu)) \simeq (H^{n-q}(N, \mathcal{O}(K_N \otimes L^{-\nu})))^* .$$

In particular, we obtain:

(3.1) PROPOSITION. If $L = K_N^{-1}$, then the above isomorphism gives us a G-equivariant identification of the $\mathbb{C}$-vector spaces $H^q(N,\mathcal{O}(L^\nu))$ and $(H^{n-q}(N,\mathcal{O}(L^{-\nu-1})))^*$, and hence we have $\rho_{q,\nu} = {}^t(\rho_{n-q,-\nu-1})^{-1}$.

4. DEFINITION OF ϕ.

Let $M = \{\mu_1, \mu_2, \ldots, \mu_{n+1}, \mu_{n+2}\}$ be an increasing sequence of (n+2) distinct integers. We then define an (n+2) by (n+1) matrix Q_M by

$$Q_M = \begin{pmatrix} 1, & \mu_1, & (\mu_1)^2, & \ldots, & (\mu_1)^n \\ 1, & \mu_2, & (\mu_2)^2, & \ldots, & (\mu_2)^n \\ \vdots & \vdots & \vdots & & \vdots \\ 1, & \mu_{n+2}, & (\mu_{n+2})^2, & \ldots, & (\mu_{n+2})^n \end{pmatrix}.$$

Let $f_M = f_M(z_1, z_2, \ldots, z_{n+2})$ be the linear function on $\mathbb{C}^{n+2}$ defined by

$$f_M(z_1, z_2, \ldots, z_{n+2}) = \det \begin{pmatrix} 1, & \mu_1, & (\mu_1)^2, & \ldots, & (\mu_1)^n, & z_1 \\ 1, & \mu_2, & (\mu_2)^2, & \ldots, & (\mu_2)^n, & z_2 \\ \vdots & \vdots & \vdots & & \vdots & \vdots \\ 1, & \mu_{n+2}, & (\mu_{n+2})^2, & \ldots, & (\mu_{n+2})^n, & z_{n+2} \end{pmatrix}.$$

Recall that the integer $\Delta_0(M) := f_M((\mu_1)^{n+1}, (\mu_2)^{n+1}, \ldots, (\mu_{n+2})^{n+1})$ is nothing but

$$\prod_{i>j} (\mu_i - \mu_j),$$

where the product is taken over all integers i and j with $n+2 \geqq i > j \geqq 1$. Furthermore, the coefficient $\Delta_k(M)$ of f_M in z_k $(1 \leqq k \leqq n+2)$ is the integer

$$(-1)^{n+k} \prod_{k \neq i > j \neq k} (\mu_i - \mu_j).$$

We now define integers $\Delta(M)$, $d(M;\mu_i)$, $d_0(M)$ by

$$\Delta(M) := \text{G.C.D.}(\Delta_1(M), \Delta_2(M), \ldots, \Delta_{n+2}(M)),$$

$$d(M;\mu_i) := \Delta_i(M)/\Delta(M), \qquad 1 \leqq i \leqq n+2,$$

$$d_0(M) := \Delta_0(M)/\Delta(M).$$

Using the notation in Section 2, we set

$$W_M := \bigoplus_{i=1}^{n+2} d(M;\mu_i)\, X(L^{\mu_i}),$$

$$\rho_M := \bigoplus_{i=1}^{n+2} d(M;\mu_i)\, \rho(L^{\mu_i}).$$

Then ρ_M: $G \to GL_{\mathbb{C}}(W_M)$ is a Lie group homomorphism which is not necessarily holomorphic. Note, however, that Theorem (2.1) implies

(4.1) PROPOSITION. *If* $c_1(L^{\mu_i} \otimes K_N^{-1}) > 0$ *for all* $i \in \{1,2,\ldots,n+2\}$, *then* W_M *coincides with*

$$\bigoplus_{i=1}^{n+2} d(M;\mu_i)\, H^0(N, \mathcal{O}(L^{\mu_i}))$$

and therefore ρ_M: $G \to GL_{\mathbb{C}}(W_M)$ *is an algebraic group homomorphism* (*defined over* $\mathbb{C}$).

Until the end of this section, we assume $L = K_N^{-1}$ and $G = \mathrm{Aut}(N)$, where the G-action on L is the natural one of Aut(N) on K_N^{-1}. The definition of ϕ is divided into two cases:

Case 1: n is odd (i.e., $n = 2p+1$ for some $p \in \mathbb{Z}$). We then set $M_n = (-(p+1), -p,\ldots,-1,0,1,\ldots,p,p+1)$. Furthermore, put

$$(4.2)\qquad \begin{aligned} V &:= d(M_n;p+1)\, X(L^{p+1}) \oplus \Big(\bigoplus_{\nu=1}^{p} (d(M_n;\nu)+d(M_n;\nu+1))\, X(L^{\nu}) \Big), \\ \phi &:= d(M_n;p+1)\, \rho(X^{p+1}) \oplus \Big(\bigoplus_{\nu=1}^{p} (d(M_n;\nu)+d(M_n;\nu+1))\, \rho(L^{\nu}) \Big). \end{aligned}$$

Case 2: n is even (i.e., $n = 2p$ for some $p \in \mathbb{Z}$). In this case, we set $M_n = (-(p+1),-p,\ldots,-1,0,1,\ldots,p-1,p)$. Furthermore, put

$$(4.3)\qquad \begin{aligned} V &:= \bigoplus_{\nu=1}^{p} d(M_n;\nu)\, X(L^{\nu}), \\ \phi &:= \bigoplus_{\nu=1}^{p} d(M_n;\nu)\, \rho(L^{\nu}). \end{aligned}$$

The Lie group homomorphism ϕ: $\mathrm{Aut}(N) \to GL_{\mathbb{C}}(V)$ thus defined satisfies the following:

(4.4) THEOREM. *Assume that* $c_1(N) > 0$. *Then* $X(L^{\nu}) = H^0(N, \mathcal{O}(K_N^{-\nu}))$ *for all nonnegative integers* ν. *Hence* ϕ: $\mathrm{Aut}(N) \to GL_{\mathbb{C}}(V)$ *is an algebraic group homomorphism* (*defined over* $\mathbb{C}$) *induced by the natural action of* Aut(N) *on* V. *Moreover for every* $g \in \mathrm{Aut}(N)$, *we have*:

(a) $\det(\rho_{M_n}(g)) = \det(\phi(g))$ _if_ n _is odd._

(b) $\det(\rho_{M_n}(g)) = (\det(\phi(g)))^2$ _if_ n _is even._

PROOF. In view of Theorem (2.1), it suffices to show (a) and (b) above. Using the notation in Section 2, we observe from Proposition (3.1) that $X(L^{-\nu-1}) = (-1)^{n+1}X(L^{\nu})$ and $\rho(L^{-\nu-1}) = (-1)^{n+1}\rho(L^{\nu})$. Note also that $\rho(L^0)$ is trivial because Aut(N) acts trivially on $X(L^0) = H^0(N,\mathcal{O}) \simeq \mathbf{C}$. Then (a) and (b) are straightforward from the fact that

$$d(M_n;-\nu-1) = \begin{cases} d(M_n;\nu+1) & \text{if } n \text{ is odd} \\ -d(M_n;\nu) & \text{if } n \text{ is even.} \end{cases}$$

(4.5) EXAMPLE. (i) Let n = 3. Then $M_3 = (-2,-1,0,1,2)$. Therefore

$$f_{M_3} = 12\,z_1 - 48\,z_2 + 72\,z_3 - 48\,z_4 + 12\,z_5\ ,$$

$$(d(M_3;-2),d(M_3;-1),d(M_3;0),d(M_3;1),d(M_3;2)) = (1,-4,6,-4,1),$$

$$d_0(M_3) = 24.$$

Hence $V = X(K_N^{-2}) \oplus -3\,X(K_N^{-1})$ and $\phi = \rho(K_N^{-2}) \oplus -3\,\rho(K_N^{-1})$. Thus, if $c_1(N) > 0$, then V is a direct sum of $H^0(N,\mathcal{O}(K_N^{-2}))$ and three copies of $H^0(N,\mathcal{O}(K_N^{-1}))^*$.

(ii) Let n = 2. Then $M_2 = (-2,-1,0,1)$. Hence

$$f_{M_2} = -2\,z_1 + 6\,z_2 - 6\,z_3 + 2\,z_4\ ,$$

$$(d(M_2;-2),d(M_2;-1),d(M_2;0),d(M_2;1)) = (-1,3,-3,1),$$

$$d_0(M_2) = 6.$$

In particular, $V = X(K_N^{-1})$ and $\phi = \rho(K_N^{-1})$. Thus, if $c_1(N) > 0$, then V is nothing but $H^0(N,\mathcal{O}(K_N^{-1}))$.

5. MAIN THEOREM.

Let $\tilde{\alpha}$ be as in the introduction, and put $G_1 := G \cap (\mathrm{Ker}\ \tilde{\alpha})$. If N is algebraic, then by our assumption in Section 1, G is also algebraic and in particular, the Chevalley decomposition allows us to express G_1 as a semidirect product $S_1 \ltimes U_1$ of a reductive algebraic subgroup S_1 of G_1 and the unipotent radical U_1 of G_1 . Now, the purpose of this section is to show the following:

(5.1) THEOREM. Let $M = \{\mu_1, \mu_2, \dots, \mu_{n+2}\}$ be an increasing sequence of $(n+2)$ distinct integers such that $c_1(L^{\mu_i} \otimes K_N^{-1}) > 0$ for all i. Then by setting $\beta := n!/d_0(M)$, we have:

$$(*) \qquad \psi_L(s) = |\det(\rho_M(s))|^{\beta},$$

for all $s \in S_1$ (see Section 1 for the definition of ψ_L).

PROOF. Step 1. First observe that, for every element s in S_1, there exists a positive integer m such that s^m belongs to the identity component $(S_1)^0$ of S_1. Hence, it suffices to show (*) for s in $(S_1)^0$. Next, by choosing a maximal compact subgroup K_1 of $(S_1)^0$, we can regard $(S_1)^0$ as complexification of K_1, i.e., $\mathrm{Lie}(S_1) = \mathrm{Lie}(K_1) \oplus \sqrt{-1}\,\mathrm{Lie}(K_1)$ (direct sum of vector spaces). Moreover, both $\{\psi_L(k); k \in K_1\}$ and $\{|\det(\rho_M(k))|^{\beta}; k \in K_1\}$ are compact subgroups of the multiplicative group $\mathbb{R}_+$, and therefore

$$\psi_L(k) = 1 = |\det(\rho_M(k))|^{\beta}$$

whenever k is in K_1. Hence the proof is reduced to showing

$$(1) \qquad (\psi_L)_*(Y) = \mathrm{Re}\{\beta(\det \circ \rho_M)_*(Y)\}$$

for all $Y \in \sqrt{-1}\,\mathrm{Lie}(K_1)$, where $\mathrm{Re}\{..\}$ denotes the real part of $\{..\}$.

Step 2. Fix a K_1-invariant Kähler metric ω on N with the corresponding Hermitian (1,0)-connection ∇^{ω} on TN. Furthermore, choose a K_1-invariant metric $h \in \mathcal{H}$ with the corresponding Hermitian (1,0)-connection ∇^h on L. Let Y' be an arbitrary element of $\mathrm{Lie}(K_1)$, and put $Y := \sqrt{-1}\,Y'$. Note that the Lie differentiation $\mathcal{L}_{Y'}$, $s \mapsto \frac{d}{dt}\big|_{t=0} \exp(-tY')\cdot s$, can act on both of the spaces $\Gamma(TN)$ and $\Gamma(L)$, where for any E, $\Gamma(E)$ denotes the space of C^{∞} sections of E. Then, $A^{\omega}(Y') := \nabla^{\omega}_{Y'} - \mathcal{L}_{Y'}$ and $A^h(Y') := \nabla^h_{Y'} - \mathcal{L}_{Y'}$ are regarded as elements of $\Gamma(\mathrm{End}(TN))$ and $\Gamma(\mathrm{End}(L))$, respectively. Since the line bundle $\mathrm{End}(L)$ is trivial, by using the notation in Section 1, we can write

$$(2) \qquad A^h(Y') = -\frac{1}{2}\sqrt{-1}\,h^{-1}(Y_{\mathbb{R}}h).$$

Let Θ_{ω} be the curvature form for the Kähler metric ω, and $\mathrm{Td} = \mathrm{Td}(Z)$ the invariant polynomial (associated with the total Todd class) defined by $\mathrm{Td}(Z) = \det((I-\exp(-Z))^{-1}Z)$. Then by a theorem of [FT] (see also [BV1], [BV2]),

$$\frac{d}{dt}\Big|_{t=0} \Lambda_{\exp(tY')}(L^{\nu}) = \frac{d}{dt}\Big|_{t=0} \int_N \mathrm{ch}(tY'; L^{\nu})\,\mathrm{Todd}(tY'; TN),$$

where $\mathrm{ch}(tY'; L^{\nu})$ and $\mathrm{Todd}(tY'; TN)$ are defined by

$$\mathrm{ch}(tY';L^\nu) := \exp(t\nu A^h(Y')+\nu c_1(L;h)),$$
$$\mathrm{Todd}(tY';TN) := \mathrm{Td}(tA^\omega(Y')+(\sqrt{-1}/2\pi)\Theta_\omega).$$

We now put $\theta := c_1(L;h)$ and $\eta := \frac{d}{dt}\big|_{t=0} \mathrm{Todd}(tY';TN)$. Recall that the total Todd form $\mathrm{Td}((\sqrt{-1}/2\pi)\Theta_\omega)$ associated with the Kähler metric ω is written as $1+\tau$, where τ is a sum of differential forms of degree $\geqq 2$. By setting $\lambda(\nu) := \frac{d}{dt}\big|_{t=0} \Lambda_{\exp(tY')}(L^\nu)$, $\nu \in \mathbb{Z}$, we have:

$$\lambda(\nu) = \int_N \exp(\nu\theta)(\nu A^h(Y')(1+\tau) + \eta).$$

Put $b_0 := \int_N \eta$ and $b_j := \int_N\{(\theta^j \wedge \eta/j!)+(\theta^{j-1}\wedge A^h(Y')(1+\tau)/(j-1)!)\}$, $j=1,2,\ldots,n$. Then in view of (2) above and (1.1) in Section 1,

$$(3) \qquad \Sigma_{j=0}^n b_j\nu^j = \lambda(\nu) - \nu^{n+1}\int_N A^h(Y')\theta^n/n! = \lambda(\nu) + \sqrt{-1}\,\nu^{n+1}(\psi_L)_*(Y)/n!\ .$$

For simplicity, put $r_i := \lambda(\mu_i) + \sqrt{-1}(\mu_i)^{n+1}(\psi_L)_*(Y)/n!$. We define column vectors $\mathbb{b}$ and $\mathbb{r}$ by

$$\mathbb{b} := {}^t(b_0, b_1, \ldots, b_n),$$
$$\mathbb{r} := {}^t(r_1, r_2, \ldots, r_{n+2}).$$

By (3) applied to $\nu = \mu_i$ $(i=1,2,..,n+2)$,

$$Q_M\,\mathbb{b} = \mathbb{r},$$

where we used the notation in Section 4. Hence, $f_M(r_1,r_2,\ldots,r_{n+2}) = 0$, i.e.,

$$f_M(\lambda(\mu_1),\lambda(\mu_2),\ldots,\lambda(\mu_{n+2})) = -\sqrt{-1}\,\Delta_0(M)(\psi_L)_*(Y)/n!\ .$$

Dividing both sides by $\Delta(M)$, we have:

$$\Sigma_{i=1}^{n+2} d(M;\mu_i)\,\lambda(\mu_i) = (\sqrt{-1}\,\beta)^{-1}(\psi_L)_*(Y).$$

Together with (2.2) applied to $\xi = \rho_M$, it follows that:

$$(\det\circ\rho_M)_*(Y') = (\sqrt{-1}\,\beta)^{-1}(\psi_L)_*(Y).$$

Now by (4.1), the homomorphism ρ_M is algebraic and in particular holomorphic. Hence, in view of the identity $Y = \sqrt{-1}\,Y'$, we obtain:

$$(\psi_L)_*(Y) = \beta(\det\circ\rho_M)_*(Y).$$

Since the left-hand side is a real number (cf. Section 1), this gives the identity (1), as required.

As an immediate consequence of Theorems (4.4) and (5.1), we obtain the following:

(5.2) COROLLARY. Assume that $c_1(N) > 0$. Let $\gamma = n!/d_0(M_n)$ if n is odd. Moreover, let $\gamma = 2(n!/d_0(M_n))$ if n is even. Then,

$$\psi(s) = |\det \phi(s)|^{\gamma}$$

for all $s \in S$ (see the introduction and (1.2) for the definition of S and ψ).

(5.3) REMARK. Recall that the image in $\mathbb{G}_m (= GL(1,\mathbb{C}))$ of a unipotent group under an algebraic group homomorphism is trivial. Hence, in Theorem (5.1), the restriction of $\det\circ\rho_M$ to U_1 is trivial. Let U be as in the introduction. Then we also see that, in (5.2), $\det\circ\phi$ is trivial when restricted to U.

(5.4) REMARK. (i) In Theorem (5.1), let $\iota_0\colon \mathbb{G}_m (= \{t \in \mathbb{C}^*\}) \hookrightarrow G$ be an arbitrary algebraic torus subgroup (if any). Put $Y_0 := (\iota_0)_*(t\partial/\partial t)$. Then by $\exp(2\pi\sqrt{-1}\, t\partial/\partial t) = 1$, we have:

$$\exp\{2\pi\sqrt{-1}(\det\circ\rho_M)_*(Y_0)\} = (\det\circ\rho_M)(\iota_0(\exp(2\pi\sqrt{-1}\, t\partial/\partial t))) = 1,$$

i.e., $(\det\circ\rho_M)_*(Y_0) \in \mathbb{Z}$. Thus, (1) in the proof of Theorem (5.1) yields

(a) $\quad (\psi_L)_*(Y_0)/\beta = (\det\circ\rho_M)_*(Y_0) \in \mathbb{Z}.$

(ii) Similarly, in (5.2), let $\iota_1\colon \mathbb{G}_m (= \{t \in \mathbb{C}^*\}) \hookrightarrow \mathrm{Aut}(N)$ be an arbitrary algebraic torus subgroup. Then by setting $Y_1 := (\iota_1)_*(t\partial/\partial t)$, we have:

(b) $\quad \psi_*(Y_1)/\gamma = (\det\circ\phi)_*(Y_1) \in \mathbb{Z}.$

(5.5) REMARK. (0.1) $\sim$ (0.3) in the introduction are straightforward from (4.2), (4.3), (4.4), (5.2), (5.3) above.

Recall that, if $c_1(N) > 0$ and furthermore N admits an Einstein-Kähler metric, then $\psi_* = 0$. Hence, (5.2) above immediately implies

(5.6) COROLLARY. If $c_1(N) > 0$ and if N admits an Einstein-Kähler metric, then the image of $\det\circ\phi\colon \mathrm{Aut}(N) \to \mathbb{G}_m$ is a finite group.

6. STUDIES OF ϕ FOR CONCRETE EXAMPLES.

In this section, we shall study ϕ for several Del pezzo surfaces. By writing out the character $\det\circ\phi$ explicitly for the Fermat cubic in $\mathbb{P}^3(\mathbb{C})$, we can show that even if N admits an Einstein-Kähler metric, the image of $\det\circ\phi$ is not necessarily trivial.

(6.1) We first consider the case where N is the blowing-up of $\mathbb{P}^2(\mathbb{C}) = \{(z_0:z_1:z_2)\}$ at the point (1:0:0) with the corresponding exceptional curve E. In view of the identification: $N - E = \mathbb{P}^2(\mathbb{C}) - \{(1:0:0)\}$, the action of $G:=(\mathbb{C}^*)^2$ on $\mathbb{P}^2(\mathbb{C}) - \{(1:0:0)\}$ defined by

$$\begin{array}{ccccc} G\ (= \mathbb{C}^*\times\mathbb{C}^*) & \times & \mathbb{P}^2(\mathbb{C}) - \{(1:0:0)\} & \to & \mathbb{P}^2(\mathbb{C}) - \{(1:0:0)\} \\ (t_1,t_2), & & (z_0:z_1:z_2) & \mapsto & (z_0:t_1z_1:t_2z_2) \end{array}$$

extends to a holomorphic G-action on N. In terms of this action, Lie(G) $(= \mathbb{C}t_1\partial/\partial t_1 + \mathbb{C}t_2\partial/\partial t_2)$ is regarded as a subspace of $H^0(N,\mathcal{O}(TN))$. Put

$$\Gamma = \{(1,-2),(1,-1),(1,0),(0,1),(-1,1),(-2,1),(-1,0),(0,-1),(0,0)\}$$

and consider the natural G-action on $H^0(N,\mathcal{O}(K_N^{-1}))$. Then, to each element $\chi = (\chi',\chi'')$ of Γ, we can associate a nonzero section $s_\chi \in H^0(N,\mathcal{O}(K_N^{-1}))$ such that

i) $\mathbf{t}\cdot s_\chi = t_1^{\chi'}t_2^{\chi''}s_\chi$ for all $\mathbf{t} = (t_1,t_2) \in G$;

ii) $\{s_\chi;\chi\in\Gamma\}$ forms a $\mathbb{C}$-basis for $H^0(N,\mathcal{O}(K_N^{-1}))$.

Let P_Γ be the polygon in $\mathbb{R}^2$ obtained as the convex hull of the points in Γ. Put $K:= \{(t_1,t_2)\in G; |t_1|=|t_2|=1\}$. Note that, for any K-invariant Kähler metric ω in the class $2\pi c_1(N)_{\mathbb{R}}$, the associated moment map $m_\omega: N \to \mathbb{R}^2$ satisfies

$$P_\Gamma = \text{Image } m_\omega \qquad \text{(cf. [M2]; see also [O1])},$$

where we identified $\mathbb{R}^2$ with $\mathbb{R}dt_1/(\sqrt{-1}\,t_1) + \mathbb{R}dt_2/(\sqrt{-1}\,t_2)$. Since the barycentre of P_Γ is $(-1/12,-1/12)$, a theorem of [M2] shows that

(1) $(\psi_*(t_1\partial/\partial t_1),\ \psi_*(t_2\partial/\partial t_2)) = c_1(N)^2[N]\cdot(-1/12,-1/12) = (-2/3,-2/3)$.

On the other hand, by $\sum_{\chi\in\Gamma}\chi = (-1,-1)$, we have $(\det\circ\phi)(\mathbf{t}) = (t_1t_2)^{-1}$ for all $\mathbf{t} = (t_1,t_2)\in G$. In particular, the corresponding Lie algebra homomorphism $(\det\circ\phi)_*: \mathrm{Lie}(G) \to \mathbb{C}$ satisfies

(2) $((\det\circ\phi)_*(t_1\partial/\partial t_1),\ (\det\circ\phi)_*(t_2\partial/\partial t_2)) = (-1,-1)$.

Now by (5.2), $\gamma = 2(2!/d_0(M_2)) = 2/3$ (cf. (ii) of (4.5)). Hence, (1) and (2) above actually satisfy the identity in (b) of (5.4).

(6.2) We next consider the case where N is the blowing-up of $\mathbb{P}^2(\mathbb{C})$ at the points (0:1:0) and (0:0:1) with the corresponding exceptional curves E' and E''. Then the action of $G:=(\mathbb{C}^*)^2$ on $\mathbb{P}^2(\mathbb{C}) - \{(0:1:0),(0:0:1)\}$ $(= N - E'\cup E'')$ defined by

$$G(=\mathbb{C}^*\times\mathbb{C}^*)\times\mathbb{P}^2(\mathbb{C})-\{(0:1:0),(0:0:1)\}\to\mathbb{P}^2(\mathbb{C})-\{(0:1:0),(0:0:1)\}$$
$$(t_1,t_2),\quad (z_0:z_1:z_2)\quad\mapsto\quad (z_0:t_1z_1:t_2z_2)$$

extends naturally to a holomorphic G-action on N. This time, we have to put

$$\Gamma=\{(1,-1),(1,0),(1,1),(0,1),(-1,1),(-1,0),(0,-1),(0,0)\}$$

and to each element $\chi=(\chi',\chi'')$ of Γ, we can again associate $s_\chi\in H^0(N,\mathcal{O}(K_N^{-1}))$ in such a way that the conditions i) and ii) in (6.1) are satisfied. Then the barycentre of the convex hull P_Γ in $\mathbb{R}^2$ of all points of Γ is (2/21,2/21) and therefore

(3) $$(\psi_*(t_1\partial/\partial t_1),\ \psi_*(t_2\partial/\partial t_2)) = c_1(N)^2[N]\cdot(2/21,2/21) = (2/3,2/3).$$

On the other hand, by $\sum_{\chi\in\Gamma}\chi=(1,1)$, the Lie algebra homomorphism $(\det\circ\phi)_*$: $\mathrm{Lie}(G)\to\mathbb{C}$ satisfies

(4) $$((\det\circ\phi)_*(t_1\partial/\partial t_1),\ (\det\circ\phi)_*(t_2\partial/\partial t_2)) = (1,1).$$

In view of $\gamma=2/3$, the above (3) and (4) again satisfy the identity in (b) of (5.4).

(6.3) Now, let N be the blowing-up of $\mathbb{P}^2(\mathbb{C})=\{(z_0:z_1:z_2)\}$ at the points (0:0:1), (1:0:0), (0:1:0) with the corresponding exceptional curves E_1, E_3, E_5, respectively. Furthermore, for each $i\in\{0,1,2\}$, let E_{2i} denote the exceptional curve in N obtained as the proper transform of the hyperplane $\{z_i=0\}$. We can then identify $G(:=(\mathbb{C}^*)^2)$ with $N-\bigcup_{i=0}^5 E_i$ by

$$G\ (=\mathbb{C}^*\times\mathbb{C}^*)\ni(t_1,t_2)\longleftrightarrow(1:t_1:t_2)\in N-\bigcup_{i=0}^5 E_i\ .$$

Hence, N is regarded as a G-equivariant compactification of G itself, since the group multiplication of G naturally extends to a holomorphic G-action on N. Define $\sigma,\ \tau\in\mathrm{Aut}(G)$ by setting

$$\sigma((t_1,t_2))=(t_2,t_1)\quad\text{and}\quad\tau((t_1,t_2))=(t_2,t_2/t_1)$$

for each $(t_1,t_2)\in G$. Since these σ and τ extend to holomorphic automorphisms of N, we obtain a finite subgroup F of Aut(N) generated by σ and τ. Note that F is a dihedral group of order 12. Moreover, Aut(N) is generated by G and F. Recall in our present case that the Futaki invariant vanishes, i.e., $\psi_*=0$. Therefore, in view of Corollary (5.2), we have $(\det\circ\phi)(G)=\{1\}$. Put

$$s_0:=(t_1\partial/\partial t_1)\wedge(t_2\partial/\partial t_2),$$
$$s_1:=(t_1\partial/\partial t_1)\wedge(\partial/\partial t_2).$$

Then both s_0 and s_1 extend to holomorphic sections $\in H^0(N,\mathcal{O}(K_N^{-1}))$. It is now easy to check that $\{s_0, s_1, \tau\cdot s_1, \tau^2\cdot s_1, \dots, \tau^5\cdot s_1\}$ forms a $\mathbb{C}$-basis for $H^0(N,\mathcal{O}(K_N^{-1}))$. We observe that $\sigma\cdot s_0 = -s_0$ and $\tau\cdot s_0 = s_0$. Furthermore, τ (resp. $-\sigma$) induces an even (resp. odd) permutation of $\{s_1, \tau\cdot s_1, \tau^2\cdot s_1, \dots, \tau^5\cdot s_1\}$. Hence

$$(\det\circ\phi)(\sigma) = (\det\circ\phi)(\tau) = 1.$$

Thus, the image of $\det\circ\phi$: $\mathrm{Aut}(N) \to \mathbb{C}^*$ is trivial. Note here that, by [S1] and [TY], our present N admits an Einstein-Kähler metric.

(6.4) We finally assume that N is the Fermat cubic surface $\{(z_1:z_2:z_3:z_4) \in \mathbb{P}^3(\mathbb{C});\ z_1^3 + z_2^3 + z_3^3 + z_4^3 = 0\}$. Obviously, the symmetric group S_4 of degree 4 is regarded as a subgroup of Aut(N) by setting

$$\sigma((z_1:z_2:z_3:z_4)) := (z_{\sigma(1)}:z_{\sigma(2)}:z_{\sigma(3)}:z_{\sigma(4)})$$

for all $\sigma\in S_4$ and $(z_1:z_2:z_3:z_4)\in N$. Recall that Aut(N) is a finite group of order $3^3\cdot 4!$ and is generated by S_4 and the group G_1 of multiplication of the coordinates by the cubic roots of unity, where multiplication of all the coordinates at the same time by the same cubic root of unity is regarded as the unit element of G_1. Note that S_4 acts on $H^0(N,\mathcal{O}(K_N^{-1}))$ as a subgroup of Aut(N). Then there exists a $\mathbb{C}$-basis $\{s_1,s_2,s_3,s_4\}$ for $H^0(N,\mathcal{O}(K_N^{-1}))$ such that

i) $(s_1(z):s_2(z):s_3(z):s_4(z)) = (z_1:z_2:z_3:z_4)$ for all $z = (z_1:z_2:z_3:z_4)\in N$;

ii) $\sigma\cdot s_i = \xi_\sigma\, s_{\sigma(i)}$ $(\sigma\in S_4)$ for some $\xi_\sigma\in\mathbb{C}^*$ independent of $i\in\{1,2,3,4\}$.

Let $\tau\in S_4$ be the transposition (1,2) of two letters 1 and 2. Obviously, τ fixes the point $p = (0:0:-1:1)\in N$. By setting $x_i := z_i/z_4$ $(i=1,2)$, we have a system (x_1,x_2) of holomorphic local coordinates of N around p. Since the automorphism τ_* of the tangent space TN_p induced by τ permutes $\partial/\partial x_1$ and $\partial/\partial x_2$, the identity

$$\tau_*((\partial/\partial x_1)\wedge(\partial/\partial x_2)) = -(\partial/\partial x_1)\wedge(\partial/\partial x_2)$$

together with $\tau\cdot s_4 = \xi_\tau\, s_4$ implies $\xi_\tau = -1 = \mathrm{sgn}(\tau)$. Similarly, we have $\xi_\sigma = \mathrm{sgn}(\sigma)$ whenever $\sigma\in S_4$ is a transposition. Therefore, the group homomorphism $S_4 \ni \sigma \mapsto \xi_\sigma\in\mathbb{C}^*$ satisfies $\xi_\sigma = \mathrm{sgn}(\sigma)$ $(\sigma\in S_4)$, i.e.,

$$\sigma\cdot s_i = \mathrm{sgn}(\sigma)\, s_{\sigma(i)} \qquad \text{for all } \sigma\in S_4 \text{ and } i\in\{1,2,3,4\}.$$

In particular, $(\det\circ\phi)(\sigma) = \mathrm{sgn}(\sigma)$ for all $\sigma\in S_4$. On the other hand, it is not so hard to check that $(\det\circ\phi)(g) = 1$ for all $g \in G_1$. Thus, the image of the character $\det\circ\phi$: $\mathrm{Aut}(N) \to \mathbb{C}^*$ is exactly $\{\pm 1\}$, though our N is known to admit an Einstein-Kähler metric (cf. [S1], [TY]).

COLLEGE OF GENERAL EDUCATION, CHIBA UNIVERSITY, 260 CHIBA, JAPAN

COLLEGE OF GENERAL EDUCATION, OSAKA UNIVERSITY, 560 OSAKA, JAPAN

(December, 1987)

REFERENCES

[BV1] N. Berline and M. Vergne: Zeros d'un champ de vecteurs et classes characteristiques equivariantes, Duke Math. J. 50 (1983), 539-549.

[BV2] N. Berline and M. Vergne: The equivariant index and Kirillov's character formula, Amer. J. Math. 107 (1985), 1159-1190.

[F1] A. Futaki: On a character of the automorphism group of a compact complex manifold, Invent. Math. 87 (1987), 655-660.

[FT] A. Futaki and K. Tsuboi: On some integral invariants, Lefschetz numbers and induction maps, preprint.

[M1] T. Mabuchi: K-energy maps integrating Futaki invariants, Tôhoku Math. J. 38 (1986), 575-593.

[M2] T. Mabuchi: Einstein-Kähler forms, Futaki invariants and convex geometry on toric Fano varieties, to appear in Osaka J. Math.

[O1] T. Oda: Convex polytopes and algebraic geometry (Japanese), Math. ser. 24, Kinokuniya, 1985, 1-229.

[S1] Y. T. Siu: The existence of Kähler-Einstein metrics on manifolds with positive anticanonical line bundle and a suitable finite symmetry group, to appear.

[TY] G. Tian and S. T. Yau: Kähler-Einstein metrics on complex surfaces with $c_1 > 0$, Commun. Math. Phys. 112 (1987), 175-203.

TENSORIAL ERGODICITY OF GEODESIC FLOWS

Masahiko KANAI

Department of Mathematics
Keio University
Yokohama 223, Japan

Introduction

In the 1930's, Birkhoff and von Neumann clarified the mathematical meaning of ergodicity. By their definition, a dynamical system with an invariant measure is said to be ergodic if it is metrically transitive, or equivalently if any L^2-integrable invariant function on the phase space is constant almost everywhere. Soon after, following their work, Hopf and Hedlund actually demonstrated the ergodicity of the geodesic flows on closed surfaces of constant negative curvature. Furthermore their result was generalized by Anosov to the geodesic flows of arbitrary closed riemannian manifolds of variable negative curvature. The significance of the geodesic flows has been recognized through these works, for they are typical examples of ergodic systems. Now, since the geodesic flow is a differentiable dynamical system, it makes sense to consider the action of the flow on the tensor fields defined on the phase space as well as the action on the functions. In particular, it seems to be reasonable to ask whether the geodesic flow possesses the "tensorial ergodicity"; that is, whether every L^2-integrable tensor field on the phase space which is invariant under the action of the geodesic flow is "constant" almost everywhere. The purpose of the present note is to show this phenomenon of the geodesic flows on certain negatively curved manifolds.

To be more precise, suppose that M is a closed riemannian manifold of negative sectional curvature. The geodesic flow φ_t of M is then defined as a smooth flow on the unit tangent bundle $V = \{v \in TM : |v| = 1\}$ of M. We now restrict ourselves to either of the following two cases: (i) M is of dimension two; (ii) M is locally symmetric. In both cases the unit tangent bundle V has a canonically defined affine connection ∇ as we will see later, and in terms of it, we can define "constant", or more precisely, parallel tensor fields on V. In fact, we say that a differentiable tensor field f on V is *parallel* if its covariant derivative ∇f identically vanishes, and a measurable tensor field on V is said to be *parallel almost everywhere* if it coincides with a certain parallel tensor field almost everywhere. Also a measurable tensor field f on V is said to be *L^2-integrable* if its norm $|f|$ with respect to the canonical riemannian metric of V is L^2-integrable over V relative to the Liouville measure of V. Our main result here is

Theorem. *Every φ_t-invariant L^2-integrable tensor field on V is parallel almost everywhere.*

The proof of the theorem will be given in §1 for locally symmetric spaces, and in §2 for surfaces. Furthermore, in the last section, we will give a reformulation of the theorem and consider a related problem concerned with φ_t-invariant differentiable tensor fields on V.

The author wishes to thank Professor S. Kaneyuki whose suggestion made our description of the canonical connection made in §2 simpler.

1. Locally Symmetric Spaces

In this section we prove the tensorial ergodicity for the geodesic flows of locally symmetric spaces, and we begin it with algebraic description of the unit tangent bundles and the geodesic flows of these spaces in order to introduce the canonical affine connections on the unit tangent bundles. Suppose first that M is a noncompact symmetric space of rank one: Namely, by multiplying the riemannian metric of M by a suitable constant, M is isometric to one of the real hyperbolic space $M_{\mathbf{R}}$, the complex hyperbolic space $M_{\mathbf{C}}$, the quaternion hyperbolic space $M_{\mathbf{H}}$ and the Cayley hyperbolic space $M_{\mathbf{O}}$. The symmetric space M is represented as a homogeneous G-space $M = G/K$ with G being a connected simple Lie group acting on M isometrically, and K a maximal compact subgroup of G. Associated with the representation of M as a homogeneous G-space, the Lie algebra $\mathbf{g}$ of G carries the Cartan decomposition

$$\mathbf{g} = \mathbf{k} + \mathbf{m},$$

where $\mathbf{k}$ is the Lie algebra of K, and $\mathbf{m}$ is a linear subspace of $\mathbf{g}$ that is naturally identified with the tangent space of M at the point $o = K$ of $M = G/K$. Now fix an element ι of $\mathbf{m} \cong T_oM$ of unit length. Direct computation shows that the eigenvalues of $\mathrm{ad}(-\iota) : \mathbf{g} \to \mathbf{g}$ are 0, ± 1 in the case where $M = M_{\mathbf{R}}$, and are 0, ± 1, ± 2 in the cases $M = M_{\mathbf{C}}$, $M_{\mathbf{H}}$ and $M_{\mathbf{O}}$. Denoting the eigenspace of each eigenvalue λ of $\mathrm{ad}(-\iota)$ by $\mathbf{g}^{\lambda}$, we obtain the eigenspace decomposition

$$\mathbf{g} = \mathbf{g}^{-2} + \mathbf{g}^{-1} + \mathbf{g}^{0} + \mathbf{g}^{+1} + \mathbf{g}^{+2}. \tag{1.1}$$

Hereafter we adopt the convention that $\mathbf{g}^{\lambda} = 0$ for $|\lambda| > 1$ provided $M = M_{\mathbf{R}}$, and $\mathbf{g}^{\lambda} = 0$ for $|\lambda| > 2$ otherwise. Then, by Jacobi's identity, we immediately have

$$[\mathbf{g}^{\lambda}, \mathbf{g}^{\mu}] \subset \mathbf{g}^{\lambda+\mu}; \tag{1.2}$$

that is, $\mathbf{g}$ is a graded Lie algebra. Furthermore, if $\mathbf{m}^{\perp}$ denotes the orthogonal complement of ι in $\mathbf{m}$ relative to the Killing form of $\mathbf{g}$, we have $[\mathbf{k}, \iota] = \mathbf{m}^{\perp}$ since M is a symmetric space *of rank one.* This yields the decomposition $\mathbf{g}^0 = (\mathbf{k} \cap \mathbf{g}^0) + \langle \iota \rangle$ of $\mathbf{g}^0$, where $\langle \iota \rangle$ denotes the linear subspace spanned by ι. Thus we obtain a new decomposition

$$\mathbf{g} = \mathbf{k}^0 + \mathbf{v} \tag{1.3}$$

of the Lie algebra $\mathbf{g}$ into $\mathbf{k}^0 = \mathbf{k} \cap \mathbf{g}^0$ and

$$(1.4) \qquad \begin{aligned} \mathbf{v} &= \mathbf{v}^{-2} + \mathbf{v}^{-1} + \mathbf{v}^{0} + \mathbf{v}^{+1} + \mathbf{v}^{+2} \\ &\text{with} \quad \mathbf{v}^0 = \langle \iota \rangle \quad \text{and} \quad \mathbf{v}^\lambda = \mathbf{g}^\lambda \quad (\lambda = \pm 1, \pm 2). \end{aligned}$$

Now consider the unit tangent bundle V of M. The isometric action of G on M is naturally lifted to an action on V, and it is transitive since M is a symmetric space *of rank one*. Further the Lie algebra of the isotropy subgroup K^0 of the action of G on V at the point $\iota \in V$ coincides with the Lie algebra $\mathbf{k}^0$ in (1.3). Hence the decomposition (1.3) of $\mathbf{g}$ means that $V = G/K^0$ is a *reductive* homogeneous G-space; i.e., $[\mathbf{k}^0, \mathbf{v}] \subset \mathbf{v}$. In particular, $\mathbf{v}$ is identified with the tangent space of V at ι. Moreover, since the splitting (1.4) of $\mathbf{v} \cong T_\iota V$ is $\mathrm{ad}(\mathbf{k}^0)$-invariant, it extends to a G-invariant splitting of the tangent bundle of V:

$$(1.5) \qquad TV = E^{-2} + E^{-1} + E^0 + E^{+1} + E^{+2}.$$

(Note here that $E^{\pm 2} = 0$ in the case of $M = M_{\mathbf{R}}$.) Fix a G-invariant riemannian metric of V for which the splitting (1.5) of the tangent bundle of V is orthogonal. The geodesic flow φ_t of M defined on the unit tangent bundle V of M commutes with the the action of G on V, and the orbit of φ_t passing through ι is given by $\varphi_t(\iota) = (\mathrm{Exp}\, t\iota) \cdot \iota$. This specifically implies that E^0 is spanned by the geodesic spray $\dot{\varphi} = (d/dt)|_{t=0}\varphi_t$, which is, by definition, the vector field on V generating the flow φ_t on V. In addition, it follows that $d\varphi_t \xi = d(\mathrm{Exp}\, t\iota) \circ \mathrm{Ad}(\mathrm{Exp}(-t\iota))(\xi)$ for $\xi \in \mathbf{v} \cong T_\iota V$, while it holds that $\mathrm{Ad}(\mathrm{Exp}(-t\iota))\xi^\lambda = e^{\lambda t}\xi^\lambda$ for $\xi^\lambda \in \mathbf{v}^\lambda$ since $\mathrm{ad}(-\iota)\xi^\lambda = \lambda \xi^\lambda$ by definition. Hence we have

$$(1.6) \qquad d\varphi_t \xi^\lambda = e^{\lambda t} \cdot d(\mathrm{Exp}\, t\iota)(\xi^\lambda), \qquad \xi^\lambda \in \mathbf{v}^\lambda \subset T_\iota V, \quad \lambda = 0, \pm 1, \pm 2.$$

In consequence, each subbundle E^λ in the splitting (1.5) is $d\varphi_t$-invariant, and satisfies

$$(1.7) \qquad |d\varphi_t \xi^\lambda| = e^{\lambda t}\, |\xi^\lambda|; \qquad \xi^\lambda \in E^\lambda, \quad \lambda = 0, \pm 1, \pm 2.$$

We now proceed to the definition of the canonical affine connection ∇ of the unit tangent bundle V. It requires another affine connection D of V which is defined as the canonical connection of the reductive homogeneous space $V = G/K^0$, and is described as follows (cf. [**KN**]). First extend the decomposition (1.3) of $\mathbf{g}$ to a left-invariant linear splitting $TG = C_{\mathbf{k}^0} + C_{\mathbf{v}}$ of the tangent bundle of G. The first component $C_{\mathbf{k}^0}$ is vertical with respect to the fibering of G over $V = G/K^0$, while $C_{\mathbf{v}}$ is horizontal. For each tangent vector ξ of V, let $\xi^* \in C_{\mathbf{v}}$ be the horizontal lift of ξ by the fibering of G over V. On the other hand, let D^* be the affine connection of G defined by $D^*\zeta \equiv 0$ for all left-invariant vector fields ζ on G. Then the canonical connection D of V as a homogeneous G-space is characterized by $(D_\xi \eta)^* = D^*_{\xi^*}\eta^*$ for any vector fields ξ and η on V. The connection D of V possesses the following properties. (1)

D is G-invariant, that is, $dg(D_\xi \eta) = D_{dg(\xi)} dg(\eta)$ for any $g \in G$ and vector fields ξ and η on V. (2) The torsion tensor T and the curvature tensor R of D have the representations $T(\xi, \eta) = -[\xi, \eta]_{\mathbf{v}}$ and $R(\xi, \eta)\zeta = -[[\xi, \eta]_{\mathbf{k}^0}, \zeta]$ for $\xi, \eta, \zeta \in T_\iota V \cong \mathbf{v} \subset \mathbf{g}$, where, for $\xi \in \mathbf{g} = \mathbf{k}^0 + \mathbf{v}$, $\xi_{\mathbf{k}^0}$ and $\xi_{\mathbf{v}}$ denote $\mathbf{k}^0$- and $\mathbf{v}$-components of ξ respectively. (3) Each subbundle E^λ in the splitting (1.5) is D-stable in the sense that the covariant derivative $D_\xi \eta$ is a section of E^λ whenever η is a section of E^λ. (4) The canonical contact form θ of V (cf. [**AM**]; see also §2) is G-invariant, and therefore, is parallel with respect to D. (5) D is φ_t-invariant, i.e., $d\varphi_t(D_\xi \eta) = D_{d\varphi_t \xi} d\varphi_t \eta$ for any vector fields ξ and η on V. Now define the *canonical connection* ∇ of V by

$$(1.8) \qquad \nabla_\xi \eta = \begin{cases} D_\xi \eta, & \xi \in E^{-2} + E^{-1} + E^{+1} + E^{+2}, \\ \mathcal{L}_{\dot\varphi} \eta, & \xi = \dot\varphi \in E^0, \end{cases}$$

for an arbitrary vector field η on V, where $\mathcal{L}_{\dot\varphi}\eta$ denotes the Lie derivative of η by the geodesic spray $\dot\varphi$. The reason why in the above definition we adopted the Lie derivative in the direction of the geodesic spray instead of the covariant derivative by D is that (locally defined) φ_t-invariant vector fields on V are, in general, non-parallel with respect to D; cf. (1.6). Note that (7) each subbundle E^λ of TV is again ∇-stable, that (8) the Liouville measure $\Lambda = \theta \wedge (d\theta)^n$ ($n + 1 = \dim M$) of V is parallel with respect to the canonical connection ∇, and that (9) ∇ is φ_t-invariant.

In summary, we have obtained the following things on the unit tangent bundle of each noncompact symmetric space of rank one: the geodesic flow φ_t, the splitting (1.5) of the tangent bundle, and the canonical connection ∇.

Henceforth assume that M is a closed locally symmetric riemannian manifold of negative curvature. Then the universal covering $\widetilde{M}$ of M is a noncompact symmetric space of rank one. Moreover the unit tangent bundle V of M is covered by the unit tangent bundle $\widetilde{V}$ of $\widetilde{M}$, and the deck transformations of the covering $\widetilde{V}$ of V preserve the above structures on $\widetilde{V}$. Thus they descend downstairs, and we obtain the corresponding structures on V which we denote by the same symbols: Namely, we are in the presence of the geodesic flow φ_t of M defined on V, the splitting (1.5) of the tangent bundle of V, and the canonical connection ∇ of V. Since the action of the geodesic flow φ_t on V is smooth, it is naturally lifted to an action ψ_t on the vector bundle $T^{(\ell,m)} = (\otimes_\ell TV) \otimes (\otimes_m T^*V)$ of (ℓ, m)-tensors of V as follows. First we require that the diagram

$$\begin{array}{ccc} T^{(\ell,m)} & \xrightarrow{\psi_t} & T^{(\ell,m)} \\ \downarrow & & \downarrow \\ V & \xrightarrow[\varphi_t]{} & V \end{array}$$

commutes, and that the restriction of ψ_t to each fiber of $T^{(\ell,m)}$ is a linear isomorphism. Further ψ_t is given by $\psi_t \xi = d\varphi_t \xi$ specifically for $\xi \in TV$, $\psi_t \alpha = \varphi_{-t}^* \alpha$

for $\alpha \in T^*V$ and $\psi_t(\xi_1 \otimes \cdots \xi_\ell \otimes \alpha_1 \otimes \cdots \otimes \alpha_m) = (\psi_t\xi_1) \otimes \cdots \otimes (\psi_t\xi_\ell) \otimes (\psi_t\alpha_1) \otimes \cdots \otimes (\psi_t\alpha_m)$ for $\xi_j \in TV$ and $\alpha_k \in T^*V$. Now the splitting (1.5) of the tangent bundle of V yields the splitting of $T^{(\ell,m)}$ into the subbundles

$$E^{(\lambda_1,\cdots,\lambda_\ell;\mu_1,\cdots,\mu_m)} = E^{\lambda_1} \otimes \cdots \otimes E^{\lambda_\ell} \otimes E^{\mu_1 *} \otimes \cdots \otimes E^{\mu_m *}$$

with $\lambda_1, \cdots, \lambda_\ell, \mu_1, \cdots, \mu_m = 0, \pm 1, \pm 2$. Each of these bundles is ψ_t-invariant, and satisfies

$$(1.9) \qquad |\psi_t \tau| = \exp\left(\sum_{j=1}^{\ell} \lambda_j - \sum_{k=1}^{m} \mu_k\right) t \cdot |\tau|, \qquad \tau \in E^{(\lambda_1,\cdots,\lambda_\ell;\mu_1,\cdots,\mu_m)},$$

by (1.6) and (1.7). (Remind here that we have chosen a G-invariant riemannian metric of V so that the splitting (1.5) of the tangent bundle of V is orthogonal). For a section f of $E^{(\lambda_1,\cdots,\lambda_\ell;\mu_1,\cdots,\mu_m)}$, define a new section $\psi_t f$ of $E^{(\lambda_1,\cdots,\lambda_\ell;\mu_1,\cdots,\mu_m)}$ by

$$(\psi_t f)(v) = \psi_t \circ f \circ \varphi_{-t}(v), \qquad v \in V.$$

We say that a section f of $E^{(\lambda_1,\cdots,\lambda_\ell;\mu_1,\cdots,\mu_m)}$ is *φ_t-invariant* if $\psi_t f = f$: In other words, f is ψ_t-invariant if and only if its "graph" $f(V)$ is a ψ_t-invariant subset of $E^{(\lambda_1,\cdots,\lambda_\ell;\mu_1,\cdots,\mu_m)}$. Note also that the canonical connection ∇ of V naturally induces a connection of $T^{(\ell,m)}$, which we denote by the same symbol ∇, and each subbundle $E^{(\lambda_1,\cdots,\lambda_\ell;\mu_1,\cdots,\mu_m)}$ of $T^{(\ell,m)}$ is ∇-stable. Every differentiable section f of $E^{(\lambda_1,\cdots,\lambda_\ell;\mu_1,\cdots,\mu_m)}$ should satisfy

$$(1.10) \qquad \nabla(\psi_t f) = \psi_t(\nabla f),$$

since the canonical connection ∇ of V is φ_t-invariant. To prove the tensorial ergodicity, it is now sufficient to show

(1.11) **Proposition.** *An L^2-integrable section f of $E^{(\lambda_1,\cdots,\lambda_\ell;\mu_1,\cdots,\mu_m)}$ is parallel almost everywhere with respect to ∇ whenever f is ψ_t-invariant.*

The proof of the proposition is divided into two cases according to whether $\sum_{j=1}^{\ell} \lambda_j - \sum_{k=1}^{m} \mu_k = 0$ or not. We first consider the easier one.

Case 1. Suppose that $\Sigma\lambda_j - \Sigma\mu_k \neq 0$. For a measurable ψ_t-invariant section f of $E^{(\lambda_1,\cdots,\lambda_\ell;\mu_1,\cdots,\mu_m)}$, we obtain

$$(1.12) \qquad |f(v)| = |\psi_{-t} \circ f \circ \varphi_t(v)| = e^{-(\Sigma\lambda_j - \Sigma\mu_k)t} |f \circ \varphi_t(v)|$$

from (1.9). Now choose a constant $c > 0$ so that $\Lambda(W) > 0$ for $W = \{v \in V : |f(v)| \leq c\}$, where Λ denotes the Liouville measure of the unit tangent bundle V. Since the Liouville measure is invariant under the the geodesic flow, Birkhoff's individual ergodic theorem (see e.g. [**P**]), together with the ergodicity of the geodesic flow, implies that $t^{-1} \int_0^t \chi_W \circ \varphi_s(v)\, ds$ converges to $\Lambda(W)/\Lambda(V) > 0$ as t goes to infinity for almost all $v \in V$, where χ_W denotes the characteristic function of W. In particular, for almost all $v \in V$ there exists a diverging sequence $\{t_k\}$ such that $\varphi_{t_k}(v) \in W$, i.e., $|f|(\varphi_{t_k}(v)) \le c$. Thus, in the case where $\Sigma\lambda_j - \Sigma\mu_k > 0$, we can obtain $f \equiv 0$ almost everywhere by letting $t = t_k \to \infty$ in (1.12). The case of $\Sigma\lambda_j - \Sigma\mu_k < 0$ can be treated in a similar way.

Case 2. Next we consider the remaining case $\Sigma\lambda_j - \Sigma\mu_k = 0$, which requires two preliminary notions — the covariant derivative in distribution sense, and an averaging operator. We have first to explain the former one. For a riemannian vector bundle F over V, denote, by $L^2(F)$, the Hilbert space of L^2 sections of F equipped with L^2-norm $\|\cdot\|_{L^2}$, and by $C^k(F)$, the Banach space of C^k sections of F equipped with the C^k-norm $\|\cdot\|_{C^k}$. Define the contraction of L^2 sections of F and of its dual bundle F^* by

$$(f\,|\,\sigma) = \int_V \langle f(v), \sigma(v)\rangle\, d\Lambda(v) \qquad \text{for} \quad f \in L^2(F), \quad \sigma \in L^2(F^*),$$

where the integrand $\langle f(v), \sigma(v)\rangle$ denotes the contraction of $f(v) \in F_v$ and $\sigma(v) \in F_v^*$. It satisfies $|(f\,|\,\sigma)| \le \|f\|_{L^2} \cdot \|\sigma\|_{L^2}$, and the correspondence $f \in L^2(F) \mapsto (f\,|\,\cdot) \in L^2(F^*)^*$ yields an isomorphism $L^2(F) \cong L^2(F^*)^*$. In the rest of this section, put $F = E^{(\lambda_1,\cdots,\lambda_\ell;\mu_1,\cdots,\mu_m)}$ for simplicity. The φ_t-invariance of the Liouville measure Λ implies

$$(1.13) \qquad (\psi_t f\,|\,\psi_t \sigma) = (f, \sigma) \qquad \text{for} \quad f \in L^2(F), \quad \sigma \in L^2(F^*).$$

Now the covariant derivative by the canonical connection ∇ is regarded as a differential operator $\nabla : C^1(F) \to C^0(T^*V \otimes F)$, and the adjoint operator $\nabla^* : C^1(TV \otimes F^*) \to C^0(F^*)$ of ∇ is again realized as a first order differentiable operator. In fact, for $\tau \in C^1(TV \otimes F^*)$, define $\nabla^*\tau \in C^0(F^*)$ by contracting T^*V- and TV-components of $-\nabla\tau \in C^0(T^*V \otimes TV \otimes F^*)$. Then we actually have

$$(1.14) \qquad (\nabla f\,|\,\tau) = (f\,|\,\nabla^*\tau), \qquad \text{for} \quad f \in C^1(F), \quad \tau \in C^1(TV \otimes F^*),$$

since the Liouville measure Λ (regarded as a volume form of V) is parallel with respect to the canonical connection ∇. With regard to (1.14), it is possible to introduce a *covariant derivative in distribution sense* as an operator $\nabla : L^2(F) \to C^1(TV \otimes F^*)^*$ by

$$(\nabla f\,|\,\tau) = (f\,|\,\nabla^*\tau) \qquad \text{for} \quad f \in L^2(F), \quad \tau \in C^1(TV \otimes F^*).$$

In particular, we say that $\nabla f = g$ *in distribution sense* for $f, g \in L^2(F)$, if $(\nabla f\,|\,\tau) = (g\,|\,\tau)$ for all $\tau \in C^1(TV \otimes F^*)$. It is not hard to see that for $f \in L^2(F)$, $\nabla f = 0$ in distribution sense if and only if f is parallel almost everywhere with respect to the canonical connection ∇ (cf. [**S**;§II.6]). Hence, to prove the proposition, it suffices to show $\nabla f = 0$ in distribution sense for all ψ_t-invariant L^2 sections f of F. Now recall that the tangent bundle of V has the s littin $TV = E^- + E^0 + E^+$ with $E^- = E^{-1} + E^{-2}$ and $E^+ = E^{+1} + E^{+2}$. This also gives rise to the decomposition of the covariant derivative in distribution sense into the "restricted derivatives" $\nabla^\nu : L^2(F) \to C^1(E^\nu \otimes F^*)^*$ ($\nu = 0, \pm$). In particular, we can easily prove that $\nabla^0 f = 0$ in distribution sense if $f \in L^2(F)$ is ψ_t-invariant. This can be seen as follows. Let $f \in L^2(F)$ be ψ_t-invariant. We have to prove $(\nabla^0 f \,|\, \tau) = (f \,|\, \nabla^{0*}\tau) = 0$ for all $\tau \in C^1(E^0 \otimes F^*)$, where $\nabla^{0*} : C^1(E^0 \otimes F^*) \to C^0(F^*)$ denotes the adjoint of ∇^0. Without loss of generality, we may assume that τ is of the form $\tau = \dot\varphi \otimes \sigma$, $\sigma \in C^1(F^*)$, where $\dot\varphi$ denotes the geodesic spray as before. Then, by (1.8), we have $\nabla^{0*}\tau = \nabla^{0*}(\dot\varphi \otimes \sigma) = \mathcal{L}_{\dot\varphi}\sigma = (d/dt)|_{t=0}\psi_t\sigma$, and therefore, (1.13) and ψ_t-invariance of f imply

$$(\nabla^0 f \,|\, \tau) = (f \,|\, \frac{d}{dt}\bigg|_{t=0} \psi_t\sigma) = \frac{d}{dt}\bigg|_{t=0} (f \,|\, \psi_t\sigma) = \frac{d}{dt}\bigg|_{t=0} (f \,|\, \sigma) = 0.$$

In consequence, to prove the proposition, it is enough to show $\nabla^{\pm} f = 0$ in distribution sense for all ψ_t-invariant $f \in L^2(F)$.

We now turn to the definition of an *averaging operator* $A^+ : L^2(F) \to L^2(F)$. First let $\mathcal{B}$ be the Banach space of bounded linear operators on $L^2(F)$ whose norm is denoted by $\|\cdot\|_{\mathcal{B}}$, and define the *weak topology* of $\mathcal{B}$ so that for B, $B_k \in \mathcal{B}$ ($k = 1, 2, \cdots$),

$$\begin{gathered} B_k \to B \text{ in the weak topology if and only if} \\ (B_k f \,|\, \sigma) \to (Bf \,|\, \sigma) \text{ for any } f \in L^2(F),\ \sigma \in L^2(F^*). \end{gathered}$$

Then the unit ball $\mathcal{B}_1 = \{B \in \mathcal{B} : \|B\|_{\mathcal{B}} \le 1\}$ in $\mathcal{B}$ is sequentially compact relative to the weak topology (cf. [**R**;§10.6]). Now for each real number t, define $A_t \in \mathcal{B}$ by

$$(A_t f)(v) = \frac{1}{t}\int_0^t (\psi_s f)(v)\, ds, \qquad f \in L^2(F), \quad v \in V.$$

By (1.9), $\psi_t : L^2(F) \to L^2(F)$ is an isometry (recall that we are assuming $\Sigma\lambda_j - \Sigma\mu_k = 0$), and therefore we have $\|A_t\|_{\mathcal{B}} \le 1$. Hence, there is a sequence $\{t_k\}$ with $t_k \to \infty$ for which A_{t_k} converges to some $A^+ \in \mathcal{B}_1$ in the weak topology.

(1.15) **Lemma.** *For any $f \in L^2(F)$, $A^+ f$ is ψ_t-invariant.*

Proof. For $f \in L^2(F)$ and $T \in \mathbf{R}$, it follows that

$$\begin{aligned} \|\psi_T A_t f - A_t f\|_{L^2} &= \frac{1}{t}\left\| \int_0^t \psi_{s+T} f\, ds - \int_0^t \psi_s f\, ds \right\|_{L^2} \\ &\le \frac{1}{t}\left\{ \left\| \int_t^{t+T} \psi_s f\, ds \right\|_{L^2} + \left\| \int_0^T \psi_s f\, ds \right\|_{L^2} \right\} \le \frac{2T}{t}\,\|f\|_{L^2}. \end{aligned}$$

On the other hand, by letting $t = t_k \to \infty$, $(\psi_T A_t f \,|\, \sigma) = (A_t f \,|\, \psi_T^* \sigma)$ converges to $(A^+ f \mid \psi_T^* \sigma) = (\psi_T A^+ f \mid \sigma)$, and $(A_t f \mid \sigma)$ to $(A^+ f \mid \sigma)$ for any $\sigma \in L^2(F^*)$. Hence we can conclude that $(\psi_T A^+ f \mid \sigma) = (A^+ f \mid \sigma)$ for all $f \in L^2(F)$ and $\sigma \in L^2(F^*)$. This proves the lemma. ∎

Suppose now that $f \in L^2(F)$ is ψ_t-invariant. Then we can always take $g \in C^1(F)$ so that $\|f-g\|_{L^2}$ is sufficiently small. Since $\|A^+\|_{\mathcal{B}} \leq 1$ and $A^+ f = f$, we obtain $\|f - A^+ g\|_{L^2} = \|A^+(f-g)\|_{L^2} \leq \|f - g\|_{L^2}$. Furthermore it holds that

$$\big|(\nabla^+(f - A^+ g) \mid \tau)\big| = \big|(f - A^+ g \mid \nabla^{+*} \tau)\big| \leq \|f - A^+ g\|_{L^2} \cdot \|\tau\|_{C^1}$$

for any $\tau \in C^1(E^+ \otimes F^*)$, where $\nabla^{+*} : C^1(E^+ \otimes F^*) \to C^0(F^*)$ denotes the adjoint of ∇^+. Thus, to see that $\nabla^+ f = 0$ in distribution sense, it is sufficient to prove

(1.16) **Lemma.** $\nabla^+ A^+ g = 0$ *in distribution sense for any* $g \in C^1(F)$.

Proof. Recall first that the action of ψ_t on F is isometric on each fiber by (1.9), while $|\psi_t \alpha| \leq e^{-t} |\alpha|$ for $\alpha \in E^{+*}$ and $t \geq 0$ by (1.7). Thus, for $\nabla^+ g \in C^0(E^{+*} \otimes F)$, we obtain $\|\psi_t \nabla^+ g\|_{C^0} \leq e^{-t} \cdot \|g\|_{C^1}$, $t \geq 0$. Hence it follows from (1.10) that

$$\|\nabla^+ A_t g\|_{C^0} \leq \frac{1}{t} \int_0^t \|\nabla^+ \psi_s g\|_{C^0} ds \leq \frac{1 - e^{-t}}{t} \|g\|_{C^1}.$$

Consequently we obtain $(A_t g \mid \nabla^{+*} \tau) = (\nabla^+ A_t g \mid \tau) \to 0$ as $t \to \infty$ for $\tau \in C^1(E^+ \otimes F^*)$. Meanwhile, by the definition of A^+, $(A_t g \mid \nabla^{+*} \tau)$ converges to $(A^+ g \mid \nabla^{+*} \tau) = (\nabla^+ A^+ g \mid \tau)$ as $t = t_k$ goes to ∞. This proves $(\nabla^+ A^+ g \mid \tau) = 0$ for all $\tau \in C^1(E^+ \otimes F^*)$, and concludes the lemma. ∎

In consequence, we have $\nabla^+ f = 0$ in distribution sense for any ψ_t-invariant $f \in L^2(F)$. Of course, it is also possible to show that $\nabla^- f = 0$ in distribution sense for all ψ_t-invariant $f \in L^2(F)$, and this completes the proof of Proposition (1.11) in the case of $\Sigma \lambda_j - \Sigma \mu_k = 0$.

2. Negatively Curved Surfaces

We now proceed to the demonstration of the tensorial ergodicity for the geodesic flows on negatively curved surfaces. We first introduce a canonical connection on the unit tangent bundle of such a surface, and then show the tensorial ergodicity by modifying the arguments in the preceding section slightly. The method emploied here to construct the canonical connection is basically the same with Kanai [**K**] except the point that a suggestion given by S. Kaneyuki made our definition much simpler than [**K**] (cf. [**KK**], [**KW**]). Anyway the canonical connection is roughly speaking defined by combining contact geometry of the unit tangent bundle together with consideration on the dynamics of the geodesic flow.

First of all, we briefly review contact geometry of the unit tangent bundles of riemannian manifolds. Let M be a riemannian manifold whose unit tangent bundle is denoted by V. For a local coordinate system $\{x_i\}$ of M, every tangent vector of M is represented as $v = \sum \dot{x}_i \partial/\partial x_i$, and in consequence, we obtain the local coordinate system $\{x_i, \dot{x}_i\}$ of the tangent bundle TM. In terms of these coordinates, it is possible to define a 1-form θ_0 on TM which is expressed locally as $\theta_0 = \sum g_{ij}\dot{x}_i dx_j$, where g_{ij}'s denote the coefficients of the metric tensor of M in the coordinate system $\{x_i\}$. Now pull back θ_0 by the inclusion of the unit tangent bundle V into TM. The resulting 1-form θ of V, called the *canonical contact form*, relates to the geodesic flow φ_t of M in the following manner. (1) The canonical contact form θ and its exterior derivative $d\theta$ are φ_t-invariant (Liouville's theorem). (2) $\theta(\dot{\varphi}) = 1$ for the geodesic spray $\dot{\varphi} = (d/dt)|_{t=0}\varphi_t$. (3) $d\theta(\dot{\varphi}, \cdot) = 0$. In addition, we can easily show that (4) the 2-form $d\theta$ is nondegenerate on the subbundle $E = \{\xi \in TV : \theta(\xi) = 0\}$.

Next suppose that M is a closed riemannian manifold of negative sectional curvature. Then the geodesic flow φ_t of M is an Anosov flow; that is, the tangent bundle of V carries a unique φ_t-invariant continuous splitting $TV = E^- + E^0 + E^+$ into linear subbundles satisfying the following two conditions: (1) E^0 is spanned by the geodesic spray $\dot{\varphi}$; (2) For each $\xi^\pm \in E^\pm$, $d\varphi_t\xi^\pm$ contracts exponentially as $t \to \mp\infty$. We call the splitting $TV = E^- + E^0 + E^+$ the *Anosov splitting* of M. Fundamental relations between the Anosov splitting and the canonical contact form θ can be summarized in

(2.1) **Lemma.** *(1)* $\theta(\xi) = 0$ *whenever* $\xi \in E^- + E^+$.
(2) For any $\xi^\pm, \eta^\pm \in E^\pm$, $d\theta(\xi^-, \eta^-) = d\theta(\xi^+, \eta^+) = 0$.

Proof. Suppose that $\xi^- \in E^-$. Then Liouville's theorem implies that $|\theta(\xi^-)| = |(\varphi_t^*\theta)(\xi^-)| = |\theta(d\varphi_t\xi^-)| \leq \|\theta\| \cdot |d\varphi_t\xi^-|$. Since $|d\varphi_t\xi^-|$ tends to zero as t goes to ∞, we have $\theta(\xi^-) = 0$. Similarly we can show that $\theta(\xi^+) = 0$ for $\xi^+ \in E^+$, and this proves (1). The second assertion (2) can be proved in a similar way. ∎

The first assertion in the lemma implies that the 2-form $d\theta$ is nondegenerate when it is restricted to $E \times E$, where $E = \{\xi \in TV : \theta(\xi) = 0\} = E^- + E^+$. In other words, $d\theta$ is a symplectic structure of the vector bundle E. Then the second assertion claims that $E = E^- + E^+$ is a lagrangian splitting of E with respect to the symplectic structure $d\theta$. By virtue of these observations, we can define a continuous pseudo riemannian metric g of V in the following way. First let I be the continuous involution of E characterized by $I|E^\pm = \pm\mathrm{id}$. Then $g_0(\xi, \eta) = d\theta(\xi, I\eta)$ $(\xi, \eta \in E)$ is a pseudo riemannian structure of E, and $g = g_0 + \theta \otimes \theta$ is the desired pseudo riemannian metric of V.

To introduce the canonical connection of V, assume especially that the Anosov splitting of M is C^1-*differentiable*. This assumption is fulfilled especially if M is a surface, or if the sectional curvature of M satisfies the pinching condition $-4 < K_M \leq -1$ (see Hirsch–Pugh [**HP**$_1$], [**HP**$_2$]). Under this condition, the pseudo riemannian metric g of V we have just defined is C^1-differentiable, and has the continuous Levi-Civita connection ∇. It is not hard to see that the canonical connection ∇ possesses the following properties. (1) The subbundles E^0, $E^{\pm}$ of TV are ∇-stable. (2) $\nabla\theta = 0$, $\nabla d\theta = 0$ and $\nabla\Lambda = 0$, where $\Lambda = \theta \wedge (d\theta)^n$ $(n+1 = \dim M)$ denotes the Liouville measure. (3) ∇ is invariant under the geodesic flow φ_t. (4) $\nabla_{\dot\varphi}\xi = \mathcal{L}_{\dot\varphi}\xi$ for any vector field ξ on V.

Notice here that the Anosov splitting of a closed locally symmetric riemannian manifold of negative curvature is C^∞-differentiable. Thus its unit tangent bundle has the canonical connection as above. However it does not coincide with the canonical connection introduced in the preceding section unless M is of constant negative curvature. In fact, for the complex, quaternion and Cayley hyperbolic spaces, the canonical connections of their unit tangent bundles given in §1 have torsion, while the canonical connection defined here is torsion-free.

We are now in the position to prove the tensorial ergodicity for surfaces. Let M be a 2-dimensional closed riemannian manifold of negative curvature. Then the subbundles $E^{\pm}$ appearing in the Anosov splitting of M are both 1-dimensional, and therefore we obtain functions $h^{\pm}(v,t)$ $(v \in V,\ t \in \mathbf{R})$ such that

$$(2.2) \qquad |d\varphi_t\xi^{\pm}| = e^{\pm h^{\pm}(v,t)}|\xi^{\pm}| \quad \text{for} \quad \xi^{\pm} \in E^{\pm}_v, \quad v \in V,$$

where $E^{\pm}_v$ denotes the fiber of $E^{\pm}$ over $v \in V$. These functions should satisfy

$$(2.3) \qquad c_1^{-1}\, t \le h^{\pm}(v,t) \le c_1\, t \quad \text{and} \quad |h^+(v,t) - h^-(v,t)| \le c_2,$$

where c_1 and c_2 are positive constants. The first assertion actually follows from the assumption on the curvature of M together with the compactness of M. Meanwhile, the last inequality is a consequence of the following three facts: (1) The symplectic structure $d\theta$ of $E = E^- + E^+$ is preserved by the geodesic flow; (2) $E = E^- + E^+$ is a lagrangian splitting of the symplectic vector bundle $(E, d\theta)$; (3) The angles between the lines E^-_v and E^+_v in T_vV are uniformly bounded in $v \in V$. Here set $E^{\pm 1} = E^{\pm}$, $h^{\pm 1}(v,t) = h^{\pm}(v,t)$ and $h^0(v,t) \equiv 0$. Then, by (2.2), the natural lift ψ_t of the geodesic flow φ_t to the vector bundle

$$E^{(\lambda_1,\cdots,\lambda_\ell;\mu_1,\cdots,\mu_m)} = E^{\lambda_1} \otimes \cdots \otimes E^{\lambda_\ell} \otimes E^{\mu_1 *} \otimes \cdots \otimes E^{\mu_m *}$$

with $\lambda_1, \cdots, \lambda_\ell, \mu_1, \cdots, \mu_m = 0, \pm 1$ satisfies

$$(2.4) \qquad |\psi_t\tau| \le c_3 \cdot \exp\left\{ \sum_{j=1}^{\ell} \lambda_j h^{\lambda_j}(v,t) - \sum_{k=1}^{m} \mu_k h^{\mu_k}(v,t) \right\} \cdot |\tau|$$

for $\tau \in E^{(\lambda_1,\cdots,\lambda_\ell;\mu_1,\cdots,\mu_m)}$. Thus, in the case where $\Sigma\lambda_j - \Sigma\mu_k \ne 0$, $\psi_t\tau$, $\tau \in E^{(\lambda_1,\cdots,\lambda_\ell;\mu_1,\cdots,\mu_m)}$, contracts exponentially as t diverges to ∞ or $-\infty$ by (2.3). This proves as in §1 that each measurable ψ_t-invariant section of $E^{(\lambda_1,\cdots,\lambda_\ell;\mu_1,\cdots,\mu_m)}$ with $\Sigma\lambda_j - \Sigma\mu_k \ne 0$ vanishes almost everywhere.

Next we consider the case of $\Sigma\lambda_j - \Sigma\mu_k = 0$. Also in this case, the proof goes as in the preceding section. Notice here that (2.4) together with the last inequality in (2.3) guarantees that ψ_t is an "almost isometry" on each fiber of $E^{(\lambda_1,\cdots,\lambda_\ell;\mu_1,\cdots,\mu_m)}$; that is, there is a constant c_4 such that

$$|\psi_t\tau| \leq c_4 \cdot |\tau| \quad \text{for} \quad \tau \in E^{(\lambda_1,\cdots,\lambda_\ell;\mu_1,\cdots,\mu_m)}, \quad t \in \mathbf{R}.$$

This again makes it possible for us to follow the arguments in §1 with suitable slight modification, and we can conclude the tensorial ergodicity of the geodesic flows of negatively curved surfaces.

3. Differentiable Tensor Fields

As we have observed in the previous sections, the geodesic flows of negatively curved surfaces and locally symmetric spaces possess the tensorial ergodicity. In particular, in the former case, the tensorial ergodicity is stated in terms of the canonical connection of the unit tangent bundle, while the definition of the canonical connection requires only the C^1-differentiability of the Anosov splitting. On the other hand, we know that there are actually a number of negatively curved manifolds other than surfaces which have C^1-differentiable Anosov splittings: For example, closed riemannian manifolds whose sectional curvatures are bounded between -4 and -1 strictly have this property (remind Hirsch–Pugh [**HP**$_1$], [**HP**$_2$]). Thus the unit tangent bundles of these manifolds carry canonical connections, and it makes sense to ask whether the geodesic flows of these manifolds possess the tensorial ergodicity. Unfortunately we have no answer to the this question. However, if we restrict ourselves to considering only *differentiable* tensor fields, we can obtain a partial result, and that is the purpose of the present section.

To mention it clearly, it is convenient to reformulate our problem slightly. Suppose that M is a closed riemannian manifold of negative curvature, and let $\widetilde{M}$ be the universal covering of M. Then the unit tangent bundle $\widetilde{V}$ of $\widetilde{M}$ has two group actions: One of them is the action of $\mathbf{R}$ on $\widetilde{V}$ as the geodesic flow of $\widetilde{M}$, and the other is the action of the fundamental group $\Gamma = \pi_1(M)$ of M that is obtained by lifting the covering action of Γ on $\widetilde{M}$ to $\widetilde{V}$. Since those actions on $\widetilde{V}$ commute with each other, the orbit space of one of these actions again has an group action induced by the other. In fact, the induced action of $\mathbf{R}$ on the the orbit space $\widetilde{V}/\Gamma$, which is naturally identified with the unit tangent bundle V of M, is nothing but the geodesic flow of M. Meanwhile, Γ acts on the orbit space of $P = \widetilde{V}/\mathbf{R}$ of the geodesic flow of $\widetilde{M}$. It is easy to see that the orbit space P naturally becomes a smooth manifold, and the action of Γ on P is differentiable. These two actions, the geodesic flow of M and the action of Γ on P, are essentially equivalent in the viewpoint of ergodic theory. For instance, the ergodicity of the geodesic flow of M can be explained by the ergodicity of the Γ-action on P, and *vice versa*. Furthermore, in some cases, such as the problem we are now involved in, the description is much simpler for the Γ-action on P than for the geodesic flow of M. So, in the rest, we will discuss on the Γ-action on P instead of the geodesic flow of M.

First we restate the results we have obtained in the earlier sections in the new framework. Throughout the following discussions, let $\pi : \widetilde{V} \to P$ be the projection of $\widetilde{V}$ onto the orbit space $P = \widetilde{V}/\mathbf{R}$. Then, for the lift $T\widetilde{V} = \widetilde{E}^- + \widetilde{E}^0 + \widetilde{E}^+$ of the Anosov splitting $TV = E^- + E^0 + E^+$ of M, the subbundle $\widetilde{E} = \widetilde{E}^- + \widetilde{E}^+$ of $T\widetilde{V}$ is horizontal relative to the $\mathbf{R}$-fibering $\pi : \widetilde{V} \to P$. For each tangent vector ξ of P, let $\xi^* \in \widetilde{E}$ be the horizontal lift of ξ. On the other hand, each 1-form α of P is pulled back to the 1-form $\alpha^* = \pi^*\alpha$ of $\widetilde{V}$. Hence it is also possible to define the horizontal lift of any tensor of P: For a tensor τ of P, denote the tensor of $\widetilde{V}$ obtained by lifting τ horizontally to $\widetilde{V}$ by τ^*. A measurable tensor field τ on P is said to be *locally L^2-integrable*,or simply, L^2_{loc} if the horizontal lift τ^* is a locally L^2-integrable tensor field on $\widetilde{V}$; i.e., $\int_K |\tau^*|^2 < \infty$ for any compact subset K of $\widetilde{V}$. Notice further that for a Γ-invariant tensor field τ on P, its lift τ^* to $\widetilde{V}$ is invariant by the geodesic flow of $\widetilde{M}$ as well as the Γ-action on $\widetilde{V}$. Therefore τ^* induces a tensor field $\hat{\tau}$ on the unit tangent bundle $V = \widetilde{V}/\Gamma$ of M which is invariant under the geodesic flow of M. Of course, $\hat{\tau}$ is L^2-integrable over V provided τ is L^2_{loc}.

A. Locally Symmetric Spaces. Suppose specifically that M is locally symmetric. Then the unit tangent bundle $\widetilde{V}$ of $\widetilde{M}$ has the canonical connection ∇ as in §1, and both the geodesic flow of $\widetilde{M}$ and the Γ-action on $\widetilde{V}$ preserve the connection ∇. Thus we can define a Γ-invariant connection D on P by $(D_\xi\tau)^* = \nabla_{\xi^*}\tau^*$ for any vector field ξ and tensor field τ on P. The connection D can be also described as follows. Since the identity component G of the isometric transformation group of the symmetric space $\widetilde{M}$ acts on $\widetilde{V}$ transitively, G acts also on P transitively so that the projection $\pi : \widetilde{V} \to P$ is G-equivariant. In fact, if G^0 denotes the connected Lie subgroup of G with the Lie algebra $\mathbf{g}^0$ in the gradation (1.1) of the Lie algebra $\mathbf{g}$ of G, then P is represented as the homogeneous space G/G^0. Further, (1.2) guarantees that $P = G/G^0$ is a reductive homogeneous G-space, and D is just the canonical connection of P as a homogeneous G-space (cf. [**KN**]). It is easy to show that a tensor field on P is parallel with respect to D if and only if it is invariant under the action of G on P. Thus the tensorial ergodicity of the geodesic flow of M we proved in §1 implies

(3.1) **Corollary.** *Every Γ-invariant L^2_{loc} tensor field on P is G-invariant almost everywhere.*

B. Negatively Curved Surfaces. Next consider a 2-dimensional closed riemannian manifold M of negative curvature. The canonical connection ∇ of the unit tangent bundle V of M which we constructed in §2 is lifted to a connection of $\widetilde{V}$, and the lifted connection on $\widetilde{V}$, which we denote by the same symbol ∇, is invariant under both the Γ-action on $\widetilde{V}$ and the geodesic flow of $\widetilde{M}$. Thus we can again define a continuous Γ-invariant connection D of P by $(D_\xi \tau)^* = \nabla_{\xi^*}\tau^*$ for any vector field ξ and tensor field τ on P. The following assertion is an immediate consequence of the tensorial ergodicity for the geodesic flows of surfaces.

(3.2) **Corollary.** *A Γ-invariant L^2_{loc} tensor field on P is parallel almost everywhere with respect to D.*

C. Manifolds with Pinched Negative Curvature. We now proceed to the study of continuous or differentiable tensor fields on P which are invariant under the action of Γ on P. Suppose that M is a closed riemannian manifold whose sectional curvature satisfies the pinching condition

$$-\left(\frac{c+1}{c}\right)^2 < K_M \leq -1 \tag{3.3}$$

with a nonnegative integer c. Then we first have

(3.4) **Proposition.** *Under the condition (3.3), every Γ-invariant continuous tensor field on P of odd degree $\leq 2c+1$ vanishes.*

Proof. Recall here that the exterior derivative $d\tilde{\theta}$ of the canonical contact form $\tilde{\theta}$ of $\widetilde{V}$ is invariant under the geodesic flow of $\widetilde{M}$ (Liouville's theorem). Thus it induces a 2-form ω on P which is characterized by the condition $\pi^*\omega = d\tilde{\theta}$, and it is easily seen to be a Γ-invariant smooth symplectic form of P. Especially it yields the Γ-equivariant isomorphism $TP \cong T^*P$ defined by $\xi \mapsto \omega(\xi, \cdot)$. This guarantees that it is enough to consider only covariant tensor fields. Let g be a continuous covariant tensor field on P of degree $2d+1 \leq 2c+1$ which is invariant under the action of Γ on P. Then $f = \hat{g}$ is a continuous covariant tensor field on the unit tangent bundle V of M which is invariant under the geodesic flow φ_t of M, and to see that $g \equiv 0$, it is sufficient to prove $f|E^{\lambda_1} \times \cdots \times E^{\lambda_{2d+1}} \equiv 0$ for $\lambda_1, \cdots, \lambda_{2d+1} = \pm$, where $TV = E^- + E^0 + E^+$ denotes the Anosov splitting of M. Let N^- be the number of $-$'s appearing in $\lambda_1, \cdots, \lambda_{2d+1}$, and N^+ the number of $+$'s. Note that we have only two possibilities $N^- > N^+$ and $N^- < N^+$ since $N^- + N^+ = 2d+1$ is odd. We here give the proof in the former case, as the other is treated in the same way. By the compactness of M, we can take a constant Λ so that $-\{(c+1)/c\}^2 < -\Lambda^2 \leq K_M \leq -1$ $(1 \leq \Lambda < (c+1)/c)$, and the standard comparison theorem implies the following estimates for the hyperbolicity of the geodesic flow φ_t:

$$\begin{aligned} e^t|\xi^-| &\leq |d\varphi_{-t}\xi^-| \leq e^{\Lambda t}|\xi^-|, & e^{-\Lambda t}|\xi^-| &\leq |d\varphi_t\xi^-| \leq e^{-t}|\xi^-|, \\ e^t|\xi^+| &\leq |d\varphi_t\xi^+| \leq e^{\Lambda t}|\xi^+|, & e^{-\Lambda t}|\xi^+| &\leq |d\varphi_{-t}\xi^+| \leq e^{-t}|\xi^+|, \end{aligned}$$

where $\xi^\pm \in E^\pm$ and $t \geq 0$. Thus we have the following for any $\xi^{\lambda_j} \in E^{\lambda_j}$ and $t > 0$ since f is φ_t-invariant:

$$|f(\xi^{\lambda_1}, \cdots, \xi^{\lambda_{2d+1}})| = |(\varphi_t^* f)(\xi^{\lambda_1}, \cdots, \xi^{\lambda_{2d+1}})| \tag{3.5}$$

$$
\begin{aligned}
&= \left| f(d\varphi_t \xi^{\lambda_1}, \cdots, d\varphi_t \xi^{\lambda_{2d+1}}) \right| \\
&\leq \|f\| \, |d\varphi_t \xi^{\lambda_1}| \cdots |d\varphi_t \xi^{\lambda_{2d+1}}| \\
&\leq \|f\| \, |\xi^{\lambda_1}| \cdots |\xi^{\lambda_{2d+1}}| \cdot e^{(\Lambda N^+ - N^-)t}.
\end{aligned}
$$

Notice here that $N^- \geq d+1$ and $N^+ \leq d$ since $N^- > N^+$ with $N^- + N^+ = 2d+1$. On the other hand, we have been assuming $\Lambda < (c+1)/c \leq (d+1)/d$. Thus we obtain $\Lambda N^+ - N^- < 0$, and, in consequence, $f(\xi^{\lambda_1}, \cdots, \xi^{\lambda_{2d+1}}) = 0$ by letting $t \to \infty$ in (3.5). ∎

Now assume especially that $c \geq 1$ in (3.3), that is, the curvature of M lies between -4 and -1 strictly. Then, by the theorem of Hirsch–Pugh [**HP**$_1$], [**HP**$_2$], the Anosov splitting of M is C^1-differentiable, and therefore the unit tangent bundle V of M has a canonical connection ∇ (recall §2), which again gives rise to a Γ-invariant continuous connection D on P as in the previous case. The connection D of P has another simpler description. As we have already mentioned in the proof of Proposition (3.4), P has a canonically defined Γ-invariant symplectic form ω. In addition, the Anosov splitting $T\widetilde{V} = \widetilde{E}^- + \widetilde{E}^0 + \widetilde{E}^+$ of $\widetilde{M}$ is also invariant under the geodesic flow of $\widetilde{M}$, and the subbundles $\widetilde{E}^\pm$ are transverse to the fibers of the $\mathbf{R}$-fibering of $\widetilde{V}$ over P. Thus they induce the C^1 subbundles $F^\pm$ of TP such that $TP = F^- + F^+$ is a lagrangian splitting of the tangent bundle of P; i.e., $\dim F^\pm = \dim P/2$ and $\omega(\xi^-, \eta^-) = \omega(\xi^+, \eta^+) = 0$ for any $\xi^\pm, \eta^\pm \in F^\pm$ (cf. Lemma (2.1)). Furthermore, the subbundles $F^\pm$ are integrable since so are $E^\pm$, and, in consequence, we obtain Γ-invariant lagrangian foliations $\mathcal{F}^\pm$ of the symplectic manifold (P, ω) which integrate $F^\pm$. Now define a Γ-invariant C^1 pseudo riemannian metric g of P by $g(\xi, \eta) = \omega(\xi, I\eta)$ for $\xi, \eta \in TP$, where I denotes the involution of TP given by $I|_{F^\pm} = \pm \mathrm{id}$. Then D coincides with the Levi–Civita connection of the pseudo riemannian metric g of P. It is obvious that for any Γ-invariant differentiable tensor field f on P, the covariant derivative Df is again Γ-invariant. Thus Proposition (3.4) has

(3.6) **Corollary.** *Under the assumption (3.3) with $c \geq 1$, we always have $Df \equiv 0$ for any Γ-invariant C^1 tensor field f on P of even degree $\leq 2c$.*

An application of Corollary (3.6) was already made by author [**K**] concerned with the smoothness problem of the Anosov splitting. It proceeds as follows. Suppose that M is a closed riemannian manifold of negative curvature whose Anosov splitting is C^∞. Then the canonical connection ∇ on the unit tangent bundle V of M is C^∞, and therefore so is the connection D of P. In particular the curvature tensor R of D is defined as a Γ-invariant smooth tensor field on P of degree four, and we can apply Corollary (3.6) to the curvature tensor R with $c = 2$:

$$
(3.7) \qquad DR \equiv 0
$$

provided $-9/4 < K_M \leq -1$. This equation means that the affine connection D on P is locally symmetric, and has a quite strong implication. In fact, we proved in [**K**] that under (3.7) the geodesic flow φ_t of M is completely isomorphic to the geodesic flow of a certain closed riemannian manifold of *constant* negative curvature. Hence we obtain

(3.8) **Theorem** ([**K**]). *Suppose that* M *is a closed riemannian manifold with sectional curvature* $-9/4 < K_M \leq -1$. *If the Anosov splitting of* M *is* C^∞, *then the geodesic flow of* M *is isomorphic to the geodesic flow of a certain closed riemannian manifold of constant negative curvature.*

In the case of dimension two, a much stronger result had been obtained by Ghys [**G**]. He proved that a 2-dimensional closed riemannian manifold M of negative curvature should be of constant curvature provided that the Anosov splitting of M is C^2-differentiable (cf. Hurder–Katok [**HK**]). Thus Theorem (3.8) is a partial generalization of the theorem of Ghys. Furthermore Katok and Feres recently improved our arguments extensively, and showed that in Theorem (3.8) we can replace the pinching condition $-9/4 < K_M \leq -1$ by $K_M < 0$ in the case of $\dim M = 3$ (Katok [**Kt**]), and by $-4 < K_M \leq -1$ in the case of $\dim M = 4$ (Feres [**F**]). Actually they proved the identity (3.7) under their conditions by the aid of the multiplicative ergodic theorem of Oseledec. This may suggest the possibility that the multiplicative ergodic theorem would sometimes strengthen our results (3.4) and (3.6).

References

[**AM**] R. Abraham and J. E. Marsden, "Foundations of Mechanics," 2nd ed., Benjamin, Reading, 1978.

[**A**] D. V. Anosov, *Geodesic flows on closed riemannian manifolds with negative curvature* (English translation), Proc. Steklov Inst. Math. **90** (1969).

[**AS**] D. V. Anosov and Ya. G. Sinai, *Some smooth ergodic systems* (English translation), Russian Math. Surveys **22** (1967), 103–167.

[**F**] R. Feres, *Rigidity of geodesic flows on negatively curved* 4*-manifolds*, to appear.

[**G**] E. Ghys, *Flots d'Anosov dont les feuilletages stables sont différentiables*, Ann. scient. Éc. Norm. Sup. **20** (1987), 251–270.

[**H**] S. Helgason, "Differential Geometry, Lie Groups, and Symmetric Spaces," Academic Press, New York, 1978.

[**HP$_1$**] M. W. Hirsch and C. C. Pugh, *Stable manifolds and hyperbolic sets*, in "Proc. Sympos. Pure Math. vol.14," Amer. Math. Soc., Providence, 1970, pp. 133–163.

[**HP$_2$**] M. W. Hirsch and C. C. Pugh, *Smoothness of horocycle foliations*, J. Diff. Geom. **10** (1975), 225–238.

[**HK**] S. Hurder and A. Katok, *Differentiability, rigidity and Godbillon–Vey classes for Anosov flows*, to appear.

[**K**] M. Kanai, *Geodesic flows of negatively curved manifolds with smooth stable and unstable foliations*, to appear in Ergod. Th. & Dynam. Sys..

[**KK**] S. Kaneyuki and M. Kozai, *Paracomplex structures and affine symmetric spaces*, Tokyo J. Math. **8** (1985), 81–98.

[KW] S. Kaneyuki and F. L. Williams, *Almost paracontact and parahodge structures on manifolds*, Nagoya Math. J. **99** (1985), 173–187.

[Kt] A. Katok, *Rigidity of geodesic flows on negatively curved* 3*-manifolds*, to appear.

[KN] S. Kobayashi and K. Nomizu, "Foundations of Differential Geometry, Vol. II," Interscience, New York, 1969.

[P] W. Parry, "Topics in Ergodic Theory," Cambridge Univ. Press, Cambridge, 1981.

[R] H. L. Royden, "Real Analysis," 2nd ed., MacMillan, New York, 1968.

[S] L. Schwartz, "Théorie des Distributions," 3rd ed., Hermann, Paris, 1966.

HARMONIC FUNCTIONS WITH GROWTH CONDITIONS ON A MANIFOLD OF ASYMPTOTICALLY NONNEGATIVE CURVATURE I

Atsushi Kasue *)

Department of Mathematics, Osaka University

Toyonaka 560, JAPAN

0. Let M be a complete, connected, noncompact Riemannian manifold of dimension m. According to a theorem due to Greene and Wu [7], M abounds harmonic functions so that M can be properly embedded into Euclidean space $\mathbb{R}^N$ by them. However, the question of whether M admits nonconstant bounded or positive harmonic functions arises in connection with the geometry of M.

On the one hand, Anderson [1] and Sullivan [14] independently proved that a simply connected, complete Riemannian manifold with negatively pinched curvature admits plentifully bounded harmonic functions and in fact, the Dirichlet problem with continuous boundary data at infinity in the geometric boundary (the Eberlain-O'Neill boundary) can be solved. Later, Anderson and Schoen [2] described the geometric boundary for such a manifold as its Martin boundary.

On the other hand, a theorem by Yau [17] states that if the Ricci curvature of M is nonnegative everywhere, then any positive harmonic function must be constant. Recently, Donnelly [5] showed that the space of bounded harmonic functions is finite, if M has nonnegative Ricci curvature outside a compact set. Moreover, Li and Tam [12] have obtained significant results in this direction. They study a manifold M of non-

---

*) supported partly by Grant-in-Aid for Sientific Research (No. 62740034), Ministry of Education

negative sectional curvature outside a compact set and have described the set of bounded or positive harmonic functions on such a manifold M. The central results in [12] are concerning the construction of, so called, barriers at infinity. However from the nature of the problem and the fact that the above condition on the curvature of M would not be kept in general under slight perturbations of the metric, it would be desirable to extend their results to a (open) class of manifolds which contains the above M.

In this note, we introduce a class of manifolds with asymptotically nonnegative curvature and discuss some properties of harmonic functions on such manifolds.

0-1. We call a complete, connected, noncompact Riemannian manifold M of *asymptotically nonnegative curvature* if the sectional curvature K_M of M satisfies:

(H.1) $$K_M \geq -k\circ r,$$

where r denotes the distance to a fixed point o of M and $k(t)$ is a nonnegative, monotone nonincreasing continuous function on $[0,\infty)$ such that *the integral* $\int^{\infty} tk(t)dt$ *is finite*. In [8] (cf. [9]), we have constructed a metric space $M(\infty)$ associated with a manifold M of asymptotically nonnegative curvature. Let us here explain it briefly (see [8] or [9] for details). We call two rays σ and γ of M equivalent if $dis_M(\sigma(t),\gamma(t))/t$ goes to zero as $t \to \infty$. Define a distance δ_∞ on the equivalence classes by $\delta_\infty([\sigma],[\gamma]) := \lim_{t\to\infty} d_t(\sigma\cap S(t),\gamma\cap S(t))/t$, where $S(t)$ denotes the metric sphere around a fixed point of radius t and d_t stands for the inner (or intrinsic) distance on $S(t)$ induced from the distance $dis_M(\ ,\ )$ on M. Then we have a metric space $M(\infty)$ of the equivalence classes of rays with the distance δ_∞ which is independent of the choice of the fixed point (i.e., the center of $S(t)$) and to which a family of scaled metric spheres $\{\frac{1}{t}S(t)\}$ converges with respect to the Hausdorff distance as t goes to infinity. We note that the complement $M\sim B(a)$ of a metric ball $B(a)$ with sufficiently large radius a is homeomorphic to $S(a)\times(a,\infty)$. For simplicity, we call a connected component of $M\sim B(a)$ (for large a) *an end* $\mathcal{E}$ *of* M. We write $M_{\mathcal{E}}(\infty)$ for the connected component of $M(\infty)$ corresponding to $\mathcal{E}$, so that $\{\frac{1}{t}\ S(t)\cap\mathcal{E}\}$ converges to $M_{\mathcal{E}}(\infty)$ with respect to the Hausdorff distance as $t\to\infty$, and then $M_{\mathcal{E}}(\infty)$ turns out to be a compact inner metric space. Since $\mathrm{Vol}_{m-1}(S(t)\cap\mathcal{E}))/t^{m-1}$ ($m = \dim M$) tends to a nonnegative constant as $t\to\infty$, let us denote the limit by $\mathcal{V}ol_{m-1}(M_{\mathcal{E}}(\infty))$. According to Li and Tam [12], we call an end $\mathcal{E}$ of M *large* (resp. *small*) if the integral $\int^{\infty} tV_{\mathcal{E}}(t)^{-1}dt$ is finite (resp. divergent), where $V_{\mathcal{E}}(t) := \mathrm{Vol}_m(B(t)\cap\mathcal{E})$.

0-2. We are now in the position to give a theorem on the construction of barriers at infinity, which was proved by Li and Tam [12] for a manifold with nonnegative sectional curvature outside a compact set.

Theorem A. Let M be a manifold of asymptotically nonnegative curvature and $\mathcal{E}$ an end of M.

(I) Suppose $\mathcal{E}$ is large. Then for any large b ($> a$), there exists a unique harmonic function f defined on $\mathcal{E}\sim B(b)$ satisfying $f_{|\partial B(b)\cap\mathcal{E}} = 1$ and $\lim_{x\in\mathcal{E}\to\infty} f(x) = 0$. Moreover one can find a constant $c_1 > 0$ such that

$$f(x) \le c_1 \int_{r(x)}^{\infty} \frac{t}{V_{\mathcal{E}}(t)}\, dt.$$

(II) Suppose $\mathcal{E}$ is small. Then for any large b ($> a$), there exists a harmonic function g defined on $\mathcal{E}\sim B(b)$ satisfying $g_{|\partial B(b)\cap\mathcal{E}} = 0$ and $\lim_{x\in\mathcal{E}\to\infty} g(x) = \infty$. Moreover one can find positive constants c_2,

$c_3, c_4 > b$ such that

$$c_2 \int_b^{r(x)} \frac{t}{V_{\mathcal{E}}(t)}\, dt \le g(x) \le c_3 \int_b^{r(x)} \frac{t}{V_{\mathcal{E}}(t)}\, dt \quad \text{if } r(x) > c_4.$$

Theorem A will be verified in Section 2. Based on Theorem A and the same argument as in [12: Theorems 6.1, 6.3, 7.2, 7.3], we can construct canonical bases for the space of bounded harmonic functions and for the positive cone of positive harmonic functions on a manifold of asymptotically nonnegative curvature (cf. Theorem B below), and then describe the space and the cone themselves (cf. Theorem C below).

Theorem B. Let M and $\mathcal{E}$ be as in Theorem A.

(I) Suppose $\mathcal{E}$ is large. Then there exists a unique positive harmonic function $f_{\mathcal{E}}$ on M such that

$$\lim_{x \in \mathcal{E} \to \infty} f_{\mathcal{E}}(x) = 1 \quad \text{and} \quad \lim_{y \in \mathcal{E}' \to \infty} f_{\mathcal{E}'}(y) = 0$$

for any other large end $\mathcal{E}'$ (if any).

(II) Suppose M has at least one large end and one small end. Suppose $\mathcal{E}$ is small. Then there exists a positive harmonic function $g_{\mathcal{E}}$ on M such that

$$\lim_{x \in \mathcal{E} \to \infty} g_{\mathcal{E}}(x) = \infty \quad \text{and} \quad \lim_{y \in \mathcal{E} \to \infty} g_{\mathcal{E}'}(y) = 0$$

for any large end $\mathcal{E}'$, and $g_{\mathcal{E}}$ is bounded on every other small end (if any). Moreover $g_{\mathcal{E}}$ is unique up to a positive scalar multiple.

Theorem C. Let M be as above.

(I) Let H_∞ be the space of all bounded harmonic functions on M.

(i) If M has only small ends, then $\dim H_\infty = 1$ and all bounded harmonic functions are constant.

(ii) Let $\mathcal{E}_i$ $(i=1,\ldots,k)$ be the large ends of M and $f_i (= f_{\mathcal{E}_i})$ the unique harmonic functions described in Theorem B(I). Then $\{f_i\}$ forms a bases for H_∞. In particular, $\dim H_\infty = k$.

(II) Let H_+ be the positive cone of all positive harmonic functions on M.

(iii) If M has only small ends, then H_+ consists only of constants functions.

(iv) If M has only large ends, then $H_+ \subset H_\infty$ and every positive harmonic function is a nonnegative linear combination of $\{f_i\}$.

(v) If M has at least one small end and one large end, then $\dim H_+$ is equal to the numbers of the ends. More precisely, let $\{\mathcal{E}_i\}_{i=1,\ldots,k}$ be the set of all large ends and $\{\mathcal{D}_j\}_{j=1,\ldots,\ell}$ the set of all small ends. Let $g_j (= g_{\mathcal{D}_j})$ be the positive harmonic functions described in Theorem B(II) such that $g_j(p) = 1$, where p is a fixed point. Then for any $u \in H_+$, u is a nonnegative linear combination of the f_i's and g_j's.

We remark here that in Theorem B(I), the value $f_{\mathcal{E}}(x)$ of $f_{\mathcal{E}}$ at $x \in M$ is equal to the hitting probability of the paths starting at x to the large end $\mathcal{E}$.

Making use of the above results, we can describe the asymptotic behaviour of the symmetric Green functions and then the Martin boundary of a manifold M of asymptotically nonnegative curvature. See Li and Tam [13] for details. Actually, all of the results in [13: Sections 2 , 3] are valid for our manifold M. For example, suppose M has at least one large end. Let $\{\mathcal{E}_i\}_{i=1,\ldots,k}$ and $\{\mathcal{D}_j\}_{j=0,\ldots,\ell}$ be as in Theorem C (II). Then there exists a unique minimal positive symmetric Green function $G(x,y)$ on $M\times M$ such that $G(x,y) \le c(x) \int_{dis_M(x,y)}^\infty t V_i(t)^{-1} dt$ for all $y \in \mathcal{E}_i \sim B(R(x))$ $(V_i(t) = Vol_m(B(t)\cap\mathcal{E}_i))$, and $G(x,y) \to C(x,\mathcal{D}_j)$ as

$y \in \mathcal{D}_j \to \infty$. The constants $R(x)$, $c(x)$, and $C(x,\mathcal{D}_j)$ are positive constants depending on the quantities in parentheses (cf. [13: Theorem 3]). We note that for a large end $\mathcal{E}_i$, if $G(x,y)(\int_{dis_M(x,y)}^{\infty} m\, tV_i(t)^{-1}dt)^{-1}$ converges to $f_i(x)$ as $y \in \mathcal{E}_i \to \infty$ for *some* $x \in M$, then this holds for *all* $x \in M$, where $m = \dim M$ and $f_i (= f_{\mathcal{E}_i})$ is the function as in Theorem B(I) (cf. [13: Corollary 2]). It is unclear whether the limit should exist and be equal to $f_i(x)$ for some $x \in M$. However we see that this is true if a large end $\mathcal{E}$ satisfies some additional conditions. Namely, we can prove the following

Proposition D. Let M be an m-dimensional manifold of asymptotically nonnegative curvature which has at least one large end $\mathcal{E}$. Let $G(x,y)$ be a unique, positive minimal symmetric Green function of M. Then for any $x \in M$,

$$G(x,y)\left(\int_{dis_M(x,y)}^{\infty} m \frac{t}{V_{\mathcal{E}}(t)}\, dt\right)^{-1} \longrightarrow f_{\mathcal{E}}(x)$$

as $y \in \mathcal{E} \to \infty$, if the scalar curvature ρ_M of M satisfies

(H.2) $$\rho_M(y) \le \frac{c}{r^2(y)} \qquad \text{for } y \in \mathcal{E}, \text{ and}$$

(H.3) $$Vol_{m-1}(M_{\mathcal{E}}(\infty)) > 0,$$

where c is a positive constant. In particular, in this case, for any $x \in M$, one has

$$G(x,y)\, dis_M(x,y)^{m-2} \longrightarrow \frac{m\, f_{\mathcal{E}}(x)}{(m-2)Vol_{m-1}(M_{\mathcal{E}}(\infty))}$$

as $y \in \mathcal{E} \to \infty$.

This proposition can be derived from the following

Fact E ([10: Lemma 2.3]). Let M be a manifold of asymptotically nonnegative curvature. Suppose that an end $\mathcal{E}$ of M satisfies (H.2) and (H.3) as in Proposition D. Then:

(i) $M_{\mathcal{E}}(\infty)$ is a compact, connected smooth manifold with $C^{1,\alpha}$-metric g_∞ $(0 < \alpha < 1)$.

(ii) Fix two positive numbers a, b with $a > b$. Then for large t, there exists a $C^{2,\alpha}$-diffeomorphism π_t from $A_t(a,b)$ into the cone $\mathcal{C}(M_{\mathcal{E}}(\infty))$ over $M_{\mathcal{E}}(\infty)$, i.e., $\mathcal{C}(M_{\mathcal{E}}(\infty)) = [0,\infty)\times_{t^2} M_{\mathcal{E}}(\infty)$, which has the following properties: as t goes to infinity, $\pi_t(A_t(a,b))$ converges to $[a,b]\times M_{\mathcal{E}}(\infty)$ and $\pi_{t*}G_t$ converges to thge metric $dt^2+t^2g_\infty$ in $C^{2,\alpha'}$ topology $(0 < \alpha' < \alpha < 1)$, where $A_t(a,b) := \{ x \in \frac{1}{t} M\cap\mathcal{E} : b \le \frac{1}{t} \mathrm{dis}_M(x,o) \le a \}$ and G_t denotes the Riemannian metric of $\frac{1}{t} M$.

The cone $\mathcal{C}(M_{\mathcal{E}}(\infty))$ is the tangent cone at infinity for $\mathcal{E}$.

0-3. Let us now discuss harmonic functions with growth conditions. For a harmonic function h on a complete, connected, noncompact Riemannian manifold M, we denote by $m_x(|h|,t)$ the maximum of $|h|$ on the metric sphere $S_x(t)$ of radius t around a point x of M. In this note, we call a harmonic function h *of finite growth* if $m_x(|h|,t)/t^p$ is bounded as $t \to \infty$, for some positive constant p. A theorem due to Cheng [3] says that if M has nonnegative Ricci curvature, then for a harmonic function h on M, any point $x \in M$, and every $t > 0$, $|dh|(x) \le c_m\, m_x(|h|,t)/t$, where c_m is a constant depending only on $m = \dim M$, and hence h must be constant if h is of sublinear growth, i.e., $\liminf_{t\to\infty} m_x(|h|,t)/t = 0$ (cf. also [16: §6.4]). It is not hard to see that if $m = 2$ and M has nonnegative Guassian curvature, and if M admits a nonconstant harmonic function h of linear growth, i.e., $\limsup_{t \to \infty} m_x(|h|,t)/t < \infty$, then M is isometric to flat cylinder or Euclidean plane. In fact, since the Guassian curvature is nonnegative, we see that $\Delta|dh|^2 \geq |\nabla dh|^2 \geq 0$. This implies that $|dh|^2$ is a bounded subharmonic function, so that $|dh|^2$ must be constant, since M possesses no nonconstant bounded subharmonic functions. Thus ∇dh must vanish everywhere on M, and hence M turns out to be isometric to flat cylinder or Euclidean plane. In higher dimensional cases, we are led to ask whether the existence of a nonconstant harmonic function h of linear growth on a complete manifold M of nonnegative sectional curvature (or more generally , nonnegative Ricci curvature) would imply that the hessian $\nabla^2 h$ vanishes on M, and hence M contains a line, that is, M splits isometrically into $\mathbb{R}\times N$ (see [15], for related open questions). The following result answers this question under some additional conditions.

Theorem F. Let M be a manifold of asymptotically nonnegative curvature. Suppose that M has nonnegative Ricci curvature and suppose that (H.2) and (H.3) (as in Proposition D) hold on M. Then M is ismometric to Euclidean space $\mathbb{R}^m$, if M admits a nonconstant harmonic function of linear growth.

Theorem F is a consequence from the Bishop's volume comparison theorem and the following

Theorem G. Let M be a manifold of dimension $m \geq 3$ with asymptotically nonnegative curvature and $\mathcal{E}$ an end of M. Suppose that (H.2) and (H.3) hold for $\mathcal{E}$ and suppose that there exists a nonzero harmonic 1-form ω defined on $\mathcal{E}$ with $\sup V_{\mathcal{E}}(t)^{-1} \int_{B(t)\cap\mathcal{E}}|\omega|^2 < \infty$. Then $M_{\mathcal{E}}(\infty)$ is isometric to the unit sphere $S^{m-1}(1)$ of Euclidean space .

We are interested in relationships (if any) between the space of harmonic functions of finite growth on a manifold M of asymptotically nonnegative curvature and the geometry of $M(\infty)$. At this stage, we have a satisfactry result for the case of dim M = 2, but for higher dimensional cases, little is known.

Theorem H. Let M be a manifold of asymptotically nonnegative curvature. Suppose that M has only one end, namely, $M(\infty)$ is connected. Then:

(i) For a nonconstant harmonic function h on M, one has

$$\liminf_{t\to\infty} \frac{\log m(|h|,t)}{\log t} \geq \log\left[\frac{(\exp c(m)\mathrm{diam}(M(\infty))) + 1}{(\exp c(m)\mathrm{diam}(M(\infty))) - 1}\right] > 0,$$

where $c(m)$ is a positive constant depending only on $m = \dim M$. In particular, M has no nonconstant harmonic functions of finite growth, if $M(\infty)$ consists of only one point.

(ii) Suppose that $m = 2$ and $\mathrm{diam}(M(\infty)) > 0$. Then for a nonconstant harmonic function h of finite growth, $\log m(|h|,t)/\log t$ converges to a constant, say ord(h), as $t \to \infty$, and ord(h) is given by $\mathrm{ord}(h) = n\pi/\mathrm{diam}(M(\infty))$ for some positive integer n. Moreover the dimension of the space of harmonic functions h such that $\mathrm{ord}(h) \leq n\pi/\mathrm{diam}(M(\infty))$ is equal to 2n+1.

It is conjectual that for a manifold of asymptotically nonnegative curvature, the space $\mathcal{H}_p$ of harmonic functions h with $\limsup_{t\to\infty} m(|h|,t)/t^p < \infty$ would be of finite dimension for any $p > 0$.

Theorem I. Let M be a manifold of dimension $m \geq 3$ with asymptotically nonnegative curvature and $\mathcal{E}$ an end of M. Suppose (H.2) and (H.3) hold for $\mathcal{E}$. Then if there exists a harmonic function

h defined on $\mathcal{E}$ such that $0 < \limsup_{t \to \infty} m(|h|,t)/t^p < \infty$ for some positive constant p, then $p(p+m-2)$ is an eigenvalue of $M_{\mathcal{E}}(\infty)$. Moreover, $p \geq 1$ and if $p = 1$, then $M_{\mathcal{E}}(\infty)$ is isometric to the unit sphere of Euclidean space $\mathbb{R}^m$.

Remark 1. Theorem H(ii) can be derived from a result due to Finn [6]. Actually, we can prove the same assertion as in Theorem H(ii) for a complete, noncompact 2-dimensional Riemannian manifold of finite total curvature which has one end.

Remark 2. In [11], Kazdan shows an example of a complete, noncompact Riemannian manifold N such that N possesses no nonconstant positive harmonic functions, but the dimension of $\mathcal{H}_p$ is infinite for some p. The sectional curvature of his example behaves like $-1/r^2 \log r$ for large r.

Remark 3. Suppose M has nonnegative sectional curvature everywhere and M is connected at infinity. Then the diameter $\mathrm{dian}(M(\infty)) \leq \pi$ and the equality holds if and only if M contains a line (cf. [8: Proposition 2.4] or [9: Proposition 4.2]). Making use of this fact, we can prove the following

Theorem F'. Let M be a (complete, connected, open) manifold of nonnegative sectional curvature. Suppose M satisfies (H.2) (as in Proposition D). Then M splits isometrically into $\mathbb{R} \times M'$ (and M' is flat), if M possesses a nonconstant harmonic function of linear growth.

The main ingrudients to prove Proposition D, Theorems F', G, H and I are some observations given in [8,9] (cf. Section 1) and Fact E. The details will be given in the second part of this note, where some observations on weighted Sobolev spaces will be also made.

1. Preliminaries

In this section, we shall give auxiliary results for the proof of Theorem A .

Let us begin with some definitions. Let M be an m-dimensional manifold of asymptotically nonnegative curvature and let o and k be as in (H.1). We denote by $J_k(t)(t\geq 0)$ the solution of a classical Jacobi equation: $J_k'' - k J_k = 0$, with $J_k(0) = 0$ and $J_k'(0) = 1$. Then it is known that $1 \leq J_k'(t) \nearrow J_k'(\infty) \leq \exp \int_0^\infty uk(u)du$, and $t \leq J_k(t) \leq J_k'(\infty)t$. Let r be the distance to the fixed point o and set $B(t) = \{x \in M: r(x) \leq t\}$, $B(t,T):=\{x \in M: t \leq r(x) \leq T\}$, and $S(t):= \partial B(t)$. We define here a function F on M by

$$F(x):= \lim_{t\to\infty} t - dis_M(x,S(t)).$$

Then obviously $F \leq r$ on M and moreover F satisfies : $F(x) = t - dis_M(x,F^{-1}(t))$ for any $x \in M$ and $t > 0$ with $F(x) < t$. Set $C(t):= \{ x \in M: F(x) \leq t \}$ and $C(t,T):= \{x \in M: t \leq F(x) \leq T \}$.

Fact 1(cf. [8] or [9: Lemmas 1.2 and 1.4]).

(i) Let ψ be a C^2-function on $[a,\infty)$ $(a \geq 0)$ with $\psi' \geq 0$. Then in the sense of distributions, one has

$$\Delta\circ(\psi\circ r) \leq \{ \psi'' + (m-1)\psi'(\log J_k)'\}\circ r \qquad \text{on } M\sim B(a)$$

$$\Delta\circ(\psi\circ F) \geq \{ \psi'' - (m-1)\psi' \eta_k\}\circ F \qquad \text{on } M\sim C(a)$$

where $\eta_k(t):= \int_t^\infty k(u)du$.

(ii) As $x \in M$ goes to infinity, one has

$$\frac{F(x)}{r(x)} \longrightarrow 1 ,$$

$$\max \{ \measuredangle(u,v) : u, v \in \nabla\circ r(x) \} \longrightarrow 0$$

$$\max \{ \measuredangle(u,v) : u \in \nabla\circ r(x),\ v \in \nabla\circ F(x) \} \longrightarrow 0,$$

where $\nabla\circ r(x) := \{ u \in T_xM : |u| = 1,\ t + r(\exp_x -tu) = r(x),\ t \in [0,r(x)]\}$, and $\nabla\circ F(x) := \{ u \in T_xM : |u| = 1,\ F(\exp_x tu) = t + F(x),\ t \in [0,\infty)\}$.

In what follows, we take a sufficiently large a and fix an end $\mathcal{E}$ of M, a connected component of $M \sim B(a)$. Set $V(t) := \mathrm{Vol}(B(t)\cap\mathcal{E})$, $A(t) := \mathrm{Vol}(\partial B(t)\cap\mathcal{E})$, $V_F(t) := \mathrm{Vol}(C(t)\cap\mathcal{E})$, and $A_F(t) := \mathrm{Vol}(\partial C(t)\cap\mathcal{E})$. Then we have the following

Lemma 1.

(i) $$\lim_{t\to\infty} \frac{V(t)}{V_F(t)} = \lim_{t\to\infty} \frac{A(t)}{A_F(t)} = 1 ,$$

$$1 \le \liminf_{t\to\infty} \frac{tA(t)}{V(t)} = \liminf_{t\to\infty} \frac{tA_F(t)}{V_F(t)} \le \limsup_{t\to\infty} \frac{tA(t)}{V(t)} = \limsup_{t\to\infty} \frac{tA_F(t)}{V_F(t)} \le m$$

$$1 \le \frac{V(at)}{V(t)} \le c_0 \frac{\alpha^m t^m - a^m}{t^m - a^m} \qquad (\alpha > 1),$$

where $c_0 := \exp m\int_0^\infty uk(u)du$.

(ii) If we set $\lambda_k(t) := \exp -(m-1)\int_1^t \eta_k(u)du$, we have

$$\lambda_k(t)\, V_F(t) \le \int_a^t \lambda_k(u)du\, A_F(t) \qquad (a \le t),$$

$$\exp -(m-1)\int_1^\infty \eta_k(u)du \le \lambda_k \le 1 ,$$

$$\frac{\int_a^t \lambda_k(u)du}{V_F(t)} \le \frac{t}{V_F(t)} \le c_1 \frac{\int_a^t \lambda_k(u)du}{V_F(t)} \quad (2a \le t),$$

where c_1 is a constant depending only on k.

(iii) If we set $\mu_k(T;t) := \int_t^T \exp \int_t^u (m-1)\eta_k(v)dv\, du$, we have

$$\int_a^T \mu_k(2T;t)dt \le c_2\, T^2 \quad , \text{ and } \quad \mu_k(2T;T) \le c_3\, T .$$

Proof. Define first a function $\psi(t)$ by $\psi(t) := \int_a^t \lambda_k(u)^{-1} \int_a^u \lambda_k(v)dv\, du$. Then it follows from Fact 1(i) that $\Delta\, \psi \circ F \ge 1$ on $M \sim C(a)$. Therefore integrating the both sides over $C(a;t) \cap \mathcal{S}$ $(t > a)$, we have

$$\frac{\int_a^t \lambda_k(u)du}{\lambda_k(t)}\, A_F(t) \ge V_F(t) .$$

Since $\int_a^t \lambda_k(u)du / t\lambda_k(t)$ tends to 1 as $t \to \infty$, we see that

$$\liminf_{t \to \infty} \frac{tA_F(t)}{V_F(t)} \ge 1 .$$

Define a function $\phi(t)$ by $\phi(t) := \int_a^t J_k^{-m+1}(u) \int_a^u J_k^{m-1}(v)dvdu$. Then it follows from Fact 1(i) that $\Delta\, \phi \circ r \le 1$ on $M \sim B(a)$. Therefore integrating the both sides over $B(a;t) \cap \mathcal{S}$ $(t > a)$, we have

$$\text{(1.1)} \qquad \frac{\int_a^t J_k^{m-1}(u)du}{J_k^{m-1}(t)}\, A(t) \le V(t) .$$

Since $tJ_k^{m-1}(t) / \int_a^t J_k^{m-1}(u)du$ tends to m as $t \to \infty$, we see that

$$\limsup_{t \to \infty} \frac{tA(t)}{V(t)} \le m .$$

Moreover it turns out from (1.1) that for $\alpha > 1$,

$$\frac{V(\alpha t)}{V(t)} \le c_0\, \frac{\alpha^m t^m - a^m}{t^m - a^m} \quad (t > a) .$$

These observations and Fact 1(ii),(iii) prove the first two assertions of Lemma 1. The last assertion of the lemma follows from simple computation.

This completes the proof of Lemma 1.

Let us now prove a Harnack inequality.

Lemma 2. Let $\mathcal{E}$ be an end of M. Then there exists a constant C such that for any positive harmonic function h defined on $B(a,T)\cap\mathcal{E}$ with $T > 4a$,

$$\sup_{\partial B(t)\cap\mathcal{E}} h \le C \inf_{\partial B(t)\cap\mathcal{E}} h \quad \text{for } 2a < t < \frac{T}{2}.$$

Before proving Lemma 2, we recall the following

Fact 2(Cheng-Yau[4]). Let N be a compact Riemannian manifold with (possibly empty) boundary and let $B_x(t)$ be the metric ball of N around a point x with radius t such that $B_x(t)\cap\partial N = \phi$. If h is a positive harmonic function on $B_x(t)$, then the following estimate holds:

$$\max_{y\in B_x(t)} (t-\mathrm{dis}_N(x,y))|\nabla \log h|(y) \le c_m(1+ \Lambda t),$$

where c_m depends only on the dimension m of N and the Ricci curvature of $N \ge -\Lambda^2$ $(\Lambda\ge 0)$.

Proof of Lemma 2. For each t : $2a < t < T/2$, we take two points p_t, q_t $\in \partial B(t)\cap\mathcal{E}$ such that $h(p_t) = \sup_{\partial B(t)\cap\mathcal{E}} h$ and $h(q_t) = \inf_{\partial B(t)\cap\mathcal{E}} h$. Let us join p_t with q_t by a Lipschitz curve $\tau_t:[0,\ell_t] \to \partial B(t)\cap\mathcal{E}$ which realizes the distance ℓ_t in $\partial B(t)\cap\mathcal{E}$ between p_t and q_t. Fix a sufficiently large integer n so that $n > \sup_{a\le s<\infty} \mathrm{diam}(\frac{1}{s}\,\partial B(s)\cap\mathcal{E})$ and put $p_{t,i} := \tau_t(i\ell_t/3n)$ $(i=0,1,\dots,3n)$. Then $\mathrm{dis}_M(p_{t,i},p_{t,i+1}) \le \ell_t/3n \le t/3$. Applying Fact 2 to h on the metric ball around $p_{t,i}$ of radius

$t/2$, we have

$$\frac{h(p_{t,i})}{h(p_{t,i+1})} < \exp c_m(2 + \sqrt{k(t/2)}\ t)\frac{\ell_t}{nt} ,$$

and hence

$$\frac{h(p_t)}{h(q_t)} < \exp c_m(2 + \sqrt{k(t/2)}\ t)\frac{\ell_t}{t} < C,$$

for a constant $C > 1$. This proves Lemma 2.

Finally we give the following

Lemma 3. Let T_i $(> a)$ be a sequence such that $\lim_{i\to\infty} T_i = \infty$. Suppose g_i are harmonic functions defined on $C(a,T_i)$ with $g_{|\partial C(a)\cap\mathcal{E}} = 0$ and $g_i > 0$ on $\partial C(T_i)\cap\mathcal{E}$. If $\limsup_{i\to\infty} \int_{\partial C(a)\cap\mathcal{E}} \langle\nabla F, \nabla g_i\rangle < \infty$, then the g_i are uniformly bounded on compact sets of $\mathcal{E}$. More precisely, for any $T > 2a$, there is a constant $C(T)$ such that

$$\sup_{C(a,T)\cap\mathcal{E}} g_i \le C(T) \quad \text{for all } i \text{ with } T_i > 4T.$$

Proof. The same argument as in the proof of [12: Lemma 5.1] is valid for our case.

2. Proof of Theorem A

In this section, we shall prove Theorem A. we follow the argument in the proofs of Theorems 4.3 and 5.2 in [12], making use of the auxiliary results presented in Section 1. We keep the notations in the previous sections.

2-1. Proof of the first part (I) of Theorem A(the case of a large end $\mathcal{E}$). For any $T > a$, let $f_T(x)$ be the solution of Dirichlet problem:

$$\begin{cases} \Delta f_T = 0 & \text{on } C(a,T)\cap\mathcal{E} \\ f_{T|\partial C(a)\cap\mathcal{E}} = 1 \quad \text{and} & f_{T|\partial C(T)\cap\mathcal{E}} = 0 . \end{cases}$$

Then the maximum principle implies that for any $T' > T > a$, $0 \le f_T < f_{T'} \le 1$ on $C(a,T)$. Therefore $f(x) := \lim_{T\to\infty} f_T(x)$ exists for $x \in \mathcal{E}\sim C(a)$, and f satisfies: $\Delta f = 0$ on $\mathcal{E}\sim C(a)$ and $f_{|\partial C(a)\cap\mathcal{E}} = 1$. For any $T > a$ and $a \le t \le T$, we define

$$S_T(t) := \sup_{\partial C(t)\cap\mathcal{E}} f_T , \quad I_T(t) := \inf_{\partial C(t)\cap\mathcal{E}} f_T ,$$

$$\theta(T) := - \int_{\partial C(a)\cap\mathcal{E}} \partial_F\, f_T > 0 ,$$

where $\partial_F\, f_T := \langle \nabla F, \nabla f_T \rangle$ and this is defined almost everywhere on $\partial C(a)\cap\mathcal{E}$ for any $t \ge a$, since $\partial C(t)$ is a Lipschitz hypersurface with the outer unit normal ∇F. Then we have an upper bound for $\theta(T)$ as follows:

$$\text{(2.1)} \qquad \frac{1 - I_T(t)}{\int_a^t A_F(s)^{-1} ds} \ge \theta(T)$$

(cf. [12: pp.185 ~ 186]).

Let us next derive a lower bound for $\theta(T)$, following the argument of [12: pp.186 ~ 187] and making use of Lemma 1. Since $\Delta_x\, \mu_k(F(x);t) \ge 0$ on $M\sim C(a)$ and $f_T = 0$ on $\partial C(T)\cap\mathcal{E}$, we have

$$\begin{aligned} - \theta(T)(T-t) &= \int_t^T \int_{\partial C(a)\cap\mathcal{E}} \partial_F f_T \, ds \\ &= \int_t^T \int_{\partial C(s)\cap\mathcal{E}} \partial_F f_t \, ds \\ &\le \int_t^T \int_{\partial C(s)\cap\mathcal{E}} \partial_F f_T \, ds + \int_{C(t,T)} \Delta\mu_k(F;t)\, f_T \\ &= \int_t^T (1 - \frac{\partial}{\partial s}\mu_k(s,t)) \int_{\partial C(s)\cap\mathcal{E}} \partial_F f_T \, ds - \int_{\partial C(t)\cap\mathcal{E}} f_T \end{aligned}$$

$$= \theta(T) \int_t^T (\frac{\partial}{\partial s} \mu_k - 1)\, ds - \int_{\partial C(t) \cap \mathcal{S}} f_T$$

$$= \theta(T)\, \mu_k(T;t) - \theta(T)(T-t) - \int_{\partial C(t) \cap \mathcal{S}} f_T .$$

Therefore we get

$$\theta(T)\, \mu_k(T,t) \geq \int_{\partial C(t) \cap \mathcal{S}} f_T ,$$

and hence for $\overline{T} \in (a,T)$, we have

$$\int_{C(a,\overline{T}) \cap \mathcal{S}} f_T = \int_a^{\overline{T}} \Big(\int_{\partial C(s) \cap \mathcal{S}} f_T \Big)\, ds$$

$$\leq \int_a^{\overline{T}} \theta(T)\, \mu_k(T;s)\, ds$$

$$= \theta(T) \int_a^{\overline{T}} \mu_k(T;s)\, ds .$$

By the maximum principle, we obtain

$$I_T(\overline{T})\, V_F(\overline{T}) = \Big(\inf_{C(a,\overline{T}) \cap \mathcal{S}} f_T \Big)\, V_F(\overline{T})$$

$$\leq \int_{C(a,\overline{T}) \cap \mathcal{S}} f_T$$

$$\leq \theta(T) \int_a^{\overline{T}} \mu_k(T;s)\, ds .$$

Writing t instead of $\overline{T}$, we have

$$\theta(T) \geq \frac{I_T(t)\, V_F(t)}{\int_a^t \mu_k(T;s)\, ds} \tag{2.2}$$

Inequalities (2.1) and (2.2) impliy

$$1 \geq I_T(t) \Big(1 + \frac{\int_a^t A_F(s)^{-1} ds\, V_F(t)}{\int_a^t \mu_k(T;s)\, ds} \Big)$$

$$\geq I_T(t) \frac{\int_a^t A_F(s)^{-1} ds\, V_F(t)}{\int_a^t \mu_k(T;s)\, ds} .$$

Setting $t = T$ and writing T as $2T$, we get

$$1 \geq I_{2T}(T) \frac{\int_a^t A_F(s)^{-1} ds \; V_F(T)}{\int_a^T \mu_k(2T;s)\, ds} \qquad (T > a).$$

Applying Lemma 1(iii) and Lemma 2, we can find a constant c_4 which is independent of T such that if $T > 2a$, then

$$(2.3) \qquad 1 \geq c_4 \frac{V_F(T)}{T^2} S_{2T}(T).$$

Inequality (2.3) yields

$$(2.4) \qquad S_T(\overline{T}) + \sum_{i=0}^n \phi(2^i T) \geq S_{2^{n+1}T}(\overline{T})$$

for all $n \geq 0$, $T > 2a$, and $\overline{T} \in (a,T)$, where $\phi(t) := \left(c_4 V(t)/t^2\right)^{-1}$ (cf. [12: p.118]). Now we claim that

$$(2.5) \qquad \sum_{i=0}^n \phi(2^i T) \leq c_5 \int_{T/2}^{\infty} \frac{t}{V_F(t)} dt$$

for a constant c_5. Actually, noting that $1 \geq \lambda_k \geq \exp-(m-1)\int_1^\infty \eta_k(u)du$ and the function $\int_a^t \lambda_k(u)du/V_F(t)$ is nonincreasing by Lemma 1(ii), we have

$$\sum_{i=0}^n \phi(2^i T) = \frac{1}{c_1} \sum_{i=0}^n \frac{(2^i T)^2}{V_F(2^i T)}$$

$$\leq c_6 \sum_{i=0}^n \frac{\int_a^{2^i T} \lambda_k(s)ds + a}{V_F(2^i T)} \cdot 2^i T \quad (c_6 := \frac{1}{c_1}\exp(m-1)\int_1^\infty \eta_k(u)du)$$

$$\leq 2c_6 \sum_{i=0}^n \int_{2^{i-1}T}^{2^i T} \frac{\int_a^t \lambda_k(s)ds + a}{V_F(t)} dt$$

$$\leq 2c_6 \sum_{i=0}^n \int_{2^{i-1}T}^{2^i T} \frac{t}{V_F(t)} dt$$

$$= 2c_6 \int_{T/2}^{2^n T} \frac{t}{V_F(t)} dt$$

$$< 2c_6 \int_{T/2}^{\infty} \frac{t}{V_F(t)}\, dt .$$

By (2.4) and (2.5), we obtain

$$S_{2^{n+1}T}(\overline{T}) \le S_T(\overline{T}) + c_5 \int_{T/2}^{\infty} \frac{t}{V_F(t)}\, dt,$$

for all $n \ge 0$, $T > 2a$, and $\overline{T} \in (a,T)$. Letting $\overline{T} \to T$ and then $n \to \infty$, we have

$$\sup_{\partial C(T)\cap \mathcal{E}} f = \lim_{n\to\infty} S_{2^n T}(T) \le c_5 \int_T^{\infty} \frac{t}{V_F(t)}\, dt$$

for $T > 2a$. Hence by using Lemma 1(i), we can find a constant c_7 such that

$$\sup_{\partial C(T)\cap \mathcal{E}} f \le c_7 \int_T^{\infty} \frac{t}{V(t)}\, dt .$$

This proves the first part of Theorem A.

2-2. Proof of the second part (II) of Theorem A(the case of a small end $\mathcal{E}$) For $T > 4a$, let g_T be the solution of equation:

$$\begin{cases} \Delta g_T = 0 & \text{on } C(a,T)\cap \mathcal{E} \\ g_T|_{\partial C(a)\cap \mathcal{E}} = 0 \quad \text{and} & g_T|_{\partial C(T)\cap \mathcal{E}} = \int_a^T \frac{1}{A_F(t)}\, dt. \end{cases}$$

As in the proof of the first part of the theorem, we denote

$$S_T(t) := \sup_{\partial C(t)\cap \mathcal{E}} g_T , \quad I_T(t) := \inf_{\partial C(t)\cap \mathcal{E}} g_T \quad \text{and}$$

$$\theta(T) := \int_{\partial C(a)\cap \mathcal{E}} \partial_F g_T = \int_{\partial C(t)\cap \mathcal{E}} \partial_F g_T > 0$$

for $t \in (a,T)$. We set here

$$\alpha(t) := \int_a^t \frac{1}{A_F(s)}\, ds , \quad \delta(t) := \int_a^t \frac{s}{V_F(s)}\, ds, \text{ and}$$

$$\hat{\delta}(t) := \int_a^t \frac{\int_a^s \lambda_k(u)du}{V_F(s)}\, ds \qquad (t > a).$$

Then by lemma 1(ii), $\hat{\delta}(t)$ is concave and

(2.6) $$c_8^{-1}\, \hat{\delta}(t) \le \delta(t) \le c_8\, \hat{\delta}(t)$$

for a constant $c_8 > 1$. Applying Lemma 2 to the positive harmonic function $\alpha(T) - g_T$, we can find a constant $c_9 > 0$ such that

(2.7) $$\alpha(T) - I_T(t) \le c_9\, (\alpha(T) - S_T(t))$$

for $2a \le t \le T/2$. As in the proof of (2.2), by considering the function $\alpha(T) - g_T$, we can prove that for all $t \in (a,T)$,

(2.8) $$\theta(T) \ge (\alpha(T) - S_T(t))\, \frac{A_F(t)}{\mu_k(T;t)} .$$

By Lemma 1(i), (2.7) and (2.8), there exists a constant $c_{10} > 0$ such that

(2.9) $$\theta(T) \ge c_{10}\, (\alpha(T) - I_T(t))\, \frac{V_F(t)}{t\, \mu_k(T;t)}$$

for $2a \le t \le T/2$. On the other hand, by Lemma 2, we have a constant $c_{11} > 0$ such that

(2.10) $$S_T(t) \le c_{11}\, I_T(t) \qquad \text{for } 2a \le t \le T/2 .$$

Now by the same argument as in [12: pp.193 ~ 194], we get

(2.11) $$S_T(t) \ge \theta(T)\, \alpha(t) \qquad \text{for } a < t < T.$$

In particular, by letting $t = T$, we obtain

(2.12) $$\theta(T) \le 1 \qquad \text{for all } T > 4a.$$

We consider the following two cases:

Case 1. For all $T > 4a$, $I_T(T/2) \ge \alpha(T/2)$. In this case, for any integer $n \ge 0$, we have

(2.13) $$g_{2^n T} \le g_{2^{n+1} T} \qquad \text{on } C(a, 2^n T).$$

By (2.12), (2.13) and Lemma 3, we can assert that $g_{2^n T}$ converges to a harmonic function g on $\mathcal{E}\sim C(a)$ which vanishes on $\partial C(a)\cap\mathcal{E}$.

By (2.12) and the assumption that $I_{2^{n+1}T}(2^nT) \geq \alpha(2^nT)$, we have

$$\alpha(2^nT) \leq I(2^nT)$$

for all $n \geq 0$, where $I(t) := \inf_{\partial C(t)\cap\mathcal{E}} g$. Using Lemma 1(i), we can find a constant $c_{12} > 0$ such that

$$(2.14) \qquad \delta(2^nT) \leq c_{12}\ I(2^nT) .$$

Since $\hat{\delta}(t)$ is a concave function, for any $n \geq 0$, we have

$$\beta_n\ \hat{\delta}(a) + (1 - \beta_n)\ \hat{\delta}(2^{n+1}T) \leq \hat{\delta}(2^nT),$$

where $\beta_n : 0 < \beta_n < 1$ is the constant satisfying: $\beta_n a + (1-\beta_n)2^{n+1}T = 2^nT$. Observe that $\lim_{n\to\infty} \beta_n = 1/2$ and $\hat{\delta}(a) = 0$. Therefore there exists n' such that for all $n \geq n'$

$$\hat{\delta}(2^{n+1}T) \leq 3\ \hat{\delta}(2^nT) .$$

For any $t > 2^{n'}T$, we can find $n \geq n'$ such that $2^nT \leq t \leq 2^{n+1}T$. Then it follows from (2.6) and (2.14) that

$$\begin{aligned}\delta(t) \leq c_8\ \hat{\delta}(t) \leq c_8\ \hat{\delta}(2^{n+1}T) \leq 3c_8\ \hat{\delta}(2^nT) \leq 3c_8^{\,2}\ \delta(2^nT) \\ \leq 3c_8^{\,2}c_{12}I(2^nT) \leq 3c_8^{\,2}c_{12}\ I(t) .\end{aligned}$$

Thus we have obtained

$$(2.15) \qquad \delta(t) \leq c_{13}\ I(t) \quad \text{for} \quad t \geq 2^{n'}T ,$$

where $c_{13} := 3c_8^{\,2}c_{12}$.

Case 2. Suppose there is a sequence $T_i \to \infty$ such that

$$(2.16) \qquad I_{T_i}(T_i/2) < \alpha(T_i/2) .$$

We may assume that $T_i > 4a$ for all i . Then by Lemma 1(i)(iii), (2.6) (2.9) and (2.16), we have for any i

$$\theta(T_i) \geq c_{10}(\alpha(T_i) - I_{T_i}(T_i/2))\ \frac{V_F(T_i/2)}{(T_i/2)\ \mu_k(T_i;T_i/2)}$$

$$> c_{14} \, (\alpha(T_i) - \alpha(T_i/2)) \, \frac{V_F(T_i/2)}{T_i^{\,2}}$$

$$> c_{14} \int_{T_i/2}^{T_i} \frac{\int_a^t \lambda_k(u)du}{V_F(t)} \, dt \, \frac{V_F(T_i/2)}{T_i^{\,2}}$$

$$> c_{15} \frac{\int_a^{T_i} \lambda_k(u)du}{V_F(T_i)} \, \frac{V_F(T_i/2)}{T_i^{\,2}} \, (T_i - T_i/2)$$

$$> c_{16} \frac{V_F(T_i/2)}{V_F(T_i)} > c_{17} > 0 \, ,$$

for some constants $c_{14} \sim c_{17}$ which are independent of i . Combining the above inequality with (2.10) and (2.11), we get

$$I_{T_i}(t) \geq \frac{1}{c_{11}} S_{T_i}(t) \geq \frac{1}{c_{11}} \theta(T_i) \, \alpha(t)$$

$$\geq \frac{c_{17}}{c_{11}} \alpha(t) = c_{18} \, \alpha(t)$$

for all t: $2a \leq t \leq T_i/2$ $(c_{18} := c_{17}/ \, c_{11})$. Since $\lim_{i \to \infty} T_i = \infty$, for a fixed $t > 2a$, we have

$$\liminf_{i \to \infty} I_{T_i}(t) \geq c_{18} \, \alpha(t) \, . \tag{2.17}$$

On the other hand, by Lemma 3 and (2.12), we can find a subsequence of g_{T_i} , denoted also by the same letters, such that $\lim_{i \to \infty} g_{T_i} = g$ for some function g satisfying

$$\Delta g = 0 \quad \text{on} \quad \mathcal{E} \sim C(a),$$

$$g = 0 \quad \text{on} \quad \partial C(a) \cap \mathcal{E}.$$

By (2.17), we have

$$I(t) = \inf_{\partial C(t) \cap \mathcal{E}} g \geq c_{18} \, \alpha(t)$$

for all $t \geq 2a$.

In any case, there are a sequence $T_i \to$ and postive constants c_{19}, c_{20} such that the g_{T_i} converges to a harmonic function g on $\mathcal{E} \sim C(a)$

with $g_{|\partial C(a)\cap\delta} = 0$ and $I(t) \geq c_{19}\delta(t)$ for all $t \geq c_{20} > 2a$.

It remains to prove that the g tends to infinity at a rate not greater than that of $\delta(t)$. This can be done just by the same argument as in [12:pp.196 ~ 198]. This completes the proof of the second part of Theorem A.

References

[1] M.T. Anderson: The Dirichlet problem at infinity for manifolds with negative curvaturte, J. Diff. Geom. 18 (1983), 701 - 721.

[2] M.T. Anderson and R. Schoen: Positive harmonic functions on complete manifolds of negative curvature, Ann. of Math. 121 (1985), 429 - 461.

[3] S.-Y. Cheng: Liouville theorems for harmonic maps, Proc. Symp. Pure Math. 36, A.M.S. 1980, 147 - 151

[4] S.-Y. Cheng and S.T. Yau:Differential equations on Riemannian and their geometric applications, Comm. Pure Appl. Math. 28 (1975), 333 - 354.

[5] H. Donnelly: Bounded harmonic functions and positive Ricci curvature Math. Z. 191 (1986), 559 - 565.

[6] M. Finn: On a class of comformal metrics, with applications to differential geometry in the large, Comm. Math. He;v. 40 (1965), 1 - 30.

[7] R.E. Greene and H. Wu: Embedding of open Riemannian manifolds by harmonic functions, Ann. Inst. Fourier (Grenoble) 25 (1975), 215 - 235.

[8] A. Kasue: On manifolds of asymptotically nonnegative curvature, M.S.R.I. Berkeley, Calf. Preprint #09208-86, July, 1986 (unpublished)

[9] ————: A compactification of a manifold with asymptotically nonnegative curvature, to appear.

[10] ————: A convergence theorem for Riemannian manifolds and some applications, to appear.

[11] J.L. Kazdan: Parabolicity and the Liouville property on complete Riemannian manifolds, Seminar on New Results in Nonlinear Partial Differential Equations, A Publication of the Max-Plank-Inst. für Math.(1987), 153 - 166, Bonn.

[12] P. Li and L.-F. Tam: Positive harmonic functions on complete manifolds with non-negative curvature outside a compact set, Ann. of Math. **125** (1987), 171 - 207.

[13] ————————: Symmetric Green's functions on complete manifolds, to appear in Amer. J. Math.

[14] D. Sullivan: The Diricjlet problem at infinity for a negatively curved manifold, J. Diff. Geom. **18** (1983), 723 - 732.

[15] H. Wu: The Bochner Technique in Differential Geometry, to appear.

[16] ——: Some open problems in the study of noncompact Kähler manifolds, Lecture presented at the Kyoto Conference on Geometric Function Theory, Sep. 8, 1978.

[17] S.-T. Yau: Harminic functions on complete Riemannian manifolds, Comm. Pure Appl. Math. **28** (1975), 201 - 228.

DENSITY THEOREMS FOR CLOSED ORBITS

Atsushi KATSUDA

Department of Mathematics Faculty of Science
Nagoya University
Chikusa, Nagoya 464, JAPAN

This is a survey article concerning about one of the applications of number theoretic idea to geometry. I will show several geometric analogue of density theorems in number theory. Two years ago, Sunada wrote the same kind of article [43]. This note is, in some part, a continuation of it. But I emphasize the application of L-function, that is " *How to influence the pole of L-fuction to the density of prime closed orbits ?*. For the properties of L-function itself and their proof, see [43] or the original papers.

§ 1. Density theorems in number theory.

In this section, we recall well-known density theorems in number theory, which are model for our results. Set $\pi_N(x) = {}^{\#}\{p \mid p\text{:prime number}, p < u = e^x\}$.

Theorem 1.1(Prime number theorem).

$$\pi_N(x) \sim \frac{u}{\log u} \sim \frac{e^x}{x}.$$

where $f(x) \sim g(x)$ means $\lim_{x\to\infty} f(x)/g(x) = 1$ as usual.

For an integer n, let $[\cdot]:\mathbb{Z} \to \mathbb{Z}/n\mathbb{Z}$ be a natural projection and

$(\mathbb{Z}/n\mathbb{Z})^\times$ be the multiplicative group contained in $\mathbb{Z}/n\mathbb{Z}$. Set $\pi_N(x,\alpha:n) = {}^{\#}\{p \mid p\text{:prime number}, p < e^x, [p] = \alpha \in (\mathbb{Z}/n\mathbb{Z})^\times\}$.

Theorem 1.2(Dirichlet density theorem).

$$\pi_N(x,\alpha:n) \sim \frac{1}{(\mathbb{Z}/n\mathbb{Z})^\times} \frac{e^x}{x}.$$

The following conjecture is not related to the results of later section. But there seems to be some similarities to our results in the work of Heath-Brown (!?) Set $\pi_{N,2}(x) = {}^{\#}\{p \mid p, p+2\text{:prime numbers}, p < e^x\}$.

Conjecture 1.3(Prime twin conjecture; Hardy and Littlewood [16]).

$$\pi_{N,2}(x) \sim \frac{Ce^x}{x^2}, \quad C = 2 \prod_{p>2} (1-(p-1)^{-2}).$$

Theorem 1.4(Heath-Brown [17]). If χ is a Dirichlet character (mod q) and $L(s,\chi)$ is the corresponding L-function. Suppose that $L(\beta_0,\chi) = 0$ for real primitive character χ (mod q) and a real β_0 with

$$1 - \beta_0 \leq (3\log q)^{-1},$$

then conjecture 1.3 holds.

§ 2. Density theorems in geometry.

2.1. Categories.

We consider three categories. Of course, the latter one is a special case of the former one.

Axiom A flow or Anosov flow.

Definition 2.1.1. The flow (X,φ_t) is Axiom A if and only if the following two conditions hold.

(a). The nonwandering set $\Omega \equiv \{x \in X \mid$ for any neighborhood U of x and $T > 0$, there exists $t \geq T$ such that $U \cap \varphi_t(U) \neq \phi\}$ is a finite union of hyperbolic singularities and closed orbits.

(b) $\Lambda = X-\Omega$ satiesfies the following conditions.

(1) The tangent space T_xX is continuously decomposed to

$$T_xX = L_x + E_x^s + E_x^u \qquad \text{for} \quad x \in \Lambda.$$

(2) The above decomposition is φ_t- invariant. i.e.

$$(d\varphi_t)_x(E_x^s) = E_{\varphi_t(x)}^s, \quad (d\varphi_t)_x(E_x^u) = E_{\varphi_t(x)}^u.$$

(3) There exist Riemannian metric on X and constants c, $\lambda > 0$ such that, for $t > o$,

$$\|(d\varphi_t)_x(w)\| \leq ce^{-\lambda t}\|w\| \qquad \text{for} \quad w \in E_x^s,$$

$$\|(d\varphi_{-t})_x(w)\| \leq ce^{-\lambda t}\|w\| \qquad \text{for} \quad w \in E_x^u,$$

Moreover if $\Lambda = X$, we call it Anosov flow.

Main tools are symbolic dynamics due to Bowen, the Ruelle zeta function, the Ruelle operators.

Geodesic flow of Anosov type.

Main tools are the same as above. But there is a natural involution I of the unit tangent bundle UM with $\varphi_{-t} \circ I = I \circ \varphi_t$.

This implies that the maximal eigenvalue of the Ruelle operator has special properties, which is, in some sense, weak "self-adjointness"

Geodesic flow of hyperbolic manifolds.

We can use the Selberg trace formula, the Selberg zeta function, the Laplacian.

2.2. Density theorems.

2.2.1. Prime orbit theorem.

Set $\pi(x) = {}^{\#}\{p \mid p:\text{prime orbit}, \ \ell(p):\text{lenth of } p < x\}$.

Theorem 2.2.1(Parry and Pollicott [27]). Let X be a compact manifold and φ_t be mixing Axiom A flow. Then,

$$\pi(x) \sim \frac{e^{hx}}{hx}$$

where h is the topological entropy of (X,φ_t).

Definition 2.2.2. A flow (X,φ_t) is (topologically) mixing when, for any open subsets U, V in X, there exists $T > 0$ such that if $t \geq T$, then $\varphi_t(U) \cap V \neq \phi$.

Definition 2.2.3. The topological entropy h of (X,φ_t) is defined by

$$h = \sup_{\delta>0} \overline{\lim_{T\to\infty}} \frac{1}{T} \log \{\max {}^{\#}A;\ A \text{ is } (T,\delta)\text{-separating set}\},$$

where A is (T,δ)-separating if, for any y, $y' \in A$ $(y \neq y')$, there exists t with $0 \leq t \leq T$, $d(\varphi_t(y),\varphi_t(y')) \geq \delta$. In the case when (X,φ_t) is a geodesic flow of negatively curved manifold, h equals the growth rate of the volume of the geodesic ball in the universal covering.

Next we consider geometric analogues of (1.2). Up to now, there are two types of geometric analogues. We call these Bowen type and Sunada type. The Bowen type is to consider the number or density of closed orbits through the subset of X and the Sunada type is to consider the number of closed orbits which belongs to a coset of a subgroup of the fundamental group of X.

Bowen type.

Let g be a non-negative C^∞ function on X. Define

$$\ell_g(p) = \int_0^{\ell(p)} g(\varphi_t(\tau))d\tau.$$

Let m be the measure of maximal entropy for φ_t. (see [26] for precise definition.) Set $\pi_g(x) = \sum_{\ell(p)<x} \ell_g(p)/\ell(p)$. Then,

Theorem 2.2.4 (Bowen [8], Parry [26]). Let (X,φ_t) be a mixing Axiom A flow. Then,

$$\pi_g(x) \sim \frac{e^x}{x}\int g\, dm.$$

Sunada type.

Take one dimensional homology class $\alpha \in H_1(X,\mathbb{Z})$ or $H_1(M,\mathbb{Z})$. Set $\pi(x,\alpha) = {}^\#\{p \mid p\text{:prime orbit, } \ell(p) < x, [p]\text{:the homology class of } p = \alpha\}$.

<u>(a) Finite case.</u>

Theorem 2.2.5(Parry and Pollicott [28], Adachi and Sunada [5]). Let (X,φ_t) is mixing Anosov flow. Assume that $^{\#}H_1(M,\mathbb{Z}) < \infty$. Then,

$$\pi(x,\alpha) \sim \frac{1}{^{\#}H_1(M,\mathbb{Z})} \frac{e^{hx}}{hx}.$$

Remark 2.2.6. Theorem 2.2.5 holds more general context, which is a geometric analogue of Chebotarev density theorem. (see [5], [28]).

(b) Infinite case.

Theorem 2.2.7(Phillips and Sarnak [29], Katsuda and Sunada[22]). Let M be a compact Riemannian surface of genus g with constant negative curvature -1. Then, for $\alpha \in H_1(M,\mathbb{Z})$,

$$\pi(x,\alpha) \sim (g-1)^g \frac{e^x}{x^{1+g}}.$$

Remark 2.2.8. Phillips and Sarnak [29] asserts more. For compact hyperbolic manifold M and $\alpha \in H_1(M,\mathbb{Z})$, there exists a constant $c > 0$ (this can be expressed by the volume of Jacobian torus $H^1(M,\mathbb{R})/H^1(M,\mathbb{Z})$) such that

$$\pi(x,\alpha) \sim \frac{ce^{(n-1)x}}{x^{1+b/2}}\left(1 + \frac{c_1}{x} + \frac{c_2}{x^2} + \cdots\right).$$

where $n = \dim M$ and $b = \dim H_1(M,\mathbb{R})$.

Theorem 2.2.9(Epstein [11]). Let M be a Riemann surface of finite volume. Then, for $\alpha \in H_1(M,\mathbb{Z})$,

$$\pi(x,\alpha) \sim \binom{2p}{p} \frac{(2g+p-2)^{p+g}}{2^{g+2}} \frac{e^x}{x^{1+g+p}}.$$

where p is the number of cusps.

Remark 2.2.10. Epstein [11] assert that the contribution of the cusp is x when dim = 2, $(x\log(1/x))^{1/2}$ when dim = 3 and $x^{1/2}$ when dim ≥ 4. (see also [10]).

Theorem 2.2.11(Katsuda and Sunada [23]). Let M be a compact Riemannian manifold with geodesic flow of Anosov type. For $\alpha \in H_1(M,\mathbb{Z})$, there exists a constant $c > 0$ depending only on the geometry of M such that

$$\pi(x,\alpha) \sim \frac{ce^{hx}}{x^{1+b/2}}.$$

Remark 2.2.12. In the case when φ_t is not mixing, there are some result corresponding to Theorems 2.2.1, 2.2.4, 2.2.5 ([26], [27], [5]). But there are examples which do not hold Theorems 2.2.7, 2.2.8, 2.2.11. (e.g. the constant suspension flow of Anosov diffeomorphism [5]).

Remark 2.2.13. There are results which combine Bowen type and Sunada type. ([2],[23],[46],etc.)

2.3. Other results

1. In the case of manifolds with boundary, see [15].

2. In the case of manifolds with nonpositive curvature of rank one, only weaker results are known. (see [20], [24]).

3. In the case of locally symmetric space of rank ≥ 2, see [6], [7], [38].

§ 3. L-functions for dynamical systems.

In this section, we recall the properties of (dynamical) L-function, which plays the crucial role in the proof of the previous results. General references of this section are [5], [27], [33]. For a factor group G of the fundamental group Γ and an N-dimensional unitary representation $\rho: G \to U(N)$, we define L-function by

$$L(s,\rho) = \prod_{p \in P} \det(I_N - \rho([p])e^{-s\ell(p)})^{-1},$$

where P is the set of all prime closed orbits, I_N is the identity matrix, $[p]$ denotes an element p whose conjugacy class corresponds to the free homotopy class of the closed orbit p.

In the case when X is the unit tangent bundle of compact Riemann surface with constant negative curvature, we see that

$$L(s,\rho) = Z(s+1,\rho)/Z(s,\rho)$$

where $Z(s,\rho)$ is the Selberg zeta function.

This L-function has the following properies.

Theorem 3.1 (Selberg [18], [19], [36]). Let (UM, φ_t) be a geodesic flow of compact Riemann surface with constant negative curvature -1. Then,

(1) $L(s,\rho)$ converges absolutely and is holomorphic when $\mathrm{Re}\, s > 1$.

(2) $L(s,\rho)$ has a meromorphic extension to whole $\mathbb{C}$-plane.

(3) $L(s,\rho)$ has poles at $s = -n$ $(n = 1,2,\cdots)$ of order

$2(g-1)N$ and at $s = 0$ of order $(g-1)N$.

(4) The other zeros and poles are

$$\text{Poles:} \quad s = \frac{1}{2} \pm \sqrt{\frac{1}{4}-\lambda_i(\rho)} \qquad i = 0,1,2,\cdots$$

$$\text{Zeros:} \quad s = -\frac{1}{2} \pm \sqrt{\frac{1}{4}-\lambda_i(\rho)} \qquad i = 0,1,2,\cdots$$

where $\lambda_i(\rho)$ is the eigenvalue of the twisted Laplacian Δ_ρ acting on the flat vector bundle E_ρ associated to the representation ρ.

This result is partially generalized to Anosov flow.

Theorem 3.2(Parry and Pollicott (for some cases) Adachi and Sunada [5], [27], [28]). Let (X,φ_t) be a mixing Anosov flow with smooth invariant measure. Then,

(1) $L(s,\rho)$ converges absolutely and is holomorphic when $\mathrm{Re}\, s > h$.

(2) $L(s,\rho)$ has a meromorphic extension to an open neighborhood $\mathbb{D}$ of $\mathrm{Re}\, s \geq h$.

(3) If ρ is irreducible and $N \geq 2$, then $L(s,\rho)$ is holomorphic in $\mathbb{D}$.

(4) $L(s,\mathbb{1})$ has simple pole at $s = h$.

(5) For a character χ, $L(s,\chi)$ has a pole at $s = h + \sqrt{-1}t$ if and only if $\chi(p) = \exp\sqrt{-1}t\ell(p)$ for all p. In this case, $L(s,\chi) = L(s-\sqrt{-1}t,\mathbb{1})$ and every pole on $\mathrm{Re}\, s = h$ is simple.

Remark 3.3. 1. Ruelle [34] showed that if Anosov flow is real analytic (i.e. the decomposition in Definition 2.2.1.b.(1) is real analytic), then $L(s,\mathbb{1})$ has a meromorphic continuation to whole $\mathbb{C}$-plane.

2. Pollicott [30] gave another conditions for the meromorphic continuation of $L(s,\mathbb{1})$. This is generalized by Adachi [3].

L-function associated to the representation of dim ≥ 2 is necessary for the proof of Remark 2.2.6.

§ 4. Outline of the proofs.

4.1. Finite case.

To prove Theorems 2.2.1, 2.2.5, is the same as the number theory once we prove Theorem 3.2. For later purpose, we recall this step briefly. Set

$$F_\alpha(s) = \sum_\chi \chi(-\alpha)\frac{L'(s,\rho)}{L(s,\rho)}$$

where χ runs over unitary characters of $H_1(M,\mathbb{Z})$. Then, using the orthogonal relation of the characters, we have

$$F_\alpha(s) = \sum_{k=1}^{\infty} \sum_p \sum_\chi \chi(-\alpha)\chi(k[p])\ell(p)e^{-sk\ell(p)}$$

$$= {}^{\#}H_1(M,\mathbb{Z}) \sum_{\substack{k=1 \\ k[p]=\alpha}}^{\infty} \sum_p \ell(p)e^{-sk\ell(p)}.$$

Thus $f(s) = F_\alpha(s/h)$ satisfies the assumption of the following theorem for $\varphi(s) = \varphi_\alpha(s/h)$, where

$$\varphi_\alpha(s) = \sum_{\substack{k=1 \\ k\ell(p)<x \\ k[p]=\alpha}}^{\infty} \sum_p \ell(p)$$

Theorem 4.1 (Tauberian theorem [25]). Let $\varphi(x)$ be a monotone non-decresing function with $\varphi(x) = 0$ for $x \leq 0$. Define $f(s)$ via

$$f(s) = \int_0^\infty e^{-sx}d\varphi(x).$$

Suppose that $f(s)$ satisfies the following properties.

(1) The integral for $f(s)$ converges for $\mathrm{Re}\ s > 1$.

(2) One can find a positive constant c such that if, for $\varepsilon > 0$, $s = 1+\varepsilon+\sqrt{-1}t$, we put

$$j_{\varepsilon}(t) = f(s) - \frac{c}{s-1},$$

then $j(t) = \lim_{\varepsilon\to 0} j_{\varepsilon}(t)$ exists almost everywhere, and there exists a locally integrable function $h(t)$ such that $|j_{\varepsilon}(t)| \leq h(t)$.

Then

$$\varphi(x) \sim c\, e^{x}.$$

Applying this theorem, we have,

$$\varphi_{\alpha}(x) \sim \frac{1}{\#H_1(M,\mathbb{Z})} e^{hx}.$$

From this, by routine arguments of analytic number theory, we get Theorem 2.2.5. Theorem 2.2.1 is similar (for Anosov flow, this is the special case of Theorem 2.2.5).

Here, we mention the proof of the result of Bowen type. To prove this, it needs to modify L-function as

$$L_g(s,\rho) = \prod_{p} \det\,(I_N - e^{-s\ell_g(p)})^{-1},$$

and establish corresponding result for Theorem 3.2.

4.2. Infinite case.

There are two methods to prove Theorem 2.2.7. The method of [29] is using Selberg trace formula, whose generalization is used in the proof of Theorem 2.2.9. Our method is analogous to the finite case. But the direct analogy does not work and needs to modify. In both methods, it is crucial to analyze the lowest eigenvalue $\lambda_0(\chi)$ of the twisted Laplacian Δ_{χ} (cf.Theorem 3.1.(4).) near $\chi = \mathbb{1}$.

First, we identify the character group $\hat{A}$ of $H_1(M,\mathbb{Z})$ with Jacobian torus $J(M) = H^1(M,\mathbb{R})/H^1(M,\mathbb{Z})$ by the following correspondence,

$$H^1(M,\mathbb{R}) \ni \omega \longrightarrow \chi_\omega \in \hat{A},$$

$$\chi_\omega(\gamma) = \exp\left(2\pi\sqrt{-1}\int_{C(\gamma)}\omega\right),$$

where $C(\gamma)$ is a closed curve belonging to the homology class of γ. Then, the following proposition can be derived by a purterbation argument.

Proposition 4.2. (1) $\lambda_0(\chi) = 0$ if and only if $\chi = \mathbb{1}$.
(2) $\lambda_0(\chi)$ is C^∞ near $\chi = \mathbb{1}$ and satisfies

$$\lambda_0(\chi_\omega) = \frac{4\pi^2}{\mathrm{vol}(M)}\|\omega\|^2 + O(\|\omega\|^4)$$

where $\|\omega\| = \left(\int_M |\omega|^2\right)^{1/2}$.

(a) Proof of Theorem 2.2.7. (I)

For $\alpha \in H_1(M,\mathbb{Z})$, we set

$$F_\alpha(s) = -\int_{\hat{A}} \chi(-\alpha)\left(-\frac{d}{ds}\right)^g \frac{L'(s,\chi)}{L(s,\chi)}\,d\chi,$$

where $d\chi$ is the normalized Haar measure on $\hat{A}$. By the orthogonal relation of the characters, we have

$$F_\alpha(s) = \sum_{k=1}^{\infty}\sum_{p}\int_{\hat{A}}\chi(-\alpha)\chi(k[p])k^g\ell(p)^{g+1}e^{-sk\ell(p)}d\chi$$

$$= \sum_{k=1}^{\infty}\sum_{\substack{p \\ k[p]=\alpha}} k^g\ell(p)^{g+1}e^{-sk\ell(p)}$$

$$= \int_0^\infty e^{-sx}d\varphi_\alpha(x)$$

where

$$\varphi_\alpha(x) = \sum_{\substack{k=1 \\ k\ell(p)<x \\ k[p]=\alpha}}^{\infty} \sum_{p} k^g \ell(p)^{g+1}.$$

Similarly as finite case, if we have the following proposition, then an elementary argument implies Theorem 2.2.7.

Proposition 4.3. $F_\alpha(s)$ satisfies the the conditions (1), (2) in Theorem 4.1.

(Outline of the proof) (1) follows from Theorem 3.1. We prove (2). For simplicity, we write $f(s) \tilde{=} g(s)$ if $h(s) = |f(s)-g(s)|$ is estimated from the above by a locally integrable function of variable t ($t = \mathrm{Im}\, s$). By Theorem 3.1, we have

(*) $\dfrac{L'(s,\chi)}{L(s,\chi)} + \sum \dfrac{1}{s-f_k(\chi)}$ is holomorphic in $\mathrm{Re}\, s > 1/2$, where $f_k(\chi) = (1+\sqrt{1-4\lambda_k(\chi)})/2$, and the summation is taken over those k with $\lambda_k(\chi) < 1/4$.

By Proposition 4.2, if we apply the Morse lemma, one can find a local coordinate $\{x_i(\omega)\}_{i=1}^{2g}$ of a neighborhood U of $\mathbb{1} \in \hat{H}$ such that

$$f_0(\chi_\omega) = f_0(x) = 1 - |x|^2$$

(1)

$$x(\mathbb{1}) = 0$$

$$J(x) = C + O(|x|) \quad \text{:the Jacobian of coordinate change}$$

$$\chi_\omega(-\alpha) = 1 + O(|x|).$$

Take small $a > 0$ with $V = \{|x|^2 \leq a^2\} \subset U$. By Theorem 3.1, we have

$$F_\alpha(s) = -\left(\int_V + \int_{A-V}\right) \cong -\int_V \chi(-\alpha)\left(-\frac{d}{ds}\right)^g \frac{L'(s,\chi)}{L(s,\chi)} d\chi$$

$$\cong -\int_V \chi(-\alpha)\left(-\frac{d}{ds}\right)^g \frac{1}{s-f_0(\chi)} d\chi \qquad \text{(by (*))}$$

$$\cong -\int_V (1+O(|x|))\left(-\frac{d}{ds}\right)^g \frac{1}{s-1+|x|^2} d\chi \qquad \text{(by (1)).}$$

Changing to the polar coordinate, we find this equals

$$- C \int_{S^{2g-1}} d\Omega \int_0^a \left(-\frac{d}{ds}\right)^g (1+O(|x|))\left(\frac{-r^{2g-1}}{s-1+r^2}\right)(1+O(|x|))dr$$

(2)

$$= - C \int_0^a \left(-\frac{d}{ds}\right)^g \left(\frac{-r^{2g-1}}{s-1+r^2}\right) dr + \left(\frac{d}{ds}\right)^g \int_0^a \frac{\phi(r\Omega)}{s-1+r^2} dr,$$

where ϕ is a function satisfying $|\phi(r\Omega)| \le Cr^{2g}$. By an elementary calculus, we have

$$\text{(The first term of (2))} \cong \frac{C}{s-1}.$$

In the last, we find that second term is locally integrable on Re s = 1 as a function of t. It is derived from the fact that

$$\int_0^a \int_{-\mu}^{\mu} \frac{r^{2g}}{(t^2+r^4)^{(g+1)/2}} dt\, dr < \infty \qquad \text{for } \mu > 0.$$

q.e.d.

(b) Proof of Theorem 2.2.7. (II)

We recall the Selberg trace formula,

$$\sum \hat{h}(r_j(\chi)) = 2(g-1) \int_{-\infty}^{\infty} r \tanh(\pi r)\, \hat{h}(r)\, dr$$

$$+ \sum_{p} \sum_{k=1}^{\infty} \frac{\chi(k[p])\ell(p)}{\sinh(k\ell(p)/2)} h(k\ell(p)),$$

where $h(s)$ is an even C^{∞} function of compact support,

$$\hat{h}(r) = \int_{-\infty}^{\infty} e^{\sqrt{-1}\, rs} h(s)\, ds \quad \text{and} \quad \lambda_j(\chi) = \frac{1}{4} + r_j(\chi).$$

Take $h(s)$ as follows. Let $k(s)$ be a C^{∞}, compactly supported, even positive function with $\int_{-\infty}^{\infty} k(s)\, ds = 1$. Set $k_{\varepsilon}(s) = \varepsilon^{-1}k(s/\varepsilon)$ and define $h(s) = \chi_{[-T,T]}*k_{\varepsilon}(s)$ by the convolution with the characteristic fuction $\chi_{[-T,T]}$ of the interval $[-T,T]$. Then $\hat{h}(r) = (2/r)\sin(Tr)\hat{k}(\varepsilon r)$, where $\hat{k}(\varepsilon r)$ is rapidly decreasing. Multiplying through by $\chi(-\alpha) \equiv \exp(-2\pi\sqrt{-1}\langle\theta,\alpha\rangle)$ and integrating over $\hat{H}$, we get

$$\sum_j \int_{\hat{H}} \hat{h}(r_j(\chi)) e^{-2\pi\sqrt{-1}\langle\theta,\alpha\rangle} d\chi = 2(g-1)\delta_{\alpha,0} \int_{-\infty}^{\infty} r \tanh(\pi r)\hat{h}(r)\, dr$$

(3)

$$+ \sum_{\substack{p \\ k[p]=\alpha}} \sum_{k=1}^{\infty} \frac{\ell(p)h(k\ell(p))}{\sinh(k\ell(p)/2)},$$

Then,

Proposition 4.4.

(L) ≡ (The left-hand side of (3))

$$= \int_U \hat{h}(r_0(\chi)) e^{-2\pi\sqrt{-1}\langle\theta,\alpha\rangle} d\chi + O(T/\varepsilon^2 + e^{\nu T}) \quad (\nu < 1/2)$$

where U is a neighborhood of $\mathbb{1}$.

(Outline of the proof) By Theorem 3.1, only a finite number of $r_j(\chi)$ is imaginary. If $r_j(\chi)$ $(j \geq 1)$ is imaginary, then it is estimated by $e^{\nu T}$ using the fact

$$0 < \mu_1 \leq \lambda_1(\chi) \leq \lambda_j(\chi),$$

where μ_1 is the lowest non-zero eigenvalue of the Laplacian acting on the fundamental domain F of M in the universal covering with the Neumann boundry condition. The contributions from real $r_j(\chi)$ can be estimated by T^2/ε using the fact that $\hat{k}(\varepsilon r)$ is rapidly decreasing. The integral $\int_{\hat{A}-U}$ is again bounded by $e^{\nu T}$ by Proposition 4.2.

Replacing $\hat{k}(\varepsilon r_0(\chi))$ by $1 + O(\varepsilon)$ and $\sinh(-\sqrt{-1}r_0(\chi)T)$ by $e^{-\sqrt{-1}r_0(\chi)T}/2$, (L) becomes

$$(L) = e^{T/2} \int_U e^{(-\sqrt{-1}r_0(\chi) + 1/2)T} \frac{e^{-2\pi\sqrt{-1}\langle\theta,\alpha\rangle}}{-\sqrt{-1}r_0(\chi)} d\chi$$

$$+ O(\varepsilon e^{T/2} + T/\varepsilon^2 + e^{\nu T}).$$

The right-hand side (R) is expressed as follows.

Proposition 4.5.

$$(R) = \sum_{\substack{p \\ [p]=\alpha \\ \ell(p)<x}} \frac{\ell(p)}{\sinh(\ell(p)/2)} + O(T^2 + \varepsilon e^{T/2}).$$

(Proof) By Theorem 2.2.1, we have

$$\sum_{\substack{p \\ T<\ell(p)<T+\varepsilon}} 1 = O(\varepsilon e^T)$$

This is used to show that the contribution of the sum $k > 1$ is $O(T^2)$. The contribution from the ε-smoothing of k_ε is $O(\varepsilon e^{T/2})$.

Combining the above, we get

$$p(s) \equiv \sum_{\substack{p \\ [p]=\alpha \\ \ell(p)<x}} \frac{\ell(p)}{\sinh(\ell(p)/2)}$$

$$= e^{T/2} \int_U e^{(-\sqrt{-1}r_0(\chi) + 1/2)T} \frac{e^{-2\pi\sqrt{-1}\langle\theta,\alpha\rangle}}{-\sqrt{-1}r_0(\chi)} d\chi$$

$$+ O(e^{\nu T}).$$

Applying the following stationary phase lemma, Theorem 2.2.7 and Remark 2.2.8 are proved by the following.

$$q(T) \equiv \sum_{\substack{p \\ [p]=\alpha \\ \ell(p)<x}} 1 = \int_0^T \frac{\ell(p)}{\sinh(\ell(p)/2)} dp(s) = \int_0^T \frac{e^{s/2}}{2s} dp(s) + O(1)$$

$$= \frac{e^{T/2}}{2T} p(T) - \int_0^T \frac{e^{s/2}}{4s} p(s)ds + \int_0^T \frac{e^{T/2}}{2s^2} p(s)ds + O(1).$$

Lemma 4.7. Let $\rho(x)$ be a smooth function in a neighborhood $U \subset \mathbb{R}^{2g}$ of $x = 0$ such that $\rho(0) = 0$, $\rho(0) > 0$ when $x \neq 0$ and $\partial_{ij}\rho = -b_{ij}$ at $x = 0$, where $B = (b_{ij})$ is positive definite. Suppose further that $f(x)$ is also smooth in U. Then,

$$\int_U e^{T\rho(x)} f(x)\, dx \sim \frac{(2\pi)^g}{T^g\sqrt{\det B}}\left(f(0) + \frac{c_1}{T} + \frac{c_2}{T^2} + \cdots\right)$$

as $T \to \infty$.

(c) Outline of the proof of Theorem 2.2.11.

Although the approach (b) is simple and get detailed information, up to now, there is no (effective) trace formula applicable to the case of variable negative curvature. So, we follow the approach of (a) and use Theorem 3.2 etc.. According to Bowen, the flow (UM, φ_t) is analyzed by symbolic dynamics ("approximation by finite oriented graphs"). In fact, Theorem 3.2 is proved by this technic. In this case, we use the twisted Ruelle operator L_χ instead of the Laplacian Δ_χ. Although L_χ acts on Banach space and thus, not self-adjoint in general, the maximun eigenvalue $\mu(\chi)$ plays similar role to $\lambda_0(\chi)$. In particular, the existence of the involution I on UM with $\varphi_{-t} \circ I = I \circ \varphi_t$ implies that $\mu(\chi)$ is real near $\chi = \mathbb{1}$ and the hessian of $\mu(\chi)$ at $\chi = \mathbb{1}$ is negative definite. From this, we also use the argument of Proposition 4.3. But in the case when b is odd, we require a little different argument and modify the Tauberian theorem.

This note is based on joint works with T. Sunada ([22],[23]).

References

1. T. Adachi: Markov families for Anosov flows with an involutive action, Nagoya Math. J. 104 (1986) 55- 62.
2. T. Adachi: Distribution of closed geodesics with a preassigned homology class in a negatively curved manifold, preprint
3. T. Adachi: Meromorphic extension of dynamical L-functions, preprint.
4. T. Adachi and T. Sunada: Homology of closed geodesics in a negatively curved manifold, J. Diff. Geom. 26 (1987), 81-99.
5. T. Adachi and T. Sunada: Twisted Perron-Frobenius theorem and L-functions, J. Funct. Anal., 71 (1987), 1-47
6. W. Ballman: Nonpositively curved manifolds of higher rank, Ann. of Math. 122 (1985) 597-609.
7. W. Ballman, M. Brin and R. Spatzier, Structure of manifolds of

nonpositive curvature II, Ann. of Math. 122 (1985) 205-235.

8. R. Bowen: The equidistribution of closed geodesics, Amer. J. of Math. 95, (1972) 413-423.

9. R. Bowen: Symbolic dynamics for hyperbolic flows, Amer. J. Math. 95 (1973), 429-460.

10. C. Epstein: *The Spectral Theory of Geometrically Periodic Hyperbolic 3-manifold*, Memoirs of the A.M.S., No. 335,1985.

11. C. Epstein: Asymptotics for closed geodesics in a homology class - finite volume case -, preprint.

12. D. Fried: Flow equivalence, hyperbolic systems and a new zeta function for flows, Comment. Math. Helvetici 57 (1982), 237-259.

13. D. Fried: The zeta functions of Ruelle and Selberg I, preprint.

14. D. Fried: Analytic torsion and closed geodesics on hyperbolic manifolds, preprint.

15. L. Guillope: Sur la distribution des longueurs des geodesiques fermees d'une surface compacte a bord totalment geodesique, Duke Math. J. 53 (1986) 827-848.

16. G. H. Hardy and J. E. Littlewood: Some problems of "partitio numerorum";III:On the expression of a number as the sum of primes, Acta Math., 44 (1923) 1-70.

17. D. R. Heath-Brown: Prime twins and Siegel zeros, Proc. London Math. Soc. (3), 47 (1983), 193-224.

18. D. A. Hejhal: The Selberg trace formula and the Riemann zeta function, Duke Math. J. 43 (1976), 441-482.

19. D. A. Hejhal: *The Selberg Trace Formula for* PSL(2,K), I. Springer Lecture Notes 548, 1976.

20. A. Katsuda: Homology of closed geodesics in a nonpositively curved manifold of rank one, preprint.

21. A. Katsuda and T. Sunada: Homology of closed geodesics in certain Riemannian manifolds, Proc. AMS 96 (1986), 657-660

22. A. Katsuda and T. Sunada: Homology and closed geodesics in a

compact Riemannian surface, to appear in Amer. J. of Math.

23. A. Katsuda and T. Sunada: L-functions and homology of closed geodesics in a compact negatively curved manifolds, in preparation.

24. G. Knieper: Das Wachstum der Äquivalenzklassen geschlossener Geodätischer in kompacten Mannigfaltigkeiten, Arch. Math. 40 (1983), 559-568.

25. S. Lang: *Algebraic Number Theory*, Addison-Wesley, 1970.

26. W. Parry: Bowen's equidistribution theory and the Dirichlet density theorem, Ergod. Th. and Dynam. Sys. 4, 117-134.

27. W. Parry and M. Pollicott: An analogue of the prime number theorem for closed orbits of Axiom A flows, Ann. of Math. 118 (1983), 573-591.

28. W. Parry and M. Pollicott: The Chebotarev theorem for Galois coverings of Axiom A flows, Ergod. Th. and Dynam. Sys. 6 (1986), 133-148.

29. R. Phillips and P. Sarnak, Geodesics in homology classes, Duke Math. J. 55 (1987), 287-297.

30. M. Pollicott: Meromorphic extensions of generalized zeta functions, Invent. math. 85 (1986) 147-164.

31. M. Pollicott: Asymptotic distribution of closed geodesics, Isrel J. of Math. 52 (1985) 209-224.

32. M. Pollicott: A note on uniform distribution for primes and closed orbits, Israel J. of Math. 55 (1986) 199-212.

33. D. Ruelle: *Thermodynamic Formalism*, Addison-Weasley, Reading, Mass., 1978.

34. D. Ruelle: Zeta functions for expanding maps and Anosov flows, Invent. Math. 34 (1976), 231-242.

35. P. Sarnak: Prime geodesic theorems, Ph. D. dissertation, Stanford University (1980).

36. A. Selberg: Harmonic analysis and discontinuous subgroups in weak-

ly symmetric Riemannian spaces with applications to Dirichlet series, J. Indian Math. Soc. 20 (1956), 47-87.

37. S. Smale: Differentiable dynamical systems, Bull. AMS. 73 (1967), 747-817.

38. R. Spatzier, dynamical properties of algebraic systems, Thesis, Warwick Univ. (1983)

39. T. Sunada: Geodesic flows and geodesic random walks, Advanced Studies in Pure Math. 3 (Geometry of Geodesics and Related Topics) (1984), 47-86.

40. T. Sunada: Tchebotarev's density theorem for closed geodesics in a compact locally symmetric space of negative curvature, preprint.

41. T. Sunada: Riemannian coverings and isospectral manifolds, Ann. of Math. 121 (1985), 169-186.

42. T. Sunada: Number theoretic methods in spectral geometry, to appear in Proc. The 6th Symp. on Differential Geometry and Differential Equations held at Shanghai, China 1985.

43. T. Sunada: L-function in geometry and some applications, Curvature and Topology of Riemannian Manifolds, Lect. note in Math. 1201, (1986) 266-284.

44. S. Zelditch: Trace formula for compact $\Gamma/PSL_2(\mathbb{R})$ and the equidistribution theory of closed geodesics, preprint

45. S. Zelditch: Selberg trace formulae, pseudo-differential operators, and geodesic periods of automorphic forms, to appear in Duke Math. J.

46. S. Zelditch: Geodesics in homology classes and Periods of automorforms, preprint

47. S. Zelditch: Splitting geodesics in homology classes, preprint

L^2-INDEX AND RESONANCES

Werner Müller
Akademie der Wissenschaften der DDR
Institut für Mathematik
DDR-1086 Berlin, Mohrenstr. 39
German Democratic Republic

1. Introduction

Let M be a compact C^∞ manifold with boundary N and let

$$\tilde{D}: C^\infty(M,\tilde{E}) \longrightarrow C^\infty(M,\tilde{F})$$

be a linear first order elliptic differential operator on M which, in a neighborhood $N \times I$ of the boundary, takes the special form

$$\tilde{D} = \sigma\left(\frac{\partial}{\partial u} + A\right) \tag{1.1}$$

where u is the inward normal coordinate, σ is a bundle isomorphism $\tilde{E}|N \longrightarrow \tilde{F}|N$ and A is a selfadjoint elliptic operator on N. Atiyah, Patodi and Singer introduced in [A-P-S] certain non-local boundary conditions which give rise to an elliptic boundary value problem for $\tilde{D}$ with finite index. The spectral boundary conditions of [A-P-S] are defined by requiring that the boundary value $\Phi|N$ of a section $\Phi \in C^\infty(M,\tilde{E})$ lies in the subspace spanned by the eigenfunctions φ_λ of A with eigenvalues $\lambda < 0$. It has been observed in [A-P-S], Corollary 3.14, that the index of the non-local boundary value problem is closely related to the index of some L^2-problem. Namely, let $X = M \cup (\mathbb{R}^+ \times N)$, and take the measure $dx = du dy$ on $\mathbb{R}^+ \times N$. The bundles $\tilde{E},\tilde{F}$ can be extended to bundles E,F on X in the obvious way. In view of (1.1), the differential operator $\tilde{D}$ has also a natural extension to an operator

$$D: C^\infty(X,E) \longrightarrow C^\infty(X,F) .$$

The space of L^2-solutions of D and D^* is finite dimensional. Let

$$L^2\text{-Index } D = \dim(\ker(D) \cap L^2) - \dim(\ker(D^*) \cap L^2) .$$

Then, by Corollary 3.14 in [A-P-S], one has

$$L^2\text{-Index } D = \text{Index } \tilde{D} + h_\infty(F) \tag{1.2}$$

where $h_\infty(F)$ is the dimension of the subspace of Ker A consisting of limiting values of extended L^2-sections Φ of F satisfying $D^*\Phi = 0$. Let $h_\infty(E)$ have the corresponding meaning with respect to E. According to (3.25) in [A-P-S], one has

$$\dim \text{Ker } A = h_\infty(E) + h_\infty(F) . \tag{1.3}$$

This suggests that every section in Ker A is uniquely expressible as a sum of limiting values coming from E and F respectively. The purpose of this paper is to give an affirmative answer to this problem. The L^2-problem on the elongated manifold X can be treated by using methods of supersymmetric scattering theory developed in [B-M-S] and the solution of the problem just mentioned is a natural consequence of this approach.

We shall now describe our method in more detail. Let H be the closure in L^2 of D^*D acting in $C_c^\infty(X,E)$. We may think of H as a perturbation of an operator H_o acting in $L^2(\mathbb{R}^+ \times N;E)$. H_o is the selfadjoint extension of D^*D acting in $C_c^\infty(\mathbb{R}^+ \times N,E)$ obtained by imposing Dirichlet boundary conditions at $\{0\}\times N$. Associated to (H,H_o) there is a scattering operator S. The corresponding on-shell scattering matrix $S(\lambda)$ acts on the sum of the eigenspaces of A^2 which correspond to eigenvalues $\mu \leq \lambda$. If Ker $A \neq 0$, the scattering matrix $S(0)$ for zero energy is a linear operator on Ker A which satisfies $S(0)^2 = \text{Id}$ and

$$\text{Tr } S(0) = h_\infty(E) - h_\infty(F) . \tag{1.4}$$

Let $\mathcal{E}^\pm \subset \text{Ker } A$ be the ± 1 eigenspaces of $S(0)$. To each $\Phi \in \text{Ker } A$ there is associated a generalized eigenfunction $E(\Phi,\lambda)$ of H. As a function of λ, $E(\Phi,\lambda)$ is a meromorphic function on a certain Riemann surface Σ. $E(\Phi,\lambda)$ is holomorphic at $\lambda = 0$ and, if $\Phi \in \mathcal{E}^+$, then one has $E(\Phi,0) \neq 0$. In this case $E(\Phi,0)$ is an extended L^2-section of E (in the sense of [A-P-S]) satisfying $DE(\Phi,0) = 0$. The same construction applied to $\Phi \in \mathcal{E}^-$ gives extended L^2-sections of F which are solutions for D^*. These sections are known in quantum mechanics as zero energy resonance states. All extended L^2-sections φ of E (resp. F) which satisfy $D\varphi = 0$ (resp. $D^*\varphi = 0$) are obtained in this way.

If we use (1.2), Theorem 3.10 in [A-P-S] and (1.4), we obtain the following formula for the L^2-index of D:

$$L^2\text{-Index } D = \int_X \alpha_o(x) - \frac{\eta(0)}{2} - \frac{1}{2}\text{Tr } S(0)$$

where $\alpha_0(x)$ is the index density for D and $\eta(0)$ is the Eta-invariant of A. Thus $\frac{1}{2}\operatorname{Tr} S(0)$ is the contribution of the zero energy resonance states to the L^2-index of D.

In the case of the signature operator it is proved in [A-P-S] that $\operatorname{Tr} S(0)$ vanishes. This will not be so in general.

2.The spectral resolution

Let X be a complete Riemannian manifold. We say that X has a cylindrical end if X has a decomposition $X = M \cup Y$ where M is a compact Riemannian manifold with boundary N and Y is isometric to the product $\mathbb{R}^+ \times N$. Let X be a manifold with a cylindrical end and let

$$\Delta : C^\infty(X,E) \longrightarrow C^\infty(X,E)$$

be a linear second order elliptic differential operator acting on sections of a vector bundle E and assume that, on $Y = \mathbb{R}^+ \times N$, Δ takes the special form

$$\Delta = -\frac{\partial^2}{\partial u^2} + B$$

where $u \in \mathbb{R}^+$ and B is a positive selfadjoint elliptic operator on N. In addition we shall assume that Δ is formally selfadjoint. If we consider Δ as a linear operator in $L^2(X,E)$ with domain $C_c^\infty(X,E)$ (the space of compactly supported C^∞-sections of E) then it is easy to see that Δ is essentially selfadjoint. We shall denote the unique selfadjoint extension of Δ by H.

In this section we shall investigate the spectral resolution of H. The point spectrum of H has been considered by Donnelly [D]. The main result is

Proposition 2.1: Let $N(\lambda)$ be the number of linearly independent eigenfunctions of H with eigenvalues $\leqq \lambda$. There exist constants $C > 0$ and $m \in \mathbb{N}$ such that

$$N(\lambda) \leqq C(1 + \lambda^m)$$

for all $\lambda \geqq 0$.

The continuous spectrum can be studied by using methods from quantum scattering theory. For this purpose we introduce an auxiliary operator H_o which corresponds to the free Hamiltonian in quantum mechanics. H_o is defined as follows. Consider $\Delta_o = -\partial^2/\partial u^2 + B$ acting in $C_c^\infty(\mathbb{R}^+ \times N, E)$ and let H_o be the selfadjoint extension in L^2 obtained by imposing Dirichlet boundary conditions at $\{0\}\times N$. Using the decomposition $L^2(X,E) = L^2(M,E) \oplus L^2(\mathbb{R}^+ \times N, E)$, we regard H_o as an operator in the Hilbert space $L^2(X,E)$. Denote by P_H^{ac} and $P_{H_o}^{ac}$ the orthogonal projections onto the absolutely continuous subspaces for H and H_o respectively. Consider the wave operators

$$W_\pm(H,H_o) = s-\lim_{t\to\pm\infty} e^{itH}e^{-itH_o}P_{H_o}^{ac}$$

If the wave operators exist and are complete, they define intertwining operators between the absolutely continuous parts of H and H_o respectively. To prove existence and completeness of the wave operators one can use the method of Enss. For details we refer to [G1], [G2] and [M,§6]. The result is

Proposition 2.2:

1) The singular continuous spectrum of H is empty.

2) The wave operators $W_\pm(H,H_o)$ exist and are complete, i.e. $W_\pm(H,H_o)$ is an intertwining operator between the absolutely continuous parts $H_{o,ac}$ and H_{ac} of H_o and H respectively.

Let $\mu_1 < \mu_2 < \cdots$ be the sequence of different eigenvalues of B and denote by $E(\mu)$ the eigenspace with eigenvalue μ. Δ_o acts on $C^\infty(\mathbb{R}^+)\otimes E(\mu_j)$ by $-\partial^2/\partial u^2 + \mu_j \mathrm{Id}$. This gives the spectral resolution for H_o. Thus the continuous spectrum of H is the interval $[\mu_1,\infty)$ and, at a given point $\lambda \in [\mu_1,\infty)$, the continuous spectrum has multiplicity $\sum_{\mu_j\le\lambda} \dim E(\mu_j)$.

3. The generalized eigenfunctions

The wave operators $W_\pm(H,H_o)$ can be described more explicitly by generalized eigenfunctions. They are defined as follows. Let Φ be an eigenfunction of B with eigenvalue μ. Denote by $\lambda \longrightarrow \sqrt{\lambda-\mu}$ the branch of the square root whose imaginary part is positive on $\mathbb{C}-[\mu_1,\infty)$ and set

$$e_\mu^\pm(\lambda,u) = e^{\pm i\sqrt{\lambda-\mu}\,u}.$$

Choose $f\in C^\infty(\mathbb{R}^+)$ so that $f(u) = 0$ for $u \leqq 1$ and $f(u) = 1$ for $u \geqq 2$. Let

$$E_\mu(\Phi,\lambda) = fe_\mu^-(\lambda)\Phi - (H-\lambda)^{-1}((\Delta-\lambda)(fe_\mu^-(\lambda)\Phi)). \qquad (3.1)$$

$E_\mu(\Phi,\lambda)$ is a smooth section of E which satisfies

$$\Delta E_\mu(\Phi,\lambda) = \lambda E_\mu(\Phi,\lambda)\ ,\ \lambda \in \mathbb{C}-[\mu_1,\infty)\ .$$

Denote by σ the set of eigenvalues of B. Let Σ be the Riemann surface which is associated to the square roots $z \longrightarrow \sqrt{z-\mu}$, $\mu \in \sigma$. The basic fact about the sections $E_\mu(\Phi,\lambda)$ is

Theorem 3.2: Each function $\lambda \longrightarrow E_\mu(\Phi,\lambda)$ has a meromorphic continuation to a function of $\Lambda \in \Sigma$.

For the proof of this Theorem see [G2] .

The restriction of $E_\mu(\Phi,\lambda)$ to $\mathbb{R}^+ \times N$ has an expansion of the form

$$e_\mu^-(\lambda)\Phi + \sum_{\mu' \in \sigma} e_{\mu'}^+(\lambda)T_{\mu'\mu}(\lambda)\Phi \qquad (3.3)$$

where $T_{\mu'\mu}(\lambda) : E(\mu) \longrightarrow E(\mu')$ is a meromorphic function of λ . The $T_{\mu'\mu}$ are closely related to the scattering matrix $S(\lambda)$ of (H,H_0).

4. Zero energy resonance states

In this section we shall assume that $\operatorname{Ker} B \neq 0$, i.e., the smallest eigenvalue of B is $\mu_1=0$. For $\Phi \in \operatorname{Ker} B$ we denote by $E(\Phi,\lambda)$ the generalized eigenfunction associated to Φ . Our purpose is to study $E(\Phi,\lambda)$ in a neighborhood of $\lambda=0$. Let μ_2 be the smallest positive eigenvalue of B and $\Sigma_2 = \pi^{-1}(\mathbb{C}-[\mu_2,\infty))$ where $\pi : \Sigma \longrightarrow \mathbb{C}$ is the covering map of the spectral surface Σ . Σ_2 can be identified with $\mathbb{C}-([\sqrt{\mu_2},\infty) \cup [-\sqrt{\mu_2},-\infty))$ and the covering map $\Sigma_2 \longrightarrow \mathbb{C}-[\mu_2,\infty)$ is the map $s \longrightarrow s^2$. Thus, by Theorem 3.2 , $E(\Phi,\lambda)$ has a continuation to a meromorphic function $E(\Phi,s)$ of $s \in \Sigma_2$. Moreover, using (3.3), it follows that there exists a linear operator $C(s)$: $\operatorname{Ker} B \longrightarrow \operatorname{Ker} B$ which is a meromorphic function of $s \in \Sigma_2$ such that, on $\mathbb{R}^+ \times N$, we have

$$E(\Phi,s) = e^{isu}\Phi + e^{-isu}C(s)\Phi + \Psi(s),\quad s \in \Sigma_2 \qquad (4.1)$$

where $\Psi(s)$ is in L^2. The operator $C(s)$ satisfies the following functional equation

$$C(s)C(-s) = \mathrm{Id},\quad s \in \Sigma_2. \qquad (4.2)$$

This follows from Corollary 5.5 in [G2] . For $s=0$ we get $C(0)^2=\mathrm{Id}$.

In particular, $C(s)$ is holomorphic at $s=0$. To prove that $E(\Phi,s)$ is also holomorphic at $s=0$, we use an analogue of the Maaß-Selberg relations for truncated Eisenstein series [H]. Put

$$\tilde{E}(\Phi,s,x) = \begin{cases} E(\Phi,s,x) \quad , \quad \text{if} \quad x\in M \\ E(\Phi,s,(u,z)) - e^{isu}\Phi(z) + e^{-isu}(C(s)\Phi)(z), \\ \qquad\qquad\qquad\qquad \text{if} \quad x=(u,z)\in \mathbb{R}^+ \times N. \end{cases}$$

It follows from (4.1) that, for all $s\in\Sigma_2$ different from poles, $\tilde{E}(\Phi,s)$ is square integrable. Let $s,s'\in\Sigma_2$ and assume that $s^2 \neq \bar{s}'^2$ and s,s' are not poles of $E(\Phi,z)$. Then

$$(s^2 - \bar{s}'^2)(\tilde{E}(\Phi,s),\tilde{E}(\Phi,s')) =$$
$$= (\Delta\tilde{E}(\Phi,s),\tilde{E}(\Phi,s')) - (\tilde{E}(\Phi,s),\Delta\tilde{E}(\Phi,s')).$$

Applying Green's formula, we obtain

$$(\tilde{E}(\Phi,s),\tilde{E}(\Phi,s')) = \frac{i}{s-\bar{s}'}\{\ \|\Phi\|^2 - (C(s)\Phi,C(s')\Phi)\ \} \ +$$
$$+ \frac{i}{s+\bar{s}'}\{(\Phi,C(s')\Phi) - (C(s)\Phi,\Phi)\ \}\ .$$

Now set $s'=0$ and let $s\longrightarrow 0$. Since $C(s)$ is holomorphic at $s=0$, the right hand side stays bounded. Thus $E(\Phi,s)$ is holomorphic at $s=0$ and

$$\|E(\Phi,0)\|^2 = -i(\tfrac{d}{ds}C(0)\Phi,C(0)\Phi) - i(\tfrac{d}{ds}C(0)\Phi,\Phi). \qquad (4.3)$$

This implies

<u>Lemma 4.4:</u> For each $\Phi\in\operatorname{Ker} B$, $E(\Phi,s)$ is holomorphic at $s=0$.

Let $\Phi\in\operatorname{Ker} B$. By Lemma 4.4, $E(\Phi,0)$ is well defined. Moreover, $E(\Phi,0)\in C^\infty(X,E)$ and it satisfies $\Delta E(\Phi,0) = 0$. To see when $E(\Phi,0) \neq 0$, we consider $C(0)$. By (4.2) we have

$$C(0)^2 = \mathrm{Id}.$$

Let $\mathcal{E}^\pm \subset \operatorname{Ker} B$ be the ± 1 eigenspaces of $C(0)$. Then we have an orthogonal sum decomposition

$$\operatorname{Ker} B = \mathcal{E}^+ \oplus \mathcal{E}^-\ .$$

Let $\Phi\in\mathcal{E}^+$. Using (4.1), it follows that, on $\mathbb{R}^+ \times N$, we have

$$E(\Phi,0) = 2\Phi + \Psi\ , \qquad (4.5)$$

where $\Psi \in L^2$. This implies that $E(\Phi,0) \neq 0$, if $\Phi \neq 0$. Thus $\frac{1}{2}E(\Phi,0)$ is a zero energy resonance state with limiting value Φ. On the other hand, if $\Psi \in E^-$, then we have $E(\Psi,0) = \tilde{E}(\Psi,0)$ and it follows from (4.3) that $\tilde{E}(\Psi,0) = 0$. This shows that the number of zero energy resonance states is equal to $\dim E^+$.

5. The index formula

Now let

$$D: C^\infty(X,E^+) \longrightarrow C^\infty(X,E^-)$$

be a linear first order elliptic differential operator and assume that on $\mathbb{R}^+ \times N$, it takes the special form

$$D = \sigma\left(\frac{\partial}{\partial u} + A\right) \tag{5.1}$$

where σ is a bundle isomorphism and A is an elliptic selfadjoint operator on N. Let $\Delta^+ = D^*D$ and $\Delta^- = DD^*$ where D^* denotes the formal adjoint operator. If we use σ to identify $E^+|N$ and $E^-|N$, then, on $\mathbb{R}^+ \times N$, both Δ^+ and Δ^- take the special form

$$-\frac{\partial^2}{\partial u^2} + A^2$$

and we can apply the results of §4. Since A is selfadjoint, we have $\operatorname{Ker} A^2 = \operatorname{Ker} A$. Given $\Phi \in \operatorname{Ker} A$, we denote by $E^\pm(\Phi,s) \in C^\infty(X,E^\pm)$ the associated generalized eigenfunctions. They give rise to two operators

$$C^\pm(0): \operatorname{Ker} A \longrightarrow \operatorname{Ker} A$$

- the on-shell scattering matrix for energy zero. We have the following relation between $C^+(0)$ and $C^-(0)$:

<u>Lemma 5.2:</u> $C^+(0) = -C^-(0)$.

<u>Proof:</u> Let $\Phi \in \operatorname{Ker} A$. It follows from (3.1) that $E^-(\Phi,s)$ is uniquely determined by the following two properties

1) $E^-(\Phi,s) - e^{isu}\Phi \in L^2(\mathbb{R}^+ \times N, E^-)$ for all s with $\operatorname{Im}(s) < 0$.

2) $\Delta^- E^-(\Phi,s) = s^2 E^-(\Phi,s)$, $\operatorname{Im}(s) < 0$.

Indeed, if we have any two C^∞ sections of E^- satisfying 1) and 2), then their difference is a square integrable solution of $(\Delta^- - s^2)\Psi = 0$ which has to be zero by the selfadjointness of H^-. Now consider $DE^+(\Phi,s)$. Using $D\Delta^+ = \Delta^- D$ and (5.1), it follows that $DE^+(\Phi,s)$

satisfies 2) and $DE^+(\Phi,s) - ise^{isu}\Phi$ is square integrable on $\mathbb{R}^+ \times N$. Thus

$$DE^+(\Phi,s) = isE^-(\Phi,s). \tag{5.3}$$

If we apply (4.1) to $E^{\pm}(\Phi,s)$ and compare both sides, we obtain

$$-isC^+(s) = isC^-(s).$$

Thus $C^+(0) = -C^-(0)$. Q.E.D.

Put

$$S(s) = C^+(s).$$

Then $S(0)$ is a linear operator on $\operatorname{Ker} A$ which satisfies $S(0)^2 = \mathrm{Id}$. Let

$$\operatorname{Ker} A = \mathcal{E}^+ \oplus \mathcal{E}^-$$

be the decomposition of $\operatorname{Ker} A$ into the ± 1 eigenspaces of $S(0)$.

Recall from [A-P-S], p.58, that an extended L^2-section of E^+ (resp. E^-) is a section φ of E^+ (resp. E^-) which is locally in L^2 and such that, on $\mathbb{R}^+ \times N$ and for large u,

$$\varphi(u,x) = \Phi(x) + \Psi(u,x)$$

where Ψ is in L^2 and $\Phi \in \operatorname{Ker} A$. Φ is called the limiting value of φ. Let $\tilde{\mathcal{E}}^+$ (resp. $\tilde{\mathcal{E}}^-$) be the subspace of $\operatorname{Ker} A$ consisting of limiting values of extended L^2-sections φ of E^+ (resp. E^-) satisfying $D\varphi = 0$ (resp. $D^*\varphi = 0$). Following [A-P-S], we put

$$h_\infty(E^\pm) = \dim \tilde{\mathcal{E}}^\pm.$$

<u>Theorem 5.4:</u> We have $\mathcal{E}^\pm = \tilde{\mathcal{E}}^\pm$.

<u>Proof:</u> Let $\Phi \in \mathcal{E}^+$ and set $\varphi = \frac{1}{2}E^+(\Phi,0)$. We know from §4 that $\varphi \neq 0$. By (5.3), we have $D\varphi = 0$ and, using (4.5), it follows that φ has the limiting value $\Phi \in \mathcal{E}^+$. Thus $\mathcal{E}^+ \subset \tilde{\mathcal{E}}^+$. Now let $\Psi \in \mathcal{E}^-$, i.e., $C^+(0)\Psi = -\Psi$. By Lemma 5.2 we have then $C^-(0)\Psi = \Psi$. Put $\psi = \frac{1}{2}E^-(\Psi,0)$. Again by §4, we have $\psi \neq 0$. It follows as above that $D^*\psi = 0$ and Ψ is the limiting value of ψ. Thus $\mathcal{E}^- \subset \tilde{\mathcal{E}}^-$. By (3.25) in [A-P-S] we have $\dim \tilde{\mathcal{E}}^+ + \dim \tilde{\mathcal{E}}^- = \dim \operatorname{Ker} A$. This implies $\mathcal{E}^\pm = \tilde{\mathcal{E}}^\pm$. Q.E.D.

Theorem 5.4 shows that each $\Phi \in \operatorname{Ker} A$ is uniquely expressible as a sum of limiting values coming from E^+ and E^- respectively. This gives an affirmative answer to the question raised in [A-P-S], p.60. Furthermore, the Theorem implies

$$\mathrm{Tr}\ S(0) = h_\infty(E^+) - h_\infty(E^-) \ . \qquad (5.5)$$

As mentioned in the introduction, D has a well defined L^2-index and the corresponding index formula can be written as

$$L^2\text{-Index } D = \int_X \alpha_0(x) - \frac{\eta(0)}{2} - \frac{1}{2}\mathrm{Tr}\ S(0) \qquad (5.6)$$

where $\alpha_0(x)$ and $\eta(0)$ have the same meaning as in the introduction. This index formula has been deduced from Theorem 3.10 in [A-P-S]. We observe that (5.6) can be also proved using [B-M-S]. The proof is similar to the proof of the index formula in [M].

REFERENCES

[A-P-S] Atiyah, M.F., Patodi, V.K., Singer, I.M.: Spectral asymmetry and Riemannian geometry I. Math. Proc. Camb. Phil. Soc. **77**, 43-69 (1975).

[B-M-S] Borisov, N., Müller, W., Schrader, R.: Relative index theorems and supersymmetric scattering theory. Preprint 1986, to appear in CMP.

[C] Colin de Verdiere, Y.: Une nouvelle démonstration du prolongement méromorphe de séries d'Eisenstein. C.R. Acad. Sci. Paris Ser. I, Math., **293**, 361-363 (1981).

[D] Donnelly, H.: Eigenvalue estimates for certain noncompact manifolds. Michigan Math. J. **31**, 349-357 (1984).

[G1] Guillopé, L.: Sur la théorie spectral de quelques variétés non compactes. Séminaire de théorie spectrale et géometrie Chambéry -Grenoble, 1985-1986, 115-126.

[G2] Guillopé, L.: Théorie spectrale de quelques variétés à bouts. Preprint, Grenoble 1987.

[H] Harish-Chandra: Automorphic forms on semisimple Lie groups. Lecture Notes in Math. **62**, Berlin-Heidelberg-New York, Springer-Verlag, 1968.

[K] Kato, T.: Perturbation theory for linear operators. Berlin-Heidelberg-New York, Springer, 1966.

[M] Müller, W.: Manifolds with cusps of rank one. Lecture Notes in Math. **1244**, Berlin-Heidelberg-New York, Springer, 1987.

APPROXIMATION OF GREEN'S FUNCTION IN A REGION WITH MANY OBSTACLES

Shin Ozawa

Department of Mathematics, Tokyo Institute of Technology, O-okayama, Megro, Tokyo, 152

One of the strongest method in calculating asymptotic behaviour of the eigenvalues of the Laplacian in region with many holes (with the Dirichlet boundary condition) is to construct an approximate Green's function. On this standpoint precise treatment of the Lenz shift phenomena is discussed recently. See Figari-Orlandi-Teta [4] , Ozawa [8] , [10] and the references therein. In the present paper we will concentrate our effort to construct and to estimate approximate Green's function. Even if we do not touch with an analysis on the Lenz shift by using approximate Green's function, we recall the lenz shift phenomena and we offer some open problems.

Lenz shift. Let M be a bounded domain in R^3 with smooth boundary. Fix $\alpha > 0$. Fix $\beta \in [1,3)$. Let $m = 1,2,3 \cdots$ be a parameter. We put $n = [m^\beta]$. We remove n balls of centers $w(m) = (w_1, \cdots, w_n) \in M^n$ with radius α/m from M and we get $M_{w(m)} = M \setminus$ n-balls. Remark that $M_{w(m)}$ may not be connected. Let M^o be a connected component of $M_{w(m)}$. Let $\mu_k(M^o)$ be the k-th eigenvalue of the Laplacian in M^o under the Dirichlet condition on ∂M^o. We arrange all $\mu_k(M^o)$ in line (all $M^o \subset M_{w(k)}$, $k = 1,2, \cdots$), then we have the j-th eigenvalue $\mu_j(w(m))$ of the Laplacian in $M_{w(m)}$ under the Dirichlet condition.

We consider M as probability space by fixing a positive continuous function V on M satisfying

$$\int_M V(x)dx = 1$$

so that $P(x \in A) = \int_A V(x)dx$. Let M^n be the product probability space. All configuration of the centers of balls $w(m)$ can be considered as a probability space M^n by the statistical law stated above. Hereafter $\mu_j(w(m))$ is considered as a random variable on M^n.

There are several papers concerning asymptotic behaviour of $\mu_j(w(m))$. The first step is made by Kac [5] and Huruslov-Marchenko [3] in 1974, when $\beta = 1$. They obtained convergence

(1.1) $$\mu_j(w(m)) - \mu_j^V \longrightarrow 0$$

in probability as $m \to \infty$, where μ_j^V is the j-th eigenvalue of the Schrodinger operator $-\Delta + 4\pi\alpha V(x)$ in M under the Dirichlet condition on ∂M. Kac [5] gave (1.1) when $V \equiv \text{const.}\ (= |M|^{-1})$. Kac used the notion of Wiener sausage to give (1.1). See Rauch-Taylor [11], Chavel-Feldman [1], Sznitman [12], Weinryb [13] etc for related interesting topics and results.

Kac [5] gave a negative opinion on perturbative calculus. It is now recognized that perturbative calculus is very nice in the sense that we have more precise information on the convergence rate of $\mu_j(w(m)) - \mu_j^V$ as m tends to infinity. Central limit theorem is discussed in [4], [8]. It is stated as follows :

$$m^{1-(\beta/2)}(\mu_j(w(m)) - (\mu_j + 4\pi\alpha\, m^{\beta-1}|M|^{-1}))$$

tends in distribution to Gaussian random variable S_j of mean $E(S_j) = 0$ and variance

$$E(S_j^2) = 4\pi\alpha(\int_M \phi_j(x)^4 dx - |M|^{-1})|M|^{-1}$$

for $V \equiv |M|^{-1}$, $\beta \in [1,12/11)$. Here ϕ_j is the normalized eigenfunction associated to μ_j.

Some readers may ask the following question. "Why do you want to try to study the case $\beta > 1$? Everything is included in the study of the case $\beta = 1$. " The reason why I emphasize consideration of the case $\beta > 1$ lies in the following fact. There occurrs a phenomena of <u>transition of fluctuation of spectra</u> in random media at $\beta = 2 + 0$. We will show below

$(1.2)_j$ $$\mu_j(w(m)) = \mu_j + 4\pi\alpha m^{\beta-1}|M|^{-1} + m^{(\beta/2)-1}S_j + \text{lower term}$$

does not hold for $\beta > 2$. Here we assume $(1.2)_j$ and $(1.2)_{j+1}$. Then, $K_j(c;m)$ determined by

$$P((\mu_j(w(m)) - 4\pi\alpha m^{\beta-1}|M|^{-1}) < c\, m^{(\beta/2)-1} + \mu_j)$$

satisfies

$$K_{j+1}((1+\varepsilon)^{-1}c;m) \leq K_j(c;m)$$

for any fixed $\varepsilon > 0$, $c > 0$, $m \gg 1$, and

$$K_{j+1}((1-\varepsilon)^{-1}c;m) \geq K_j(c;m)$$

for any fixed $\varepsilon > 0$, $c < 0$, $m \gg 1$. Then, by a simple observation we have $E(S_j^2) = E(S_{j+1}^2)$ which does not hold for generic M.

It is of great interest to the author to find a complete statistical property of $\mu_j(w(m))$ for the case $1 < \beta < 3$. For $\beta = 3$, it is well known that $(\mu_j(w(m)) - \mu_j)m^{-2}$ has a distribution which is not concentrated on $4\pi\alpha|M|^{-1}$. See [8], etc.

In Ozawa [10], the reason why $(1.2)_j$ holds for lower $\beta - 1$ is stated. Term

$$\mu_j + 4\pi\alpha\, m^{\beta-1}|M|^{-1} + m^{(\beta/2)-1}S_j$$

comes from sums of independent identically distributed random variables

$$(1.3)_j \qquad \mu_j + (4\pi\alpha/m)\sum_{k=1} \phi_j(w_k)^2$$

for lower $\beta - 1$. When $\beta > 2$, then $(1.3)_j > (1.3)_{j+1}$ may happen. Thus, we need order statistics for this process. See, for example Feller [2] for order statistics. Rearrange $(1.3)_1, (1.3)_2, \cdots$ in order that these are increasing. Therefore, if $\mu_j(w(m))$ is approximated by one of $(1.3)_t$ fairly well, then statistics of $\mu_j(w(m))$ is reduces to an analysis of order statistics of $\{(1.3)_k\}_{k=1}^{\infty}$ However, as in [10] we have other reason why we must modify $(1.3)_j$ to other one.

Finally, we offer an open problem. Prove CLT for $\mu_j(w(m))$ by using the theory of Wiener sausage.

Approximate Green's function. Construction of an approximate Green's function of $-\Delta + T\,m^{\beta-1}$ $(T = \text{const.} > 0)$ in $M \setminus$ n-balls under the Dirichlet condition on $\partial(M \setminus \text{n-balls})$ by using Green's function of $-\Delta + T\,m^{\beta-1}$ in M under the Dirichlet condition on ∂M is a very strong tool to calculate asymptotic properties of $\mu_j(w(m))$. See Ozawa [8]. In the present note we devote ourselves to exhibit L^2 calculus by which we can estimate difference of genuine and approximate Green's function. As a result we have Theorem 1. It should be emphasized that Theorem 1 itself is weaker than Proposition 1 in Ozawa [8]. Recall that our proof of Proposition 1 in [8] need tricky L^p calculus. The author prefers L^2 calculus to L^p calculus. Moreover, we see that L^2 calculus presented in this note is nicely adapted itself to construction of an approximate eigenvalue of Green's function.

It should be remarked that $M\setminus n$ balls may not be connected. We can avoid this technical complication by the following subset $\mathcal{O}_1(m)$ on $w(m) \in M^n$ which assures that the exactly one component of $M_{w(m)}$ is a main part of $M_{w(m)}$. The components other than ω are negligible to study $\mu_j(w(m))$. We introduce $\mathcal{O}_1(m)$ in [8].

$\mathcal{O}_1(m)$: The number of α/m radius ball such that K and ball has a intersection does not exceed $(\log m)^2$ for any open ball K of radius $m^{-\beta/3}$ in R^3.

We see that the measure of the set $w(m)$ in M^n satisfying $\mathcal{O}_1(m)$ tends to 1 as m tends to infinity. $\mathcal{O}_1(m)$ implies the following property $\widetilde{\mathcal{O}}(m)$. See .

$\widetilde{\mathcal{O}}(m)$: Only the one component of $M\setminus n$ balls has a property that its diameter is greater than $3\alpha\, m^{-1} (\log m)^2$.

The eigenvalue problem of $-\Delta + Tm^{\beta-1}$ in ω is transformed into the eigenvalue problem of the Green operator. Put $Tm^{\beta-1} = \lambda$. Main idea of our calculation is to approximate $G_{w(m)}$ by construction of operator $H_{w(m)}$ which is writte by using the centers $w_1,\cdots,w_n$. We put $\tau = 4\pi\alpha \exp(\lambda^{1/2}\alpha/m)$, $m_o = (\log m)^2$. We put

$$h(x,y;w(m)) = G(x,y) + (-\tau/m)\sum_{i=1}^{n} G(x,w_i)G(w_i,y)$$

$$+ \sum_{s=2}^{m_o} (-\tau/m)^s \sum\nolimits_{(s)} G(x,w_{i_1})G_{i_1 i_2}\cdots G_{i_{s-1} i_s} G(w_{i_s},y).$$

Here $G_{ij} = G(w_i,w_j)$ and the indices in $\sum_{(s)}$ run over all $1 \leq i_1, i_2, \cdots, i_s \leq n$ such that $i_\nu \neq i_\mu$ when $\nu \neq \mu$. The sum $\sum_{(s)}$ is called <u>self-avoiding sum</u>. We put

$$H_{w(m)}f(x) = \int_\omega h(x,y;w(m))f(y)dy, \qquad x \in \omega.$$

Our proof of the following Theorem 1 is one of our main theme. Basic importance lies in a way to get Theorem 1.

Theorem 1. Fix T in λ sufficiently large. Fix $\sigma > (2/3)\beta$, $\varepsilon > 0$. Then, the measure of the set satisfying

$$\| H_{w(m)} - G_{w(m)} \|_{2,\omega} \leq C\, m^p,$$

where $p = (3/4)\beta - (7/4) + (\sigma/2) + \varepsilon$ tends to 1 as m tends to infinity.

Some Lemmas. We put

$$J_\xi(\lambda, x) = \sum_r \exp(-\lambda^{1/2}|w_r - x|)|w_r - x|^{-(1-\xi)}.$$

Then, we have the following.

Lemma 1. Fix $\xi > -2$. We assume that $w(m)$ satisfies $\mathcal{O}_1(m)$. Then there exists a constant C independent of m satisfying

$$\sup_{x\in\omega} J_\xi(\lambda, x) \leq C(\log m)^2(\lambda^{-(2+\xi)/2} m^\beta + m^{1-\xi}).$$

Lemma 2. Under the same assumption as in Lemma 1 we have

$$\sup_{x\in M} J_1(\lambda, x) \leq C(\log m)^2(\lambda^{-3/2} m^\beta + 1).$$

Proof of Lemma 1. We put

$$A(k) = \left\{ z\ ;\ k\, m^{-\beta/3} \leq |z - x| \leq (k+1)m^{-\beta/3} \right\}.$$

If $w(m)$ satisfies $\mathcal{O}_1(m)$, then $J(\lambda, x)$ does not exceed

$$C(m^{1-\xi}(\text{the number of } w_r \text{ in } B(m^{-\beta/3}; x))$$

$$+ \sum_{k=1} \sum_{w_r \in A(k)} \exp(-\lambda^{1/2}|w_r - x|)|w_r - x|^{-(1-\xi)})$$

$$\leq C''(\log m)^2(m^{1-\xi} + m^{(\beta/3)(1-\xi)} \sum_{k=1} k^{1+\xi} \exp(-\lambda^{1/2} k m^{-\beta/3}/2))$$

noticing

$$|w_r - x| \geq \alpha/m$$

for any $x \in \omega$. Since

$$\sum_{k=1}^{n} k^{1+\xi} e^{-kt} \leq C\, t^{-2-\xi}$$

for large t, we get the desired result.

We put

$$F_\theta(\lambda,x,y) = |x - y|^{-\theta} \exp(-\lambda^{1/2}|x - y|)$$

and

$$F_\theta^\circ(\lambda,w_r,w_s) = \int_{M\setminus(B_r\cup B_s)} F_\theta(\lambda,x,w_r)F_\theta(\lambda,x,w_s)dx.$$

Here B_r denotes the ball of radius α/m with the center w_r.

Lemma 3. We have $F_1^\circ(\lambda,w_r,w_s) \leq C\lambda^{-1/2}\exp(-\lambda^{1/2}|w_r - w_s|/2)$ and $F_2^\circ(\lambda,w_r,w_s) \leq C|\log(\lambda^{1/2}|w_r - w_s|)|\, F_1(\lambda/4,w_r,w_s)$.

We write $\|f\|_{L^p(A)}$ as $\|f\|_{p,A}$.

Lemma 4. If $w(m)$ satisfies $\mathcal{O}_1(m)$, then

$$\text{(2.2)} \qquad \sum_{i=1}^n Gf(w_i)^2 \leq C\, m^{2-\beta}(\log m)^2 \|f\|_{2,\omega}^2$$

holds for a constant C independent of m, $w(m)$ and $f \in C_0^\infty(\omega)$.

Proof. We see that (2.2) does not exceed

$$\sum_{i=1}^n \int_M G(w_i,y)f(y)^2 dy \int_M G(w_i,y)dy$$

$$\leq C\lambda^{-1} \max_{y\in\omega} \sum_{i=1}^n G(w_i,y)\, \|f\|_{2,\omega}^2.$$

We get the desired result.

Remark. The law of large numbers tells us

$$\frac{1}{n}\sum_{i=1}^n Gf(w_i)^2 \longrightarrow \|Gf\|_{2,M}^2 \leq C\lambda^{-2}\|f\|_{2,M}^2$$

for fixed f. It should be remarked that (2.2) is an estimate for every running $f \in C_0^\infty(\omega)$.

Lemma 5. If $u \in C^\infty(\omega)\cap C^0(\overline{\omega})$ satisfies $(-\Delta + \lambda)u(x) = 0$ for $x \in \omega$, $u(x) = 0$ for $x \in \partial M \cap \partial\omega$ and $M_r = \max_{T_r} |u(x)|$, $r = 1,2,\dots,n$. Here $T_r = B_r \cap \partial\omega$. Here we put $M_r = 0$, $\partial B_r \cap \partial\omega = \phi$.

Assume that $w(m)$ satisfies $\mathcal{O}_1(m)$. Then, there exists a constant C independent of m such that

$$\|u\|_{2,\omega} \leq C m^{-\beta/2}(\log m)(\sum_r M_r^2)^{1/2}.$$

<u>Proof</u>. By the Hopf maximum principle we have

$$|u(x)| \leq C(\alpha/m) \sum_r F_1(\lambda, x, w_r) M_r.$$

Thus, $\|u\|^2_{2,\omega}$ does not exceed

$$C^2(\alpha/m)^2 (\lambda^{-1} \sum_r M_r^2 + \sum_{r,s} M_r M_s K_{rs})$$

by Lemma 3. Here $K_{rs} = F_1^o(\lambda, w_r, w_s)$. Put

$$L_o = \max_r (\sum_s K_{rs}) = \max_s (\sum_r K_{rs}).$$

Then,

$$(\sum_{r,s} M_r M_s K_{rs})^2 \leq (\sum_r M_r^2)(\sum_r (\sum_s M_s K_{rs})^2).$$

Now,

$$\sum_r (\sum_s M_s K_{rs})^2 \leq \sum_r (\sum_s M_s^2 K_{rs} \sum_s K_{rs})$$

$$\leq L_o \sum_{r,s} M_s^2 K_{rs}$$

$$\leq L_o^2 \sum_s M_s^2.$$

Therefore,

$$\sum_{r,s} M_r M_s K_{rs} \leq L_o \sum_r M_r^2.$$

We get the desired result.

<u>Rearrangement of Green's function</u>. We put $Q_{w(m)} = G_{w(m)} - H_{w(m)}$. Then, we have $(-\Delta + \lambda) Q_{w(m)} f(x) = 0$ for $x \in \omega$ and $Q_{w(m)} f(x) = 0$ for $x \in \partial\omega \cap \partial M$ when f belongs to $C_o^\infty(\omega)$. Notice that $Q_{w(m)} f(x) = -H_{w(m)} f(x)$ on T_r.

We put

$$I_r^o f(x) = Gf(x) - (\tau/m) G(x, w_r) Gf(w_r).$$

We abbreviate $G(w_{i_k}, w_{i_h})$ as $G_{i_k i_h}$ for the sake of convienience. We put $I_r^s f(x)$ as

$$\sum\nolimits'_{(s)} G(x, w_{i_1}) G_{i_1 i_2} \cdots G_{i_{s-1} i_s} Gf(w_{i_s})$$

$$- (\tau/m) \sum\nolimits'_{(s)} G(x, w_r) G(w_r, w_{i_1}) \cdots G_{i_{s-1} i_s} Gf(w_{i_s}).$$

Here the indices $i_1, \cdots, i_s$ in $\sum'_{(s)}$ run over all $i_1, \cdots, i_s$ which satisfies $i_\nu \neq i_\mu$ for $\nu \neq \mu$ and $i_\nu \neq r$ for $\nu = 1,\cdots,s$. The following rearrangement of $H_{w(m)}$ is very useful to estimate $Q_{w(m)}$.

$$H_{w(m)}f(x) = \sum_{s=0}^{m_o} (-\tau/m)^s I_r^s f(x) \quad + \quad (-\tau/m)^{m_o} Z^r_{w(m)},$$

where

$$Z^r_{w(m)} = \sum\nolimits'_{(m_o)} G(x,w_{i_1})\cdots Gf(w_{i_{m_o}}).$$

Lemma 6. We have

$$\|Q_{w(m)}f\|_{2,\omega}$$

$$\leq C\, m_o^{1/2} m^{-\beta/2}(\log m)\left\{ \sum_{s=0}^{m_o} (\tau/m)^{2s} \sum_r \max_{T_r} |I_r^s f(x)|^2 \right.$$

$$\left. + (\tau/m)^{2m_o} \sum_r \max_{T_r} (Z^r_{w(m)})^2 \right\}^{1/2}.$$

The following Lemma 7 is given in [6].

Lemma 7. Fix $\beta \in [1,3)$. Assume that $w(m)$ satisfies $\mathcal{O}_1(m)$. Then, there exists a constant C independent of m such that

$$\max_{T_r} |G(x,w_i) - G(w_r,w_i)| \leq C(\alpha/m)|w_i - w_r|^{-2} e^{-\lambda^{1/2}|w_i - w_r|/C}$$

and

$$(\#) \qquad \max_{T_r} |S(x,w_r)G(w_r,w_i)| \leq C(\log m)^2 |w_i - w_r|^{-2} e^{-\lambda^{1/2}|w_i - w_r|/C}$$

hold.

Remark. We do not have any assumption on $|w_i - w_r|$. Thus, w_i and w_r can be close each other.

Lemma 8. In (#) w_i can be replaced by any $y \in M \setminus \overline{B}_r$.

Analysis of $I^o f$. We have the following :

Proposition 1. Assume that $w(m) \in \mathcal{O}_1(m)$. Fix an arbitrary positive constant $\varepsilon > 0$. Then,

$$\sum_r \|I_r^o f\|^2_{\infty,\partial B_r} \leq C\, m^{-1+\varepsilon} \|f\|^2_{2,\omega}$$

holds for a constant C independent of m, f.

Proof. We put $I_r^{oo}f(x) = Gf(x) - Gf(w_r)$, $I_r^{o1}f(x) = (-\tau/m)S(x,w_r) \times Gf(w_r)$. Then, $I_r^{o}f(x) = I_r^{oo}f(x) + I_r^{o1}f(x)$, $x \in \partial B_r$. Thus, Proposition 1 is a consequence of the following inequality which is valid for $w(m) \in \mathcal{O}_1(m)$.

$$\sum_r \|I_r^{ok}f\|^2_{\infty,\partial B_r} \leq C\, m^{-1+\varepsilon}\, \|f\|^2_{2,\omega} \qquad (k = 0,1).$$

Take $f \in C_o^\infty(\omega)$. We put $B_r^o = B(2\alpha/m; w_r)$ and $D_r = (B_r^o)^C \cap \omega$. See Lemmas 7,8. Then, the left hand side of the above inequality does not exceed (when $k = 0$)

$$2\sum_r \max_{\overline{B}_r} G(\chi_{B_r^o} f)(x)^2 + Cm^{-2}\sum_r \Big(\int_{D_r} F_2(\lambda/4, w_r, y) f(y)dy\Big)^2.$$

Here $\chi_{B_r^o}$ is the chracteristic function of the set B_r^o. Since

$$\max_{\overline{B}_r} \int_{B_r^o} G(x,y)^2 dy \leq C\, m^{-1},$$

then the first term in the sum of above two terms does not exceed

$$C\, m^{-1} \sum_r \int f(y)^2\, \chi_{B_r^o}(y)dy.$$

By $\mathcal{O}_1(m)$,

$$\sum_r |\chi_{B_r^o}(y)| \leq C\, (\log m)^2.$$

We have only to estimate the second term to get our proof. The second term does not exceed

$$m^{2\xi-2} \sum_r \|F_{3-\xi}(\lambda/4, w_r, \cdot)\|_{1,D_r} \int_{D_r} F_{1+\xi}(\lambda/4, w_r, y) f(y)^2 dy$$

for any $\xi > 0$. Here we used the fact that $|y - w_r| \geq 2\alpha/m$. We see $F_{3-\xi}$ is integrable on D_r. Thus, we see that the above term is less than $m^{2\xi-2} \max J_{-\xi}(\lambda/4, y)\, \|f\|^2_{2,\omega}$. By Lemma 1 we have a bound for the sum of L^∞ norm term of $I_r^{oo}f$.

We have a similar estimate for $I_r^{o1}f$ term using (#). We get the desired result.

Analysis on I^1f. We put $F_2(\lambda, w_r, w_i) = F_{ri}$. Then, by Lemma 7 we see that

$$\sum_r \|I_r^1 f\|^2_{\infty,\partial B_r} \leq C\, (\log m/m)^2 \mathcal{F}_G^1.$$

where

$$\mathfrak{F}^1_G = \sum_{r=1}^{n}\Big(\sum_{i=1,\neq r}^{n} F_{ri}Gf(w_i)\Big)^2.$$

We have

$$\mathfrak{F}^1_G \leq C\, m^{2-\beta}(\log m)^2 \sum_{\substack{r,i\\ r\neq i}} F_{ri}^2 \, \|f\|^2_{2,\omega}$$

by Lemma 4. Let us introduce the following subset $S_\sigma(m)$ of $w(m)$.
$S_\sigma(m)$: For any w_i, w_j $(i \neq j)$ in $w(m)$, we have

$$|w_i - w_j| \geq C\, m^{-\sigma}.$$

It is easy to check that $\lim_{m\to\infty} P(w(m) \in M^n\ ;\ S_\sigma(m) \text{ holds}) = 1$ when $\sigma > (2/3)\beta$. If we assume $S_\sigma(m)$ $(\sigma > (2/3)\beta)$, then

$$\sum_{r\neq i}^{n} F_{ri}^2 \leq C\, m^{\sigma(1+\xi)} \sum_{r,i} F_{3-\xi}(\lambda, w_r, w_i).$$

Thus, the measure of the set satisfying

$$\sum_{r\neq i}^{n} F_{ri}^2 \leq C\, m^{2\beta+\sigma+\varepsilon}$$

tends to 1 as $m \longrightarrow \infty$.

Summing up these facts we have

Proposition 2. Assume that $w(m) \in \mathcal{O}_1(m)$ and $S_\sigma(m)$. Then,

$$\sum_r \|I^1_r f\|^2_{\infty,\partial B_r} \leq C\, m^{\sigma+\beta+\varepsilon}\|f\|^2_{2,\omega}$$

holds for a constant C independent of m, f, if T in λ is large enough.

Remark. We used $\mathcal{O}_1(m)$ in Lemmas 4,7.

Analysis of $I^s f$ need combinatorial argument. By Lemma 7 we see that

$$\sum_r \|I^s_r f\|^2_{\infty,\partial B_r} \leq C\,(\log m/m)^2\,\mathfrak{F}^s_G,$$

where

$$\mathfrak{F}^s_G = \sum_r \Big({\sum}'_{(s)} F_{ri_1} G_{I(s)}\Big)^2 \|Gf\|^2_{\infty,M}.$$

Here $G_{I(s)} = G_{i_1 i_2} \cdots G_{i_{s-1} i_s}$. We see that the first term in the above formula does not exceed

$$\Big(\sum_{r,i_1} F_{ri_1}^2\Big) \sum_{i_1} \Big(\sum_{i_2,\cdots,i_s} G_{I(s)}\Big)^2.$$

If we assume that $S_\sigma(m)$ holds, then

$$\sum F_{ri_1}^2 \le C\, m^{2\beta+\sigma+\varepsilon}.$$

We expand the term

$$\sum_{i_1} \Big(\sum_{i_2,\cdots} \cdots\Big)^2$$

as

$$\sum_{I,J} G_{IJ},$$

where $G_{IJ} = G_{i_1 i_2} \cdots G_{i_{s-1} i_s} G_{j_1 j_2} \cdots G_{j_{t-1} j_t}$, $t = s$. In the following we want to study a general case where s and t may be distinct.

<u>Definition</u>. If there are exactly q-cuples of $(h(k),p(k))$ ($k = 1, \cdots ,q$) such that $i_{h(k)} = j_{p(k)}$, we say that $(i_1,\cdots,i_s)$ and $(j_1, \cdots ,j_t)$ have q-intersections.

The following Lemma 9 is crucial for our study.

<u>Lemma 9</u>. Assume that $s,t \geq 2$. Assume that $i_1 = j_1$. Assume that G_{IJ} is of q-intersections $(q \geqq 2)$. Then,

$$E(G_{IJ}) \leqq (C\lambda^{-1})^{s+t-(3/2)q-(1/2)}$$

holds.

<u>Proof</u>. We assume that $i_{h(k)} = j_{p(k)}$ for $k = 1,\cdots,q$. Here $h(k)$ $(k = 1,\cdots,q)$ is a sequence satisfying $h(k) < h(k+1)$. Then, there is a permutation σ on $(1,\cdots,q)$ such that $p(\sigma(k)) < p(\sigma(k+1))$, $k=1,\cdots,q-1$. For the sake of simplicity, we write $p(\sigma(k))$ as $r(k)$. We put $d(k) = h(k) - h(k-1)$, $d_\sigma(k) = r(k) - r(k-1)$. It is convenient to write $w_{i_{h(k)}}$ as $\overline{h(k)}$ and $w_{j_{p(k)}}$ as $\underline{p(k)}$. Thus,

$$G(w_{i_1}, w_{j_{p(k)}})$$

is written as $G(\overline{1},\underline{p(k)})$. We use a notational conention $G_o(z,w) \equiv 1$.

We treate the case $q \geqq 3$. Integrate G_{IJ} by the probability measure associated with the variables other than $\overline{h(2)},\overline{h(3)},\cdots,\overline{h(q)}$ and $\underline{r(2)},\underline{r(3)},\cdots,\underline{r(q)}$. Notice that $w_{i_1} = w_{j_1}$, $\overline{h(1)} = \underline{r(1)} = 1$. Then, we get $S(1)S(2)S(3)$, where

$$S(1) = G_{h(2)+r(2)-2}(\overline{h(2)},\underline{r(2)})$$

$$S(2) = G_{s-h(q)}(\overline{h(q)},\overline{s}) \prod_{k=3}^{q} G_{d(k)}(\overline{h(k-1)},\overline{h(k)})$$

and

$$S(3) = G_{t-r(q)}(\underline{r(q)},\underline{t}) \prod_{k=3}^{q} G_{d_\sigma(k)}(\underline{r(k-1)},\underline{r(k)}).$$

We have

$$E(S(1)S(2)S(3)) \leq E(S(1)S(2)^2)^{1/2}E(S(1)S(3)^2)^{1/2}.$$

We consider $E(S(1)S(2)^2)$. We put $G_1(x,y) = G(x,y)$ and

$$G_{k+1}(x,y) = \int_M G_k(x,z)V(z)G(z,y)dz.$$

Note that

$$g_u = \max_{y \in M} \int_M G_u(x,y)^2 V(x)dx$$

satisfy $g_u \leqq (C\lambda^{-1})^{2u-(3/2)}$. We also note that

$$g_u^\circ = \max_{y \in M} \int_M G_u(x,y)V(x)dx \leqq (C\lambda^{-1})^u.$$

Therefore,

$$E(S(1)S(2)^2) \leqq (C\lambda^{-1})^{h(2)+r(2)-2}(g_{s-h(q)} \prod_{k=3}^{q} g_{d(k)}).$$

By a simple calculation we have the desired result.

We treate the case $q = 2$. We have

$$E(G_{IJ}) = E(G_{h(2)+r(2)-2}(\overline{h(2)},\underline{r(2)})G_{s-h(2)}(\overline{h(2)},\overline{s})$$
$$G_{t-r(2)}(\underline{r(2)},\underline{t})).$$

Use the inequality $\max G_p(x,y) \leqq (C\lambda^{-1})^{p-(3/2)}$ when $t = r(2)$. Then, we get the proof. We can get the same desired result when $t \neq r(2)$.

<u>Proposition 3</u>. Fix $s \geqq 2$. Then, the measure of the set satisfying

$$\sum_r ||I_r^s f||^2_{\infty,\partial B_r} \leqq C\, m^{2s+\sigma+(5/2)\beta-(7/2)+\varepsilon} ||f||^2_{2,\omega}$$

tends to 1 as m tends to infinity, if T is sufficiently large and $\sigma > (2/3)\beta$.

<u>Proof of Proposition 3</u>. It is easy to get the above Proposition 3 observing

$$||Gf||_{\infty,M} \leqq C\lambda^{-1/4} ||f||_{2,M}$$

and the fact that the indices I, J in $\sum_{I,J} G_{IJ}$ is at least of q-intersections $(q \geqq 1)$.

Proof of Theorem 1 is now easy. By Lemma 6 we get the desired result, if we take T in λ sufficiently large.

References.

[1] I. Chavel and E. A. Feldman, The Wiener sausage and a theorem of Spitzer in Riemannian manifolds, Probability and harmonic analysis, J. Chao and W. A. Woyczynski, eds., Marcel Dekker Inc., (1986), 45-60.

[2] W. Feller, An introduction to probability theory and its applications, II, John Wiley and Sons, Inc., 1966.

[3] E. Ja. Huruslov and V. A. Marchenko, Boundary value problems in regions with fine grained boundaries. (in Russian) Kiev 1974.

[4] R. Figari, E. Orlandi and S. Teta, The Laplacian in regions with many obstacles: Fluctuations around the limit operator, J. Statistical Phys., 41 (1985), 465-487.

[5] M. Kac, Probabilistic methods in some problems of scattering theory, Rocky Mountain J. Math., 4 (1974), 511-538.

[6] S. Ozawa, Random media and eigenvalues of the Laplacian, Commun. Math. Phys., 94 (1984), 421-437.

[7] ________, Point interaction approximation for $(-\Delta + U)^{-1}$ and eigenvalues of the Laplacian on wildly perturbed domain, Osaka J. Math. 20 (1983), 923-937.

[8] ________, Fluctuation of spectra in random media, Proceedings of the Taniguchi symposium "Probabilistic methods in mathematical physics" eds.,by N. Ikeda and K. Ito, Kinokuniya, (1987), 335-361.

[9] ________, Construction of approximate eigenfunction in disordered media. in prep.

[10] ________, Mathematical study of spectra in random media, in Hydrodynamic behaviour and infinite interacting particle system, IMA series in mathematics, Springer 1987.

[11] J. Rauch and M. Taylor, Potential and scattering theory on wildly perturbed domains, J. Funct. Anal. 18 (1975), 27-59.

[12] A. S. Sznitman, Some bounds and limiting results for the measure of Wiener sausage of small radius associated to elliptic diffusions, Stochastic processes and their applications., 25, (1987), 1-25.

[13] S. Weinryb, Etude asymptotique de l'image par des measures de certains ensembles aleatoires liés à la courbe Brownienne, Prob. Theory Rel. Fields 73 (1986), 135-148.

Lower bounds of the essential spectrum of the Laplace-Beltrami operator and its application to complex geometry

Ken-ichi SUGIYAMA

§0. Introduction

In the present paper, we shall investigate the relation between holomorphic vector bundle valued L^2-cohomology and the infimum of the essential spectrum of its Laplace-Beltrami operator. To be more precise, let us give some notations.

Let (M,g) be an n-dimensional non-compact complete Hermitian manifold and $\{E,a\}$ a holomorphic vector bundle over M with a Hermitian fibre metric a. Now we consider a complex;

$$(0.1) \quad 0 \longrightarrow D^{p,0}(\bar\partial) \longrightarrow D^{p,1}(\bar\partial) \longrightarrow \cdots \longrightarrow D^{p,n}(\bar\partial) \longrightarrow 0,$$

where $D^{p,q}(\bar\partial) := \{\phi \in L^2(M,\Lambda^{p,q}(E)) \mid \bar\partial\phi \in L^2(M,\Lambda^{p,q+1}(E))\}$. Here L^2 means square integrable with respect to the metrics g and a. Then <u>E-valued L^2-cohomology</u> $H_2^{p,q}(M_g,E_a)$ with respect to metrics g and a is defined to be the cohomology of the complex (0.1). Let Δ_a be the Laplace-Beltrami operator of $\{E,a\}$. When the dimension of $H_2^{p,q}(M_g,E_a)$ is finite, it is known that $H_2^{p,q}(M_g,E_a)$ is isomorphic to $\mathrm{Ker}\{\Delta_a : L^2(M,\Lambda^{p,q}(E)) \longrightarrow L^2(M,\Lambda^{p,q}(E))\}$. Therefore, in the case $\dim H_2^{p,q}(M_g,E_a)$ is finite, we will be able to obtain various informations about E-valued L^2-cohomology just as we obtained much results using the harmonic integral theory in case that the base manifold M is compact. Then, what is a sufficient condition for $\dim H_2^{p,q}(M_g,E_a)$ to be finite?

One of the answers to this question is as follows. Let

$$\gamma_{p,q} := \inf \sigma_{ess}\{\Delta_a : L^2(M,\Lambda^{p,q}(E)) \longrightarrow L^2(M,\Lambda^{p,q}(E))\}.$$

and assume that there exists an integer m, and that $\gamma_{p,0},\dots,\gamma_{p,m}$ are positive. Then, approximating the complex(0.1) by certain finite complex, we shall show that $\dim H_2^{p,q}(M_g,E_a)$ is finite for $0 \leqq q \leqq m-1$.(See Theorem(1.22)). Therefore the problem becomes when $\gamma_{p,q}$ is positive. In section 2, we shall prove the following formula (See Theorem(2.1));

$$(0.2)\quad \gamma_{p,q} = \sup_{K\subset\subset M} \inf_{\phi\in C_0^\infty(M\setminus K,\Lambda^{p,q}(E)),\|\phi\|_2=1} \langle\phi,\Delta_a\phi\rangle.$$

This is an analogue of the Persson's formula for Schrödinger operator. When E is a holomorphic line bundle, observe that, using Bochner-Weizenböck formula, the right hand side of (0.2) is written in terms of the Ricci form of $\{E,a\}$. In fact, in section 3, due to this observation, we shall show certain criterions so that the dimesion of E-valued L^2-cohomology should be finite, or should vanish. In section 4, we shall generalize the Demailly's results to the case the base manifold is non-compact complete Kähler manifold.

§1. L^2-cohomology and the essential spectrum of Δ_a

Let (M,g) be an n-dimensional complex manifold with a Hermitian metric g, and let $\{E,a\}$ a holomorphic vector bundle with a Hermitian fibre metric a. In this section, we always assume that the metric g is complete. Firstly, we give some notations which will be used all in this paper.

(1,1)Notations.

$\Omega^{p,q}(M,E) := \{$ all E-valued C^∞ (p,q)-forms over $M \}$,

$\Omega^{p,q}(M,E)_0 := \{ u \in \Omega^{p,q}(M,E) \mid u$ is compactly supported $\}$,

$\Omega^{p,q}_2(M_g,E_a) := \{$ all E-valued L^2 (p,q)-forms over M with respect to the metrics g and $a \}$.

(1,2)Definition. We extend the differential operator $\bar\partial$ to the operator from $\Omega^{p,q}_2(M_g,E_a)$ to $\Omega^{p,q+1}_2(M_g,E_a)$ with its domain $D^{p,q}(\bar\partial) := \{ u \in \Omega^{p,q}_2(M_g,E_a) \mid \bar\partial u \in \Omega^{p,q+1}_2(M_g,E_a) \}$, and the extension will be denoted by $\bar\partial$ again.

(1,3)Definition. Let the differential operator ϑ_a be the formally adjoint of $\bar\partial$ with respect to the metrics g and a. Then we extend ϑ_a to the operator from $\Omega^{p,q}_2(M_g,E_a)$ to $\Omega^{p,q-1}_2(M_g,E_a)$ with its domain $D^{p,q}(\vartheta_a) := \{ u \in \Omega^{p,q}_2(M_g,E_a) \mid \vartheta_a u \in \Omega^{p,q-1}_2(M_g,E_a) \}$, and the extension will be denoted by ϑ_a again.

In what follows, both $\bar\partial$ and ϑ_a always denote the extended operator. Note that the operators $\bar\partial$ and ϑ_a are closed.

Let $u \in \Omega^{p,q}(M,E)_0$. Then the Laplace-Beltrami operator Δ_a of $\{E,a\}$ is defined as

(1.4) $\quad \Delta_a u := (\vartheta_a \bar\partial + \bar\partial \vartheta_a) u.$

We extend Δ_a to the operator from $\Omega^{p,q}_2(M_g,E_a)$ to $\Omega^{p,q}_2(M_g,E_a)$ with its domain $D^{p,q}(\Delta_a) := \left\{ u \in \Omega^{p,q}_2(M_g,E_a) \mid \Delta_a u \in \Omega^{p,q}_2(M_g,E_a) \right\}$, and the extension will be denoted by Δ_a again. In what follows, Δ_a always denotes the extended operator. Now since (M,g) is complete, the following lemma(1.5), and (1.6) hold. The proof of the lemmas seem to be very standard, and maybe well-known, hence we omit it.

(1.5)Lemma. Let $u \in D^{p,q}_2(\Delta_a)$. Then $u \in D^{p,q}(\bar\partial) \cap D^{p,q}(\vartheta_a)$, and moreover $< \Delta_a u, u > = \|\bar\partial u\|_2^2 + \|\vartheta_a u\|_2^2$, where $< \, , \, >$ denotes the inner product of $\Omega^{p,q}_2(M_g,E_a)$ and $\| \;\|$ denotes its norm.

(1.6)Lemma. The operator Δ_a is self-adjoint. Moreover $u \in D^{p,q}(\Delta_a)$ implies $\bar\partial u \in D^{p,q+1}(\vartheta_a)$ and $\vartheta_a u \in D^{p,q-1}(\bar\partial)$, and the identity $\Delta_a u = (\vartheta_a \bar\partial + \bar\partial \vartheta_a) u$ holds.

Now, for any $0 \leqq p,q \leqq n$, we define a positive number $\gamma_{p,q}$ to be

(1.7) $\quad \gamma_{p,q} := \inf \sigma_{ess}\left\{\Delta_a : \Omega^{p,q}_2(M_g,E_a) \longrightarrow \Omega^{p,q}_2(M_g,E_a)\right\},$

where $\inf \sigma_{ess}$ means the infimum of the essential spectrum of the operator. When $\gamma_{p,q} > 0$, we take a positive number δ so that $0 < \delta < \lambda_1\left\{\Delta_a : \Omega^{p,q}_2(M_g,E_a) \longrightarrow \Omega^{p,q}_2(M_g,E_a)\right\}$, where λ_1 denotes the non-zero first eigenvalue of the operator, and we define the Green's operator G_a of Δ_a as

(1.8) $G_a := \int_\delta^\infty \lambda^{-1}\, dP_\lambda$

where P is the projection valued measure associated to Δ_a. Then, from the definition, G_a is a bounbed operator and its range is contained in the domain of Δ_a, and moreover $I_d = P_{p,q}(0) + \Delta_a G_a$, where $P_{p,q}(0)$ means the harmonic projection. Now we define a operator

(1.9) $G_{p,q}: D^{p,q}(\bar\partial) \longrightarrow D^{p,q}(\bar\partial)$

by the commutative diagram

$$\begin{array}{ccc} D^{p,q}(\bar\partial) & \longrightarrow & \Omega^{p,q}_2(M_g, E_a) \\ G_{p,q}\downarrow & & \downarrow G_a \\ D^{p,q}(\bar\partial) & \longleftarrow & D^{p,q}(\Delta_a) \end{array} \quad (1.10)$$

Let $\Omega \subset \mathbb{R}$ be a bounded Borel measurable set. Then we define a bounded operator

(1.11) $P_{p,q}(\Omega): D^{p,q}(\bar\partial) \longrightarrow D^{p,q}(\bar\partial)$

by the commutative diagram

$$\begin{array}{ccc} D^{p,q}(\bar\partial) & \longrightarrow & \Omega^{p,q}_2(M_g, E_a) \\ P_{p,q}(\Omega)\downarrow & & \downarrow P_\Omega \\ D^{p,q}(\bar\partial) & \longleftarrow & D^{p,q}(\Delta_a). \end{array} \quad (1.12)$$

Now, since $\Delta_a\bar\partial = \bar\partial\Delta_a$, we obtain the following lemma.

(1.13)Lemma. Assume that $\min\left\{\gamma_{p,q}, \gamma_{p,q+1}\right\} > 0$, and let $0 \leqq \lambda < \min\left\{\gamma_{p,q}, \gamma_{p,q+1}\right\}$. Then we obtain the following commutative diagram;

$$\begin{array}{ccc} D^{p,q}(\bar\partial) & \xrightarrow{\bar\partial} & D^{p,q+1}(\bar\partial) \\ P_{p,q}([0,\lambda])\downarrow & & \downarrow P_{p,q+1}([0,\lambda]) \\ D^{p,q}(\bar\partial) & \xrightarrow[\bar\partial]{} & D^{p,q+1}(\bar\partial) \end{array} \quad (1.14)$$

$$\begin{array}{ccc} D^{p,q}(\bar\partial) & \xrightarrow{\bar\partial} & D^{p,q+1}(\bar\partial) \\ G_{p,q}\downarrow & & \downarrow G_{p,q+1} \\ D^{p,q}(\bar\partial) & \xrightarrow[\bar\partial]{} & D^{p,q+1}(\bar\partial) \end{array} \tag{1.15}$$

As a consequence, we can show the following proposition.

(1.16)Proposition. Assume that there exists an integer m, and that $0 < \min\left\{\gamma_{p,0},\ldots,\gamma_{p,m}\right\}$. And choose $\lambda \in \mathbb{R}$ arbitraly so that $0 < \lambda < \min\left\{\gamma_{p,0},\ldots,\gamma_{p,m}\right\}$. Then, for any $0 \leqq q \leqq m-1$, there exists a linear operator

$$\Phi_q(\lambda):D^{p,q+1}(\bar\partial) \longrightarrow D^{p,q}(\bar\partial)$$

such that

$$\mathrm{Id} - P_{p,0}([0,\lambda]) = \Phi_0(\lambda)\bar\partial, \tag{1.17}$$

and

$$\mathrm{Id} - P_{p,q}([0,\lambda]) = \bar\partial\Phi_{q-1}(\lambda) + \Phi_q(\lambda)\bar\partial \quad \text{for } 1 \leqq q \leqq m-1. \tag{1.18}$$

Proof. Since $\mathrm{Id} = P_{p,q}(0) + \Delta_a G_{p,q}$, we obtain

$$\begin{aligned} \mathrm{Id} - P_{p,q}([0,\lambda]) &= (P_{p,q}(0)+\Delta_a G_{p,q})(\mathrm{Id}-P_{p,q}([0,\lambda])) \\ &= \Delta_a G_{p,q}(\mathrm{Id}-P_{p,q}([0,\lambda]) \\ \text{(from (1.6))} \quad &= (\bar\partial\vartheta_a+\vartheta_a\bar\partial)G_{p,q}(\mathrm{Id}-P_{p,q}([0,\lambda]) \end{aligned}$$

(from (1.14) and (1.15))

$$= \bar\partial[\vartheta_a G_{p,q}(\mathrm{Id}-P_{p,q}([0,\lambda]))] + [\vartheta_a G_{p,q+1}(\mathrm{Id}-P_{p,q+1}([0,\lambda]))]\bar\partial.$$

Now set

$$\Phi_q(\lambda) := \vartheta_a G_{p,q+1}(\mathrm{Id}-P_{p,q+1}([0,\lambda])).$$

Q.E.D.

We consider a complex

(1.19) $0 \longrightarrow D^{p,0}(\bar\partial) \xrightarrow{\bar\partial} D^{p,1}(\bar\partial) \longrightarrow \dots \xrightarrow{\bar\partial} D^{p,n}(\bar\partial) \longrightarrow 0,$

and we define the E-valued L^2-cohomology $H^{p,q}_2(M_g,E_a)$ with respect to the metrics g and a as

(1.20) $H^{p,q}_2(M_g,E_a) := H^q(D^{p,*}(\bar\partial),\bar\partial).$

Let λ be a real number in (1.16) and let $P_{p,q}(\lambda) := \mathrm{Im}\, P_{p,q}([0,\lambda])$. Then, from (1.14), $\bar\partial P_{p,q}(\lambda) \subset P_{p,q+1}(\lambda)$, and therefore we obtain a finite dimensional complex

(1.21) $0 \longrightarrow P_{p,0}(\lambda) \xrightarrow{\bar\partial} \dots \xrightarrow{\bar\partial} P_{p,m}(\lambda) \longrightarrow 0.$

Now, from (1.16), we finally obtain the following theorem.

(1.22)Theorem. Assume that there exists an integer m, and that $0 < \min\left\{\gamma_{p,0},\dots,\gamma_{p,m}\right\}$. Choose a real number λ arbitraly so that $0 < \lambda < \min\left\{\gamma_{p,0},\dots,\gamma_{p,m}\right\}$. Then

$$H^{p,q}_2(M_g,E_a) \simeq H^q(P_{p,*}(\lambda),\bar\partial)$$
$$\simeq \mathrm{Ker}\left\{\Delta_a : \Omega^{p,q}_2(M_g,E_a) \longrightarrow \Omega^{p,q}_2(M_g,E_a)\right\}$$

for $0 \leqq q \leqq m-1$. In particular, $H^{p,q}_2(M_g,E_a)$ is finite dimensional for $0 \leqq q \leqq m-1$.

§2. Persson's formula for Hermitian vector bundles.

In this section, we shall prove the following formula.

(2.1)Theorem. Let (M,g) be an n-dimensional non-compact complete Hermitian manifold and let $\{E,a\}$ be a Hermitian vector bundle over M. Then, for any $0 \leqq p,q \leqq n$, we have the following formula;

$$\gamma_{p,q} = \sup_{K \subset\subset M} \inf_{\phi \in C_0^\infty(M \setminus K, \Lambda^{p,q}(E)),\ \|\phi\|_2=1} \langle \phi, \Delta_a \phi \rangle,$$

where $\Lambda^{p,q}(E)$ denotes a vector bundle $\Lambda^{p,q} \otimes E$ over M.

In the proof of the theorem (2.1), we shall often use the following fact.

(2.2)Fact(Weyl's criterion). Let H be a Hilbert space over $\mathbb{C}$, and let $A: H \longrightarrow H$ be a self-adjoint operator. Then a real number ρ is an element of the essential spectrum of A if and only if there exists a sequence $\{\psi_m\}_{m=1}^\infty$, $\psi_m \subset D(A)$ such that $\|\psi_m\| = 1$, $\psi_m \xrightarrow{w} 0$, and that $\lim_{m\to\infty} \|(A-\rho)\psi_m\| = 0$.

Firstly, we shall prove the following lemma.

Lemma(2.3). Let $\{\psi_m\}_{m=1}^\infty \subset \Omega^{p,q}(M,E)_0$ be a sequence such that $\|\psi_m\|_2=1$ for any m, and that $\psi_m \xrightarrow{w} 0$, $\|(\Delta_a-\lambda)\psi_m\|_2 \longrightarrow 0$ as $m \longrightarrow \infty$. Then for any $f \in C_0^\infty(M,\mathbb{R})$, $\|(\Delta_a-\lambda)(f\psi_m)\|_2 \longrightarrow 0$ and $\|f\psi_m\|_2 \longrightarrow 0$ as $m \longrightarrow \infty$.

Proof. Before prooving the lemma, we prove the following claim.

(2.4)Claim. $\|f\psi_m\|_2 \longrightarrow 0$ as $m \longrightarrow \infty$.

Proof of (2.4). Let fix a positive number β, then, since Δ_a is a positive operator, there exists a bounded operator

$$(\Delta_a+\beta)^{-1}: \Omega^{p,q}_2(M_g,E_a) \longrightarrow \Omega^{p,q}_2(M_g,E_a),$$

and moreover, for any $\chi \subset C_0^\infty(M,\mathbb{R})$, $\chi(\Delta_a+\beta)^{-1}$ is a compact operator. Since

$$f\psi_m = f(\Delta_a+\beta)^{-1}(\Delta_a+\beta)\psi_m = f(\Delta_a+\beta)^{-1}(\Delta_a-\lambda)\psi_m + f(\Delta_a+\beta)^{-1}(\lambda+\beta)\psi_m,$$

we have

$$\|f\psi_m\|_2 \leqq \|f(\Delta_a+\beta)^{-1}(\Delta_a-\lambda)\psi_m\|_2 + |\lambda+\beta|\,\|f(\Delta_a+\beta)^{-1}\psi_m\|_2.$$

Now, from the assumption and the observation above, we obtain

$$\|f\psi_m\|_2 \longrightarrow 0 \quad \text{as} \quad m \longrightarrow \infty. \qquad \text{Q.E.D.}$$

We continue the proof of the lemma. Note that a simple computation yields

$$(2.5) \qquad \|\Delta_a(f\psi_m)-f(\Delta_a\psi_m)\|^2 \leqq \chi_f\left\{ \|\bar\partial\psi_m\|^2 + \|\vartheta_a\psi_m\|^2 + \|\psi_m\|^2\right\},$$

where χ_f is a compactly supported non-negative C^∞ function depending only on f. Therefore

$$(2.6) \qquad \|(\Delta_a-\lambda)(f\psi_m)\|_2 \leqq \|\Delta_a(f\psi_m)-f(\Delta_a\psi_m)\|_2 + \|f(\Delta_a-\lambda)\psi_m\|_2$$

(from (2.5))

$$\leqq \left[\int_M \chi_f\left\{ \|\bar\partial\psi_m\|^2 + \|\vartheta_a\psi_m\|^2 + \|\psi_m\|^2\right\}\right]^{\frac{1}{2}} + \sup_{x\in M}|f(x)|\; \|(\Delta_a-\lambda)\psi_m\|_2.$$

(2.7)Claim. For any non-negative $\chi \subset C_0^\infty(M,\mathbb{R})$,

$$\int_M \chi\left\{ \|\bar\partial\psi_m\|^2 + \|\vartheta_a\psi_m\|^2\right\} \longrightarrow 0 \qquad \text{as} \quad m \longrightarrow \infty.$$

Proof of (2.7). Firstly note that, since

$$\|\chi(\Delta_a-\lambda)\psi_m\|_2 \leqq \sup_{x\in M}|\chi(x)|\,\|(\Delta_a-\lambda)\psi_m\|_2,$$

$$\|\chi(\Delta_a-\lambda)\psi_m\|_2 \longrightarrow 0 \quad \text{as} \quad m \longrightarrow \infty.$$

On the other hand, we have

$$<\chi(\Delta_a-\lambda)\psi_m,\psi_m> = <(\Delta_a-\lambda)\psi_m,\chi\psi_m> = <\Delta_a\psi_m,\chi\psi_m> - \lambda<\psi_m,\chi\psi_m>$$
$$= <\bar\partial\psi_m,\bar\partial(\chi\psi_m)> + <\vartheta_a\psi_m,\vartheta_a(\chi\psi_m)> - \lambda<\psi_m,\chi\psi_m>$$
$$= <\bar\partial\psi_m,\chi\bar\partial\psi_m> + <\bar\partial\psi_m,\bar\partial\chi\wedge\psi_m> + <\vartheta_a\psi_m,\chi\vartheta_a\psi_m> + <\vartheta_a\psi_m,\iota(\bar\partial\chi)\psi_m>$$
$$- \lambda<\psi_m,\chi\psi_m>,$$

where ι denotes the inner product. Since

$$|<\bar\partial\psi_m,\bar\partial\chi\wedge\psi_m>|^2 \leqq \|\bar\partial\psi_m\|_2^2\,\|\bar\partial\chi\wedge\psi_m\|_2^2 \leqq <\Delta_a\psi_m,\psi_m>\,\|\bar\partial\chi\wedge\psi_m\|_2^2$$
$$= <(\Delta_a-\lambda)\psi_m,\psi_m>\,\|\bar\partial\chi\wedge\psi_m\|_2^2 + \lambda\|\psi_m\|_2\|\bar\partial\chi\wedge\psi_m\|_2^2,$$

and since $\|\psi_m\|_2=1$, $\|(\Delta_a-\lambda)\psi_m\|_2 \longrightarrow 0$, using (2.4), we obtain

$$\lim_{m\to\infty} |<\bar\partial\psi_m,\bar\partial\chi\wedge\psi_m>| = 0.$$

By the similar way, we obtain

$$\lim_{m\to\infty} |<\vartheta_a\psi_m,\iota(\bar\partial\chi)\psi_m>| = 0.$$

Now by the first observation and by (2.4), and by non-negativity of χ, we obtain the required claim. Q.E.D.

Now, the lemma is obvious from claim(2.4), claim(2.7), and (2.6).

Q.E.D.

(2.8)Proposition. Let $K_1 \subset\subset K_2 \subset\subset \dots \subset\subset M$ be a compact exhaustion of M. Then, for any real number λ, the following (2.9) and (2.10) are equivalent.

(2.9) $\lambda \in \sigma_{ess}\left\{\Delta_a:\ \Omega^{p,q}_2(M_g,E_a) \longrightarrow \Omega^{p,q}_2(M_g,E_a)\right\}$.

(2.10) There exists a sequence $\left\{\phi_m \subset \Omega^{p,q}(M\backslash K_m,E)_0\right\}_{m=1}^{\infty}$ such that $\|\phi_m\|_2=1$, and that $\|(\Delta_a-\lambda)\phi_m\|_2 \longrightarrow 0$ as $m \longrightarrow \infty$.

Proof. It is obvious that, from Weyl's criterion, (2.10) implies (2.9). Therefore we show that (2.9) implies (2.10). Let λ be a element of the essential spectrum. Then, by Weyl's criterion again, there exists a sequence $\left\{\psi_m \in \Omega^{p,q}(M,E)_0\right\}_{m=1}^{\infty}$ such that $\|\psi_m\|_2=1$, $\psi_m \xrightarrow{w} 0$, and that $\|(\Delta_a-\lambda)\psi_m\|_2 \longrightarrow 0$ as $m \longrightarrow \infty$. Now, for any m, we take $\chi_m \in C^{\infty}(M:[0,1])$ so that $\chi_m(x) = 1$ if $x \in M\backslash K_{m+1}$, and that $\chi_m(x) = 0$ if $x \in K_m$. Then, from the assumption of $\left\{\psi_m\right\}_{m=1}^{\infty}$ and lemma(2.3), we obtain a subsequence $\left\{\psi_{i(m)}\right\}_{m=1}^{\infty}$ of $\left\{\psi_m\right\}_{m=1}^{\infty}$ satisfying $\|(\Delta_a-\lambda)\psi_{i(m)}\|_2 \leqq m^{-1}$, $\|(\Delta_a-\lambda)(1-\chi_m)\psi_{i(m)}\|_2 \leqq m^{-1}$, and $\|(1-\chi_m)\psi_{i(m)}\|_2 \leqq m^{-1}$, and therefore

$$\begin{aligned}\|(\Delta_a-\lambda)(\chi_m\psi_{i(m)})\|_2 &\leqq \|(\Delta_a-\lambda)(1-\chi_m)\psi_{i(m)}\|_2+ \|(\Delta_a-\lambda)\psi_{i(m)}\|_2 \\ &\leqq 2m^{-1}.\end{aligned}$$

Moreover, note that

$$\|\chi_m\psi_{i(m)}\|_2 \geqq \|\psi_{i(m)}\|_2 - \|(1-\chi_m)\psi_{i(m)}\|_2 \geqq 1-m^{-1}.$$

Now we define, for $m \geqq 2$,

$$\phi_m := \left\{\chi_m\psi_{i(m)}\right\}/\|\chi_m\psi_{i(m)}\|_2.$$

Then, from the observation above, it follows that

$$\lim_{m\to\infty} \|(\Delta_a-\lambda)\phi_m\|_2 = 0, \ \phi_n \in \Omega^{p,q}(M\backslash K_m,E)_0, \text{ and that } \|\phi_m\|_2=1.$$

Q.E.D.

Now the theorem (2.1) can be proved in completely the same way as (Theorem 3.12 of [1]), hence we omit the proof.

§3. Finiteness and vanhishing theorem of L^2-cohomology of holomorphic line bundles

In this section, we shall apply the theorem(2.1) to prove finiteness and vanishing theorems of L^2-cohomology of holomorphic line bundles of certain type. Firstly, we shall prove a finiteness theorem of L^2-cohomology.

(3.1)Theorem. Let (M,g) be an n-dimesional non-compact Hermitian manifold and let $\{L,h\}$ be a holomorphic line bundle over M with a Hermtian fibre metric h. If there exists a compact subset K of M such that

(3.2) $\quad -c_1(L,h)(x) > 0 \quad$ for any $x \in M\backslash K$,

and that

(3.3) $\quad \Phi := -c_1(L,h)+\lambda\chi_K\omega_g \quad$ gives a complete Hermitian metric of M

for some positive number λ and $\chi_K \in C_0^\infty(M:[0,1])$, where ω_g is the fundamantal form of g. Then $\dim H^{p,q}_2(M_\Phi,L_h) < \infty$ for $p+q<n-1$.

Proof. Let K_1 be the support of χ_K. Note that, on $M\backslash K_1$, Φ is a Kähler metric. Then, by the Kodaira-Nakano's identity(see, for instance, [5]), we obtain

$$\langle\Delta_h\phi,\phi\rangle - \langle\Delta_h(*\phi),*\phi\rangle = \Big\{n-(p+q)\Big\}\|\phi\|_2^2 \quad \text{for any } \phi \in \Omega^{p,q}(M\backslash K_1,L)_0.$$

Therefore, in paticular, we have

$$\langle\Delta_h\phi,\phi\rangle \geqq \Big\{n-(p+q)\Big\}\|\phi\|_2^2 \quad \text{for any } \phi \in \Omega^{p,q}(M\backslash K_1,L)_0.$$

Now, from theorem (2.1),

$$\gamma_{p,q} = \sup_{K\subset\subset M}\ \inf_{\phi\in C_0^\infty(M\backslash K,\Lambda^{p,q}(L)),\ \|\phi\|_2=1} \langle\phi,\Delta_h\phi\rangle$$

$$\geq \inf_{\phi\in C_0^\infty(M\backslash K_1, \Lambda^{p,q}(L)),\ \|\phi\|_2=1} \langle\phi, \Delta_h\phi\rangle$$
$$\geq n-(p+q),$$

and therefore we obtain the theorem by (1.22). Q.E.D.

(3.4)Example. Let M be an n-dimesional non-compact complex manifold with a Hermitian metric g, and moreover assume that M is strongly pseudo-convex, namely assume that there exists a strictly plurisubharmonic exhaustion function ϕ on M. Now we consider the trivial line bundle $L:=M\times\mathbb{C}$ with a Hermitian fibre metric $h:=e^{\rho(\phi)}$ where $\rho\in C^\infty(\mathbb{R},\mathbb{R})$ is a monotone increasing, convex function. If we take ρ so that $\rho(r)$, $\rho'(r)$, and $\rho''(r)$ grow sufficiently fast to infinity as $r\uparrow\infty$, it is easy to see that (M,g) and $\{L,h\}$ satisfies the assumption of (3.3).

Assume that the first chern form $c_1(L,h)$ of a hermitian line bundle of $\{L,h\}$ is negative semi-definite everywhere and moreover $\mathrm{rk}\ c_1(L,h)(x) \geq k$ for any $x\in M$. Let $\{\gamma_i\}_{i=1}^n$ be the eigenvalues of $c_1(L,h)(x)$ and satisfy $\gamma_1\leq,\ldots,\leq\gamma_n\leq 0$. Note that, from the assumption, γ_k is negatine. Then we define a continuos function λ_k with negative values by $\lambda_k(x):= \gamma_k$. Now we have

(3.5)Vanishing theorem. Let (M,g) be an n-dimensional non-compact complete Kähler manifold and let $\{L,h\}$ be a holonorphic line bundle over M. Assume that there exists positive numbers α and β such that

(3.6) $-\alpha\omega_g(x) \leq c_1(L,h)(x) \leq 0$ for any $x\in M$, where ω_g is the fundamental form of g, and

(3.7) $\mathrm{rk}\ c_1(L,h)(x) \geq k$ and $\lambda_k(x) \leq -\beta$ for any $x\in M$.

Then we have

$$H^{p,q}_2(M_g,L_h) = 0 \qquad \text{for} \quad p+q \leqq k-2.$$

Proof. For any non-negative number μ, we set $\Phi_\mu := \omega_g - \mu c_1(L,h)$. Then, by the assumption (3.6), (3.7), and by Gigante's lemma(see, for instance [3]), we obtain that there exists a positive number μ_0 and δ such that, for any $\mu \geqq \mu_0$,

(3.8) $$\sqrt{-1} <\left\{\Lambda_\mu e(c_1(L,h)) - e(c_1(L,h))\Lambda_\mu\right\}\phi,\phi> \leqq -\delta\|\phi\|_2^2$$

for any $\phi \in \Omega^{p,q}(M,L)_0$ ($p+q \leqq k-1$), where Λ_μ is the dual of the operator $e(\Phi_\mu)$. In what follows, we consider that M carries a complete Kähler metric Φ_μ. Since, by the Bochner-Weiztenböck formula,

$$\Delta_h - *^{-1}\Delta_h * = \sqrt{-1}\left\{e(c_1(L,h))\Lambda_\mu - \Lambda_\mu e(c_1(L,h))\right\},$$

and therefore, from (3.8), we obtain

(3.9) $$<\Delta_h\phi,\phi> \geqq \delta\|\phi\|_2^2 \qquad \text{for any} \quad \phi \in \Omega^{p,q}(M,L)_0.$$

Using Theorem(2.1) and using the same argument of Theorem(3.1), we have $\gamma_{p,q} \geqq \delta$ for $p+q \leqq k-1$, and therefore, from Theorem(1.22), we obtain

$$H^{p,q}_2(M_{\Phi_\mu},L_h) \simeq \mathrm{Ker}\left\{\Delta_h : \Omega^{p,q}_2(M_{\Phi_\mu},L_h) \longrightarrow \Omega^{p,q}_2(M_{\Phi_\mu},L_h)\right\}$$

for $p+q \leqq k-2$. Let $\phi \in \mathrm{Ker}\left\{\Delta_h : \Omega^{p,q}_2(M_{\Phi_\mu},L_h) \longrightarrow \Omega^{p,q}_2(M_{\Phi_\mu},L_h)\right\}$. Then, (3.9) implies

$$0 = <\Delta_h\phi,\phi> \geqq \delta\|\phi\|_2^2,$$

and therefore $\phi = 0$. Hence $H^{p,q}_2(M_{\Phi_\mu},L_h) = 0$ for $p+q \leqq k-2$. Now, since the L^2-cohomology does not change under quasi-isomotries, we obtain $H^{p,q}_2(M_g,L_h) = 0$ for $p+q \leqq k-2$. Q.E.D.

(3.10)Example(R.Kobayashi, [4] Lemma 1). Let $\overline{M}$ be an compact projective algebraic manifold and let $D = \sum_{i=1}^{k} D_i$ be a divisor on $\bar{M}$ with only normal crossings. Now assume that $K_{\bar{M}} \otimes D$ is ample, and $\| \;\|$ denotes the Hermitian fibre metric over D, and take $\sigma_i \in H^0(\bar{M}, D_i)$ so that $D_i = \mathrm{Div}(\sigma_i)$. Since $K_{\bar{M}} \otimes D$ is ample, there exists a volume form Ω on $\bar{M}$ such that

$$-c_1(K_{\bar{M}}, \Omega) - \sum_{i=1}^{k} \sqrt{-1}\ \partial\bar{\partial} \log \|\sigma_i\|^2 > 0 \quad \text{on} \quad \bar{M}.$$

Set $\Psi := \Omega / \prod_{i=1}^{k} \|\sigma_i\|^2 (\log \|\sigma_i\|^2)^2$ and $M := \bar{M} \backslash D$. Then, it is known that $-c_1(K_M, \Psi) > 0$ on M and moreover $-c_1(K_M, \Psi)$ is a complete Kähler metric on M. Therefore, if we set $g := -c_1(K_M, \Psi)$, by the theorem (3.5), we obtain $H_2^{p,q}(M_g, K_{M_\Psi}) = 0$ for $p+q \leqq n-2$.

§4. Non-vanishing theorem.

Let (M,g) be an n-dimesional non-compact complete Kähler manifold, and let $\{L,h\}$ be a holomorphic line bundle with a Hermitian fibre metric h. Let Δ_k be the Laplace-Beltrami operator of $\{L^k,h^k\}$. Then we obtain

$$(4.1)\qquad \frac{2}{k}\int_M \langle \Delta_k\phi,\phi\rangle = \int_M \left\{\frac{1}{k}\|\nabla_k\phi\|^2-(1-\frac{1}{k})\langle V\phi,\phi\rangle\right\}$$

for any $\phi\in C_0^\infty(M,\Lambda^{0,q}(L^k))$, where $\langle V\phi,\phi\rangle=2^q\Sigma_{|J|=q}(\alpha_{CJ}-\alpha_J)|\phi_{J,1}|^2$ and ∇_k is the covariant derivative of $\Lambda^{0,q}(L^k)$. Here $\{\alpha_1,\dots\alpha_n\}$ are the eigenvalues of $c_1(L,h)(x)$ and, for $J=(j_1,\dots,j_q)$, $\alpha_J:=\alpha_{j_1}+\dots+\alpha_{j_q}$. (see, for instance, [2]).

(4.2)Definition. Let (M,g) be a n-dimensional non-compact complete Kähler manifold, and $\{L,h\}$ be a holomorphic line bundle over M with a Hermitian fibre metric h. $\{L,h\}$ is said to be of <u>asymptotically constant Ricci form</u> if there exists a real number γ such that

$$\inf_{K\subset\subset M}\ \sup_{x\in M\setminus K}\ \|c_1(L,h)(x)-\gamma\omega_g\|(x) = 0.$$

(4.3)Lemma. Let (M,g) be an n-dimensional($n\geqq5$) non-compact complete Kähler manifold, and let $\{L,h\}$ be a holomorphic line bundle with a Hermitian fibre metric which is of asymptotically constant $c_1(L,h)$ with the constant $\gamma<0$. Then there exists a positive number δ such that

$$\inf \sigma_{ess}\left\{\frac{2}{k}\Delta_k:\ \Omega_2^{0,q}(M,L^k)\longrightarrow \Omega_2^{0,q}(M,L^k)\right\} > \delta$$

for $q=0,1,2$, and for any $k \geqq 2$.

Proof. From the assumption, there exists a compact subset $K_0 \subset\subset M$ such that $\sup_{x \in M \setminus K_0} \|c_1(L,h) - \lambda\omega_g\|(x) < -\frac{\gamma}{100}$. Namely, let $\{\alpha_1, \dots \alpha_n\}$ be the eigenvalues of $c_1(L,h)(x)$ for $x \in M \setminus K_0$, then we have $\frac{\gamma}{100} < \alpha_i - \gamma < -\frac{\gamma}{100}$ for any i. In the case of $|J|=0$, $-\alpha_{CJ} = -\sum_{i=1}^{n} \alpha_i \geqq -\frac{99n}{100}\gamma$. Therefore, by (4.1), there exists a positive constant δ such that $-(1-\frac{1}{k})\langle V\phi, \phi\rangle(x) \geqq \delta\|\phi\|^2(x)$ for any $\phi \in \Lambda_x^{0,0}(L^k)$ and for any $k \geqq 2$. Note that $n \geqq 5$, and we obtain $-(1-\frac{1}{k})\langle V\phi, \phi\rangle(x) \geqq \delta\|\phi\|^2(x)$ for any $\phi \in \Lambda_x^{0,q}(L^k)(q=1,2)$ and for any $k \geqq 2$. Now using Theorem(2.1), we have proved the required lemma.

Q.E.D.

Assume that

$$\inf \sigma_{ess}\left\{\frac{2}{k}\Delta_k : \Omega^{0,q}_2(M,L^k) \longrightarrow \Omega^{0,q}_2(M,L^k)\right\} > 0.$$

Let $\Omega \subset M$ be an open subset, and let

$$\rho < \inf \sigma_{ess}\left\{\frac{2}{k}\Delta_k : \Omega^{0,q}_2(M,L^k) \longrightarrow \Omega^{0,q}_2(M,L^k)\right\}.$$

Then we define

(4.4) $N^q_{\Omega,k}(\rho) := \Big\{$ number of eigenvalues of

$$\frac{2}{k}\Delta_k : W^{1,2}_0(\Omega, \Lambda^{0,q}(L^k)) \longrightarrow W^{1,2}_0(\Omega, \Lambda^{0,q}(L^k))$$

equal or less than ρ with counting multiplicity$\Big\}$.

Then we obtain the following lemma.

(4.5)Lemma. Let (M,g) and $\{L,h\}$ be as in Lemma(4.3). Then, for any real number λ and μ such that $0<\lambda<\lambda+\mu<\delta$ and that $0<\mu<\lambda$, a relatively compact open domain $\Omega_{\lambda,\mu}$ of M exists and satisfies that $N^q_{\Omega_{\lambda,\mu},k}(\lambda) \leqq N^q_{M,k}(\lambda) \leqq N^q_{\Omega_{\lambda,\mu},k}(\lambda+\mu)$ for any $k \geqq 2$

and q=0,1,2. Here δ is the constant in Lemma (4.3).

Proof. From the proof of Lemma(4.3), there exists a compact subset K_0 of M such that

$$\mu_{K_0} := \inf_{x\in M\setminus K_0} \left\{\inf_{\phi\in\Lambda_x^{0,q}(L^k)} -(1-\frac{1}{k})\langle V\phi,\phi\rangle(x)\right\} > (\lambda+\mu)\|\phi\|^2(x)$$

for any $k\geqq 2$, and $q=0,1,2$. Next we choose J_0, $J_1\in C^\infty(M:[0,1])$ and a relatively compact open domain $\Omega=\Omega_{\lambda,\mu}$ ($K_0 \subset\subset \Omega$) satisfying the following conditions;

(4.6) $\quad J_0(x) = 1$ if $x\in K_0$, and $J_0(x) = 0$ if $x\in M\setminus\Omega$,

$\quad J_0(x)^2 + J_1(x)^2 = 1$ for any $x\in M$,

and its derivative satisfies

(4.7) $\quad \sup_{x\in M}|dJ_i(x)|^2, \sup|\nabla^2 J_i(x)| \leqq \frac{1}{100}\operatorname{Min}\left\{\lambda,\mu\right\}.$

(4.8)Claim. $N^q_{\Omega,k}(\lambda) \leqq N^q_{M,k}(\lambda)$ for any $k\geqq 2$, and $q=0,1,2$.

Proof of (4.8). Let

$$\mathcal{F}(\lambda) := \oplus\left\{\phi_\alpha\in W_0^{1,2}(\Omega,\Lambda^{0,q}(L^k)) \mid \frac{2}{k}\Delta_k\phi_\alpha=\lambda_\alpha\phi_\alpha,\ \lambda_\alpha\leqq\lambda\right\}.$$

Then $\dim\mathcal{F}(\lambda) = N^q_{\Omega,k}(\lambda)$. Now set $N:= N^q_{\Omega,k}(\lambda)$, then

$$\mu_N := \inf_{F_N\subset W}\ \sup_{f\in F_N,\ \|f\|_2=1} \langle f, \frac{2}{k}\Delta_k f\rangle$$

$$\leqq \sup_{f\in\mathcal{F},\ \|f\|_2=1} \langle f, \frac{2}{k}\Delta_k f\rangle \leqq \lambda < \delta.$$

Here $W:= W^{1,2}(M,\Lambda^{0,q}(L^k))$ and F_N denotes an N-dimensional subspace of W. Now, from Lemma(4.3) and by Min-max principle, μ_N is the N-th eigenvalue of $\frac{2}{k}\Delta_k: \Omega_2^{0,q}(M_g,L_h^k) \longrightarrow \Omega_2^{0,q}(M_g,L_h^k)$, and therefore $N^q_{\Omega,k}(\lambda) \leqq N^q_{M,k}(\lambda)$. Q.E.D.

(4.9)Claim. $N^q_{M,k}(\lambda) \leqq N^q_{\Omega,k}(\lambda+\mu)$ for any $k \geqq 2$, and $q=0,1,2$.

Proof of (4.9).

(4.10)Subclaim. For any $\psi \in C_0^\infty(M, \Lambda^{0,q}(L^k))$, we have

$$<\frac{2}{k}\Delta_k\psi,\psi> \leqq <\frac{2}{k}\Delta_k(J_0\psi),J_0\psi>+<\frac{2}{k}\Delta_k(J_1\psi),J_1\psi>$$

$$\leqq <\frac{2}{k}\Delta_k\psi,\psi>+ \frac{1}{50} \mathrm{Min}\left\{\lambda,\mu\right\} \|\psi\|_2^2$$

Proof of (4.10).

$$<\frac{2}{k}\Delta_k(J_0\psi),J_0\psi>+<\frac{2}{k}\Delta_k(J_1\psi),J_1\psi> = \int_M \frac{1}{k} \left\{\|\nabla_k(J_0\psi)\|^2+\|\nabla_k(J_1\psi)\|^2\right\}$$
$$- \int_M(1- \frac{1}{k})<V\psi,\psi>$$
$$= \frac{1}{k} \int_M\left\{\|dJ_0\wedge\psi\|^2+ 2\mathrm{Re}<J_0dJ_0\wedge\psi,\nabla_k\psi> + J_0^2\|\nabla_k\psi\|^2\right\}$$
$$+ \frac{1}{k} \int_M\left\{\|dJ_1\wedge\psi\|^2+ 2\mathrm{Re}<J_1dJ_1\wedge\psi,\nabla_k\psi> + J_1^2\|\nabla_k\psi\|^2\right\}$$
$$- \int_M(1- \frac{1}{k})<V\psi,\psi>.$$

Now, since, $J_0^2+J_1^2=1$, $J_0dJ_0+J_1dJ_1=0$, and using (4.7), we obtain (4.10). Q.E.D. of (4.10).

Let $\mathcal{F}(\lambda+\mu):= \oplus\left\{\phi_\alpha \in W_0^{1,2}(\Omega,\Lambda^{0,q}(L^k)) \mid \frac{2}{k}\Delta_k\phi_\alpha=\lambda_\alpha\phi_\alpha,\ \lambda_\alpha \leqq \lambda+\mu\right\}$, and let $\left\{\phi_1,\dots,\phi_{N'}\right\}$ be an orthonormal basis of $\mathcal{F}(\lambda+\mu)$, where $N':=N^q_{\Omega,k}(\lambda+\mu)$. The following Subclaim (4.11) can be shown in the same way as ([2], Proposition 2.6 (b)).

(4.11)Subclaim.

$$\mu'_{N'+1} := \sup_{F_{N'} \subset W} \ \inf_{\psi \perp F_{N'},\|\psi\|_2=1} <\psi,\frac{2}{k}\Delta_k\psi> > \lambda.$$

When $\mu'_{N'+1} < \inf \sigma_{ess}\left\{\frac{2}{k}\Delta_k:\Omega_2^{0,q}(M_g,L_h^k) \longrightarrow \Omega_2^{0,q}(M_g,L_h{}^k)\right\}$, Min-max principle tells us that $\mu'_{N'+1}$ is the $(N'+1)$-th eigenvalue of $\frac{2}{k}\Delta_k$: $\Omega_2^{0,q}(M_g,L_h^k) \longrightarrow \Omega_2^{0,q}(M_g,L_h^k)$, and therefore $N^q_{M,k}(\lambda) \leqq N'$. When $\mu'_{N'+1} = \inf \sigma_{ess}\left\{\frac{2}{k}\Delta_k:\Omega_2^{0,q}(M_g,L_h^k) \longrightarrow \Omega_2^{0,q}(M_g,L_h{}^k)\right\}$, by Min-max

principle again, there exist at most N'-eigenvalues of
$\frac{2}{k}\Delta_k : \Omega_2^{0,q}(M_g, L_h^k) \longrightarrow \Omega_2^{0,q}(M_g, L_h^k)$, and therefore $N_{M,k}^q(\lambda) \leqq N'$.

Q.E.D.

Now we are in position to prove the Main Theorem of this section.

(4.12)Theorem. Let (M,g) be an n-dimensional non-compact complete Kähler manifold($n \geqq 5$), and let $\{L,h\}$ be a holomorphic line bundle with a Hermitian metric h. Assume that the first chern form of $\{L,h\}$ is asymptotically constant with the constant $\gamma<0$. Then

$$\liminf_{k\to\infty} k^{-n}\dim H_2^0(M_g, L_h^k) \geqq \sum_{q=0}^{1} \frac{1}{n!}\int_{M_h(q)} c_1(L,h)^n,$$

where $M_h(q) := \{x \in M \mid$ q-eigenvalues of $c_1(L,h)(x)$ are negative and the other (n-q)-eigenvalues are positive$\}$, and
$H_2^0(M_g, L_h^k) := \{$holomorphic sections of the line bundle L^k whose L^2-norm with respect to the metric g and h are bounded$\}$.

Proof. From (4.3) and (1.22), and by the Strong Morse's inequality, we obtain

$$\dim H_2^0(M_g, L_h^k) = \dim H^0(P_{0,*}(\delta), \bar\partial) \geqq \dim P_{0,0}(\tfrac{\delta}{3}) - \dim P_{0,1}(\tfrac{\delta}{3}).$$

where δ is the constant in (4.3). Now take a constant λ, μ in Lemma(4.5) so that $\lambda = \frac{\delta}{3}$, $\mu = \frac{\delta}{6}$. Then there exists a relatively compact open domain Ω_δ of M such that

$N_{\Omega_\delta,k}^q(\frac{\delta}{3}) \leqq N_{M,k}^q(\frac{\delta}{3}) \leqq N_{\Omega_\delta,k}^q(\frac{\delta}{2})$ for any $k \geqq 2$, and $q=0,1,2$.

Therefore it follows that

$$\liminf_{k\to\infty} k^{-n}\dim H_2^0(M_g, L_h^k)$$

$$\geqq \liminf_{k\to\infty} k^{-n} N_{\Omega_\delta,k}^0(\tfrac{\delta}{3}) - \limsup_{k\to\infty} k^{-n} N_{\Omega_\delta,k}^1(\tfrac{\delta}{2}).$$

Using ([2], Theorem 2.16), we obtain

$$\geqq \int_{\Omega_\delta} \nu_B(\tfrac{2}{3}\delta + \textstyle\sum_{i=1}^n \alpha_i) - \sum_{j=1}^n \int_{\Omega_\delta} \bar\nu_B(\delta - \alpha_j + \sum_{i\neq j}\alpha_i).$$

Here note that ν_B is a non-negative function (see (0.5) of [2] for the definition of ν_B), and that, from the proof of (4.3), a compact subset K_0 of Ω_δ exists and satisfies

$$\inf_{x\in M\setminus K_0} \left\{\inf_{\phi\in\Lambda_x^{0,q}(L^k)} -\langle V\phi,\phi\rangle(x)\right\} > \frac{|\gamma|}{2} \|\phi\|^2(x) \quad \text{for} \quad q=0,1,2.$$

In paiticular, $\sum_{i=1}^n \alpha_i < -\frac{|\gamma|}{2}$ on $M\setminus K_0$. Therefore, by the definition of ν_B,

$$\liminf_{k\to\infty} k^{-n}\dim H_2^0(M_g, L_h^k)$$

$$\geqq \int_{K_0} \nu_B(\tfrac{2}{3}\delta + \textstyle\sum_{i=1}^n \alpha_i) - \sum_{j=1}^n \int_M \bar{\nu}_B(\delta - \alpha_j + \sum_{i\neq j}\alpha_j),$$

for sufficiently small $\delta>0$. Since ν_B is monotone increasing, using Lebesgue's covergence theorem and the computation of ([2], p224), we obtain the estimation;

$$\liminf_{k\to\infty} k^{-n}\dim H_2^0(M_g, L_h^k) \geqq \frac{1}{n!}\sum_{q=0}^1 \int_{M_h(q)} c_1(L,h)^n,$$

by $\delta\downarrow 0$. Q.E.D.

References

[1]H.L.Cycon-R.G.Froese-W.Kirsh-B.Simon, Schrödinger operators, Texts and monographs in physics, Springer-Verlag, 1986.

[2]J.P.Demailly, Champs magnetiques et inégalité de Morse pour la d''-cohomology, Ann. Inst. Fourier 35(1985), 189-229.

[3]G.Gigante, Vector bundle with semidefinite curvature and cohomology vanishing theorem, Adv. in Math. 41(1981), 40-56.

[4]R.Kobayashi, Kähler-Einstein metric on an open algebraic manifold, Osaka J. Math. 21(1984), 399-418.

[5]S.Nakano, On the inverse of monoidal transformations, Publ. RIMS, Kyoto Univ. 6(1970/1971), 483-502.

Department of Mathematics
Faculty of Science
University of Tokyo
Hongo, Tokyo, 113
Japan

FUNDAMENTAL GROUPS AND LAPLACIANS

Toshikazu Sunada*
Department of Mathematics, Nagoya University,
Nagoya 464, Japan

This lecture is primarily concerned with the spectral theory of the Laplacian acting in the L^2-space of functions on a non-compact Riemannian manifold with compact quotient. Before going to my thesis, I wish to start the lecture with a brief explanation of background and motive of my problems. For some relevant works, see R. Brooks [3] and M. Burger [5] .

The celebrated Riemann hypothesis, which has been neither proved nor disproved, asserts that, in the region $0 < \mathrm{Re}\, s < 1$, all the zeros of the Riemann zeta function

$$\zeta(s) = \sum_{n=1}^{\infty} n^{-s}$$

(continued meromorphically to the whole s-plane) lie on the line $\mathrm{Re}\, s = 1/2$. Since 1859 (the year Riemann's paper on his zeta function appeared), many programs have been proposed towards the proof of the hypothesis. Among others, the most interesting program, due originally to Hilbert, is to relate the zeros of $\zeta(s)$ to eigenvalues of certain self-adjoint operators acting in a Hilbert space, although nobody yet succeeded in finding such an authentic operator.

One may ask why it is expected that the eigenvalue problem of self-adjoint operators may come up in connection with the Riemann hypothesis. To explain the mysterious reason we should first point out that the Riemann zeta function has been conceptually generalized to various zeta functions in algebraic geometry and differential geometry. Some of those zeta functions turn out to be closely related to self-adjoint operators whose eigenvalues determine completely the zeros of the zeta functions. The "Riemann hypothesis" for those zeta-functions, therefore, results in some estimates of the eigenvalues. To convince the reader, we shall take a look at two examples.

I. Ihara zeta functions. Let M be a regular graph, that is, M is a countable set with a map

* supported by The Ishida Foundation

$$M \longrightarrow 2^M \quad (= \text{the set of subsets in } M)$$

$$x \longmapsto V(x)$$

satisfying the conditions:

1) $y \in V(x)$ if and only if $x \in V(y)$,

2) $x \notin V(x)$,

3) $\# V(x)$, the number of elements in $V(x)$, is finite, and does not depend on x (and will be written $q+1$ conventionally).

An element in M is called a vertex, and a pair $\{x,y\}$ is called an (unoriented) edge if $y \in V(x)$. We regard M, in a natural manner, as a one dimensional simplicial complex. A sequence $c = (x_0,\dots,x_n)$ is called a closed geodesic if $\{x_i, x_{i+1}\}$ are edges, $x_0 = x_n$, and $x_{i-1} \neq x_{i+1}$ for all $i \in \mathbb{Z}/n\mathbb{Z}$ For a closed geodesic $c = (x_0,\dots,x_n)$, the k-multiple c^k is a closed geodesic, given by

$$(\underbrace{x_0,\dots,x_n}_{1},\ \underbrace{x_0,\dots,x_n}_{2},\ \dots,\ \underbrace{x_0,\dots,x_n}_{k})$$

If a closed geodesic c is not a k-multiple of another one $(k \geq 2)$, then c is called prime. Two closed geodesics $c_1 = (x_0,\dots,x_n)$ and $c_2 = (y_1,\dots,y_n)$ are said to be equivalent if there exists an integer d such that $y_k = x_{k+d}$ $(k \in \mathbb{Z}/n\mathbb{Z})$. An equivalence class of a prime closed geodesic will be called a prime geodesic cycle.

We now define the (Ihara) zeta function of a finite regular graph M by

$$Z(s) = \prod_p (1 - u^{\ell(p)})^{-1}, \quad (u = q^{-s}),$$

where p runs over all prime geodesic cycles in M, and $\ell(p)$ is the length of p, that is , $\ell(p) = n$ if p is a cycle represented by $(x_0,\dots,x_n)$. It is known that $Z(s)$ converges absolutely in $\operatorname{Re} s > 1$ $(|u| < q^{-1})$.

On the other hand, given a regular graph M (not assumed to be finite), we set

$$L^2(M) = \{\phi: M \to \mathbb{C}\ ;\ \sum_{x \in M} |\phi(x)|^2 < \infty\},$$

which is a Hilbert space with the standard scalar product

$$\langle\phi,\psi\rangle = \sum_{x \in M} \phi(x)\,\overline{\psi(x)}.$$

The adjacency operator $A = A_M$ is a bounded self-adjoint operator acting in $L^2(M)$, defined by

$$(A\phi)(x) = \sum_{y \in V(x)} \phi(y).$$

If M is a finite graph, then $q+1$ is a simple eigenvalue of A with constant eigenfunctions.

The relationship between the zeta function $Z(s)$ and the adjacent operator A is now stated in:

<u>Proposition</u> 1. (Y.Ihara [9] and [24]). Let M be a finite regular graph, then

$$Z(s) = (1-u^2)^{-g} \det(I - Au + qu^2)^{-1} ,$$

where $u = q^{-s}$ and $g = (q-1) \# M/2$. In particular, $Z(s)$ is a rational function of u .

<u>Remark</u>. Originally, Ihara zeta functions are defined as the zeta functions associated with discrete subgroups in $PGL_2(K_p)$, K_p being a p-adic number field, which turns out to be described in terms of regular graphs (J.P.Serre [18]) .

One now has, in view of the above proposition,

<u>Corollary</u> 1.1. The zeta function $Z(s)$ satisfies the "Riemann hypothesis" (i.e. all the singular points (poles in this case) of $Z(s)$ in $0 < \mathrm{Re}\, s < 1$ lie on the line $\mathrm{Re}\, s = 1/2$) if and only if the eigenvalues λ except for $\pm(q+1)$ satisfy the estimate

$$|\lambda| \leqq 2q^{1/2} .$$

<u>Remark</u>. Generally, $-(q+1)$ is not an eigenvalue. If it is, then the graph M has a very special property; namely M is partitioned into two parts M_+ , M_- such that if $x \in M_\pm$, then $V(x) \subset M_\mp$ (so that M looks like an "expander").

II. <u>Selberg zeta functions</u>. Let M be a closed Riemann surface with a metric of constant (-1) negative curvature. The notions such as closed geodesics and prime closed geodesics are defined in the usual sense. Prime geodesic cycles mean, in this setting, oriented 1-cycles represented by prime closed geodesics in M . We denote by $\ell(p)$ the length of a prime geodesic cycle p . We set

$$Z(s) = \prod_{k=0}^{\infty} \prod_{p} (1 - e^{-(s+k)\ell(p)}) ,$$

which is what we call the Selberg zeta function of the surface M . It is known that $Z(s)$ converges absolutely in $\mathrm{Re}\, s > 1$ and extends to an entire function. Furthermore one has

Proposition 2. (A.Selberg [17]). The zeros of $Z(s)$ in $0 \leqq \mathrm{Re}\, s \leqq 1$ are given by the equation

$$s = 1/2 \pm (1/4 - \lambda_k)^{1/2} \ , \ k = 0,1,2, \ldots \ ,$$

where the sequence

$$0 = \lambda_o < \lambda_1 \leqq \lambda_2 \leqq \cdots$$

denotes the eigenvalues of the Laplacian Δ_M on M .

Since $\lambda_k \uparrow \infty$ $(k \to \infty)$, the above proposition says that the zeta function $Z(s)$ nearly satisfies the "Riemann hypothesis" in the sense that all the zeros of $Z(s)$ in $0 < \mathrm{Re}\, s < 1$ lie on the line $\mathrm{Re}\, s = 1/2$ except for finitely many zeros possibly located on the interval $(0,1)$. The zeros on $(0,1)$ correspond to eigenvalues λ_k with $\lambda_k < 1/4$. Thus one has

Corollary 2.1. The zeta function $Z(s)$ satisfies truly the "Riemann hypothesis" if and only if

$$\lambda_1 \geqq 1/4 \ .$$

Here comes a question: What on earth is the nature of those numbers $2q^{1/2}$ and $1/4$ in Corollary 1.1 and 2.1 ? The answer is given in terms of the spectrums of the corresponding operators defined on the universal covering spaces of those spaces M . We continue to look at the examples I and II .

I. Let M be a finite regular graph, and let

$$\pi : X \to M$$

be the universal covering map. The space X has a graph structure such that the map π is a morphism of graphs. Actually X is a regular tree.

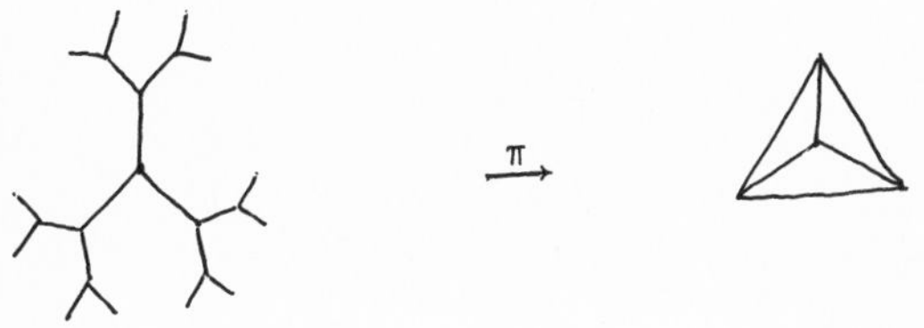

(q=2)

M

X

We now pay our attention to the adjacency operator A_X on X . In view of the theory of spherical functions on the regular tree, we may obtain a complete description of the spectral decomposition of the self-adjoint operator A_X , (see Appendix 3) , from which it follows that the spectrum of A_X is equal to the interval

$$[-2q^{1/2}, 2q^{1/2}].$$

(We shall give a definition of spectrums later).

Remark. We shall show that the spectrum of A_X is contained in the interval $[-2q^{1/2}, 2q^{1/2}]$, since it is rather elementary. It is enough to prove that

$$|\langle Af,f\rangle| \leqq 2q^{1/2} \|f\|^2, \qquad (A = A_X)$$

for all finite support functions f on X. For an edge $\{x,y\}$ in X, the ordered pair (x,y) or (y,x) is called an oriented edge. Thus each edge has just two orientations. We now employ the following notations; if $e = (x,y)$ is an oriented edge, then $O(e) = x$, and $t(e) = y$. We can assign an orientation to each edge in such a way that for each vertex x in X there exists only one oriented edge e with $O(e) = x$. This assignment is possible because there is no closed path in the tree X.

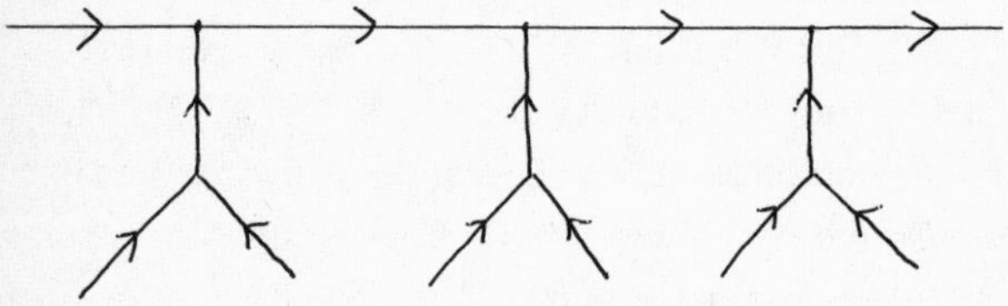

(q=2)

We denote by E a set of oriented edges with this property. We then have

$$\begin{aligned}
|\langle Af,f\rangle| &= \Big|\sum_{x\in X}\sum_{y\in V(x)} f(y)\overline{f(x)}\Big| \\
&= \Big|\sum_{e\in E}\{f(t(e))\overline{f(O(e))} + f(O(e))\overline{f(t(e))}\}\Big| \\
&= \Big|2\sum_{e\in E}\operatorname{Re} f(O(e))\overline{f(t(e))}\Big| \\
&\leqq 2\sum_{e\in E}|f(O(e))|\,|f(t(e))| \\
&\leqq 2\Big(\sum_{e\in E}|f(O(e))|^2\Big)^{1/2}\Big(\sum_{e\in E}|f(t(e))|^2\Big)^{1/2} \\
&= 2\Big(\sum_{x\in X}|f(x)|^2\Big)^{1/2} q^{1/2}\Big(\sum_{x\in X}|f(x)|^2\Big)^{1/2} \\
&= 2q^{1/2}\|f\|^2,
\end{aligned}$$

as desired, where we have used the assumption that the correspondence $e \to O(e)$ is one-to-one, and hence the correspondence $e \to t(e)$ is q-to-one.

II. Let M be a closed Riemann surface with constant (-1) negative curvature. The universal covering space of M with the metric obtained by lifting up the metric on M is identified with the upper half plane H^2 with the Poincare metric ds^2 :

$$H^2 = \{(x,y) \in \mathbb{R}^2 \ ; \ y > 0\}$$

$$ds^2 = y^{-2}(dx^2 + dy^2) .$$

Let us consider the Laplacian Δ_{H^2} acting in $L^2(H^2)$:

$$\Delta_{H^2} = - y^2 \left(\frac{\partial^2}{\partial x^2} + \frac{\partial^2}{\partial y^2} \right) ,$$

which turns out to be an (unbounded) self-adjoint operator. In the same vein as the case of regular trees, the theory of spherical functions on H^2 leads us to the conclusion that the spectrum of Δ_{H^2} coincides with the interval $[1/4, \infty)$ (see Appendix 3). (It is again easy to check that the spectrum is contained in $[1/4, \infty)$).

From the above arguments, it turns out (and is still mysterious) that those constants $2q^{1/2}$ and $1/4$ in Corollary 1.1 and 2.1 are the end points of the spectrums of the operators A_X and Δ_{H^2} on the universal covering spaces respectively. Those phenomena suggest us to study relationships of spectrums of some geometrically defined operators on base spaces and covering spaces.

We are now in a position to formulate our problems. We first establish some basic terminology. Let X be a smooth manifold with a metric $ds^2 = \Sigma g_{ij} dx_i dx_j$. The canonical measure on X is defined by the density

$$dx = dv(x) = (\det g)^{1/2} dx_1 \cdots dx_n , \quad (g = (g_{ij})) .$$

Throughout, $L^2(X)$ denotes the space of square integrable functions on X with respect to this canonical measure. The Laplacian Δ_X is defined by

$$\Delta_X = -(\det g)^{-1/2} \sum_{i,j=1}^{n} \frac{\partial}{\partial x_i} \left(g^{ij} (\det g)^{1/2} \frac{\partial}{\partial x_j} \right) ,$$

where the matrix (g^{ij}) is the inverse matrix of g . The operator Δ_X is an elliptic differential operator of second order acting on functions on X , and satisfies:

(1) If $f_1, f_2 \in C_0^\infty(X)$, then

$$\langle \Delta_X f_1, f_2 \rangle = \langle f_1, \Delta f_2 \rangle .$$

(2) $\langle \Delta_X f, f \rangle \geqq 0$ for all $f \in C_o^\infty(X)$.

Let

$$\pi: X \to M$$

be a covering map onto a compact Riemannian manifold M . As a Riemannian metric on X , we always take the pull-back of the metric on M , so that we have the following strikingly simple relation between Δ_M and Δ_H :

$$(\Delta_M f) \circ \pi = \Delta_X (f \circ \pi) , \quad f \in C^\infty(M) .$$

But an important thing to remember is that this is not a relation between those operators acting in the L^2-spaces. It makes sense, therefore, to want to know the relationship between the spectrum of Δ_X and that of Δ_M . Especially we wish to investigate the structure of the spectrum of Δ_X in the case that X is non-compact. Similar problems are also raised for the adjacency operators of graphs (see Appendix 1). But we restrict ourselves, from now on, to the case of Laplacians.

The setting of our problems is generalized in the following way. Let X be a Riemannian manifold with compact quotient in the sense that X is a branched locally isometric convering space on a compact Riemannian orbifold M . (In all the examples below, X has an isometry group, say Γ , acting discontinuously on X with a compact quotient space $\Gamma \backslash X$ (possibly with singularities)). It should be noted that a manifold with compact quotient is always complete, and hence the Laplacian Δ_X with the domain $C_o^\infty(X)$ is extended in a unique way to a self-adjoint operator acting in $L^2(X)$. Our concern is the spectrum of Δ_X , or more generally the spectrum of the Schrödinger operator

$$H = H_X = \Delta_X + q(x) ,$$

where $q(x)$ is a smooth real-valued function on X obtained as a lift of a function on M .

To make sure, let us recall the definition of self adjointness of operators and spectrum.

Let V be a Hilbert space, and $\mathcal{D}$ be a subspace of V . An operator H of V with a domain $\mathcal{D}$, which we write $H : V \to V$ by abuse of notations, is a linear operator H of $\mathcal{D}$ into V . If the domain $\mathcal{D}$ is dense in V , then H is said to be densely defined. An operator H is said to be symmetric if $\langle Hu, v \rangle = \langle u, Hv \rangle$ for every u, v in $\mathcal{D}$. Thus the operator H_X with the domain $C_o^\infty(X)$ is symmetric. A densely defined symmetric operator H with a domain $\mathcal{D}$ is called self-adjoint if the following condition is satisfied :

If, for every u in $\mathcal{D}$, $\langle Hu,v\rangle = \langle u,v^*\rangle$, then v is in $\mathcal{D}$, and $v^* = Av$.

The spectrum of a self-adjoint operator H , denoted by $\sigma(H)$, is a subset in $\mathbb{R}$, which is decomposed into two parts, say $\sigma_p(H)$ and $\sigma_c(H)$. Here, $\sigma_p(H)$ is the set of eigenvalues of H , and $\sigma_c(H)$ is the set of real numbers λ such that $(H-\lambda I)^{-1}$ exists, but not bounded. The set $\sigma_p(H)$ is called the point spectrum, and $\sigma_c(H)$ is called the continuous spectrum. The spectrum $\sigma(H)$ is a closed subset in $\mathbb{R}$. If $H = H_X$, then the bottom of the spectrum

$$\lambda_o = \lambda_o(H) = \inf \ \sigma(H)$$

is a finite value $(>-\infty)$. Especially, if $q(x) \equiv 0$, then

$$\lambda_o = \lambda_o(\Delta_X) \geqq 0 .$$

A special attention will be paid to the bottom of the spectrum later on.

It is a standard fact that if X is compact (a special case of our setting), then $\sigma(H) = \sigma_p(H)$; namely there is no continuous spectre. Moreover the multiplicity of each eigenvalue is finite and zero is always a simple eigenvalue of Δ_X . Our question is now to ask how about the case of non-compact X .

Example 1. Let X be the n-dimensional Euclidean space $\mathbb{R}^n$ with the standard metric. Since $\mathbb{R}^n$ is the universal covering space of a flat torus (say, $\mathbb{R}^n/\mathbb{Z}^n$) , $\mathbb{R}^n$ belongs to our class of manifolds. The spectral properties of the Laplacian

$$\Delta_{\mathbb{R}^n} \quad (= -(\partial^2/\partial x_1^2 + \ldots + \partial^2/\partial x_n^2))$$

are extracted from the formula

$$F\Delta_{\mathbb{R}^n} F^{-1} f(\xi) = 4\pi^2|\xi|^2 f(\xi) ,$$

where F denotes the Fourier transform:

$$F(f)(\xi) = \int_{\mathbb{R}^n} f(x)\, e^{-2\pi i\langle x,\xi\rangle} dx \quad .$$

It should be noted that F extends to a unitary transformation of $L^2(\mathbb{R}^n)$ (Parseval's identity).

It should not be a waste of time to recall the spectral properties of multiplication operators in a general measure space. Let (S,μ)

be a measure of space. We assume that μ is a σ-finite measure. Let f be a real valued measurable function on S. Define the multiplication operator $M_f : L^2(S) \to L^2(S)$ by putting $M_f(g) = f \cdot g$ The domain of M_f is the set $\{g \in L^2(S) \,;\, f \cdot g \in L^2(S)\}$. One can then easily check that

1) M_f is self-adjoint,

2) λ is an eigenvalue of M_f if and only if the inverse image $f^{-1}(\lambda)$ has positive measure, and

3) λ is in the continuous spectrum of M_f if and only if $f^{-1}(\lambda)$ has zero measure and $f^{-1}(\lambda-\varepsilon, \lambda+\varepsilon)$ has positive measure for all positive ε.

Since $\Delta_{\mathbb{R}^n}$ is unitarily equivalent to the multiplication operator $M_{4\pi^2|\cdot|^2}$, it is apparent that $\sigma(\Delta_{\mathbb{R}^n}) = \sigma_c(\Delta_{\mathbb{R}^n}) = [0,\infty)$.

It is worthwhile to note that, according to the spectral theorem, an arbitrary self-adjoint operator acting in a Hilbert space is unitarily equivalent to a multiplication operator. But there are very few cases in which the spectral theorem is available to investigate the spectrum of a given operator (adjacency operators on regular trees and Laplacians on symmetric spaces are such examples for which we may construct natural "Fourier transforms" by means of spherical functions).

Example 2. (periodic Schrödinger operators). This is an example which comes up in quantum physics in crystal, and a model in our arguments. As X, we take the Euclidean space $\mathbb{R}^n$ with the standard metric, and let Γ be a lattice in $\mathbb{R}^n$ which acts on $\mathbb{R}^n$ by parallel translations. Therefore a function $q(x)$ on $\mathbb{R}^n$ is assumed to be periodic with respect to the lattice, that is,

$$q(x+\sigma) = q(x) \qquad \text{for all } \sigma \in \Gamma \quad .$$

It is known (cf. M.C.Reed and B.Simon [15]) that $\sigma(H) = \sigma_c(H) =$ a disjoint union of closed intervals.

Much about the spectrum has been known in the one dimensional case. Let us review some of the classical results in this specific case. We may assume that H has the shape

$$-\frac{d^2}{dx^2} + q(x) \quad , \qquad q(x+1) = q(x) \quad ,$$

which is what we call a Hill's operator. Consider the following (periodic) boundary value problem:

$$\left(-\frac{d^2}{dx^2} + q(x)\right) u(x) = \lambda\, y(x)$$
$$u(x+1) = e^{\sqrt{-1}\alpha}\, u(x) \; .$$

where $0 \leqq \alpha < 2\pi$. We enumerate the eigenvalues for this problem, counting multiplicity, in nondecreasing order:

$$\lambda_o(\alpha) \leqq \lambda_1(\alpha) \leqq \ldots ,$$

so that $\lambda_k(\alpha)$ are continuous functions of α . On the other hand, we put, for each $\lambda \in \mathbb{R}$,

$$S_\lambda = \{u \in C^\infty(\mathbb{R}) \; ; \; \left(-\frac{d^2}{dx^2} + q(x)\right) u(x) = \lambda\, u(x)\} \quad ,$$

and define a linear operator $M_\lambda : S_\lambda \to S_\lambda$ by $(M_\lambda u)(x) = u(x+1)$. The trace of M_λ , say $\Delta(\lambda) = \mathrm{tr}\, M_\lambda$, is called the discriminant (note that $\dim S_\lambda = 2$) , which is an entire function of order $1/2$, and gives a complete description of the eigenvalues $\{\lambda_k(\alpha)\}$ in the following way:

$$\begin{aligned} \{\lambda_k(\alpha)\} &= \{\lambda \in \mathbb{R} \; ; \text{ there exists some } u \text{ in } S_\lambda \text{ with } M_\lambda u = e^{\sqrt{-1}\alpha} u \} \\ &= \{\lambda \in \mathbb{R} \; ; \; \det(M_\lambda - e^{\sqrt{-1}\alpha} I) = 0 \} \\ &= \{\lambda \in \mathbb{R} \; ; \; \Delta(\lambda) = 2 \cos \alpha \} \; . \end{aligned}$$

It is known that $\Delta(-\lambda)$ is asymptotically equal to $2 \cosh \sqrt{\lambda}$ as $\lambda \uparrow \infty$. The direct integral representation of H (which we mention later on in more general form) allows us to conclude that

$$\begin{aligned} \sigma(H) &= \bigcup_{0 \leqq \alpha < 2\pi} \bigcup_{k=0}^{\infty} \{\lambda_k(\alpha) \\ &= \{ \lambda \in \mathbb{R} \; , \; |\Delta(\lambda)| \leqq 2 \} \quad . \end{aligned}$$

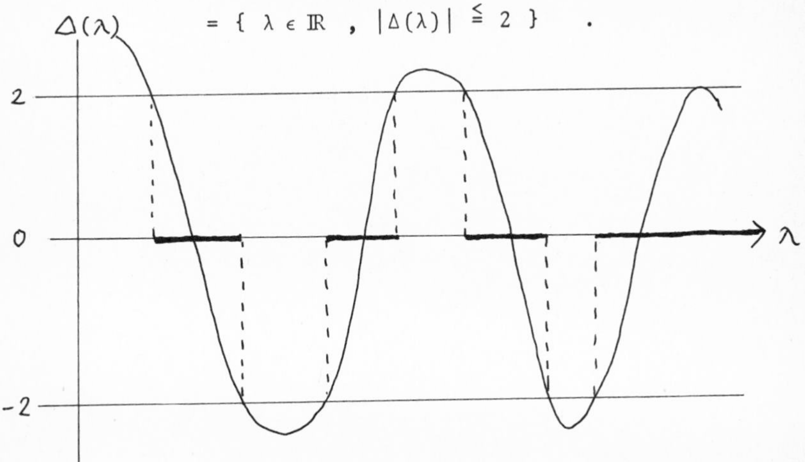

Example 3. If X is a globally symmetric space of non-compact type, (which has a compact quotient by a classical theorem of A.Borel), then $\sigma(\Delta_X) = \sigma_c(\Delta_X) = [\lambda_o,\infty)$ with a positive λ_o. Especially, if X is an n-dimensional hyperbolic space of curvature -1, then $\sigma(\Delta_X) = [(n-1)^2/4,\infty)$.

Example 4. Let $X = S^1 \times H^2$. Then Δ_X has no point spectre. This is a special case of the fact that $\Delta_{X_1 \times X_2}$ has no point spectre if one of Δ_{X_1} and Δ_{X_2} has no point spectre. We will find out later an example of a S^1-fiber bundle X over H^2 such that Δ_X has point spectra.

In general cases, we do not know much about the structure of the spectrum. But those examples above suggest us to conjecture that if X is non-compact, then $\sigma_c(H)$ constitutes a large part in $\sigma(H)$. Namely we expect that the condition on X "having compact quotient" imposes quite a bit of restriction on the existence of eigenvalues. More precisely, we propose

Conjecture A. The continuous spectrum is not bounded from above.

Conjecture B. If M has no null-homotopic closed geodesic (except for point geodesics), or equivalently if the universal covering space X has no closed geodesic, then $\sigma(\Delta_X) = \sigma_c(\Delta_X)$.

Non-positively curved manifolds satisfy the condition in Conjecture B. But even in this case, it seems rather hard to decide the structure of the spectrum. It is worthwhile to note that H.Donnelly and P.Li [28] constructed a simply connected negatively curved manifold X with $\sigma(\Delta_X)$ consisting of only eigenvalues with finite multiplicity. In their example, however, the curvature goes to $-\infty$ at infinity, so that X can never be a manifold with compact quotient. The Conjecture B is a bit related to the discussion in Appendix 2.

From now on, we confine ourselves to the special situation that X has an isometry group Γ, discontinuously acting on X, with compact quotient space $\Gamma\backslash X$. What has been known so far on the continuous spectrum is:

Proposition 3. The bottom $\lambda_o(H)$ is in $\sigma_c(H)$. Especially, $\sigma_c(H) \neq \emptyset$.

This is a consequence of the following two facts.

Fact 1 (cf. P.Sarnak [16]). If λ is an eigenvalue of H, then the multiplicity of λ is infinite.

To show this, we first note that there exists a Γ-isomorphism

$$L^2(X) \cong L^2(\Gamma) \hat{\otimes} L^*(D)$$

where D is a fundamental domain in X for the Γ-action. We regard $L^2(D)$ as a trivial Γ-module. Since the eigenspaces of H are Γ-invariant (this is only the condition we shall use), it suffices to show that an arbitrary Γ-subspace V in $L^2(\Gamma)$ is (0) or of infinite dimension. Let $\{v_1,\ldots,v_N\}$ be an orthonormal basis of V. We then have

$$v_i(\sigma\mu) = \sum_{j=1}^{N} u_{ij}(\sigma)v_j(\mu) \qquad \sigma,\mu\in\Gamma$$

for some unitary matrix $(u_{ij}(\sigma))$. We now find

$$\begin{aligned} N = \sum_{i=1}^{N} \|v_i\|^2 &= \sum_{i=1}^{N}\sum_{\sigma\in\Gamma} |v_i(\sigma\mu)|^2 \\ &= \sum_{\sigma\in\Gamma}\sum_{i=1}^{N}\sum_{j=1}^{N} u_{ij}(\sigma)\,\overline{u_{ik}(\sigma)}\, v_j(\mu)\overline{v_k(\mu)} \\ &= \sum_{\sigma\in\Gamma}\sum_{j=1}^{N} |v_j(\mu)|^2 \\ &= \#\Gamma \sum_{j=1}^{N} |v_j(\mu)|^2 \quad . \end{aligned}$$

Since we have assumed that $\#\Gamma = \infty$, it follows that $N = 0$ or ∞.

Fact 2 (cf. D.Sullivan [19]). If the bottom λ_0 is an eigenvalue of H, then the multiplicity of λ_0 is just one (this is true for arbitrary complete manifolds and bounded potential functions).

To establish this, consider the semigroup e^{-tH} acting in $L^2(X)$ with the generator $-H$. According to the Feynmann-Kac formula, the operator e^{-tH}, $t>0$, has a smooth positive kernel function, say $h(t,x,y)$. On the other hand, since $H-\lambda_0$ is positive, the operator $e^{-t(H-\lambda_0)}$ is a contraction of $L^2(X)$, that is,

$$\|e^{-t(H-\lambda_0)}f\| \leqq \|f\|.$$

We now employ the Perron-Frobenius-theorem-argument to establish Fact 2, which goes as follows. Let f be a real valued eigenfunction of H with the eigenvalue λ_0. We then have

$$e^{-tH}f = e^{-t\lambda_0}f \quad .$$

Since $h(t,x,y) > 0$, we find that

$$e^{-t\lambda_o}|f(x)| = |\int_X h(t,x,y)f(y)dy|$$
$$\leq \int_X h(t,x,y)|f(y)|dy$$
$$= e^{-tH}|f|(x) \quad ,$$

and

$$|f(x)|^2 \leq |(e^{-t(H-\lambda_o)}|f|)(x)|^2 \quad .$$

Therefore, taking account of the contraction property, we find

$$|f| = e^{-t(H-\lambda_o)}|f| \ .$$

Suppose that $f(x) > 0$ for some point x in X . Then we have

$$\int_X h(t,x,y)\, f(y)\, dy = e^{-t\lambda_o} f(x)$$
$$= e^{-t\lambda_o} |f(x)|$$
$$= \int_X h(t,x,y)|f(y)|dy \quad ,$$

from which it follows that $f(y) = |f(y)| \geq 0$ for all $y \in X$. This, in particular, implies that for every eigenfunction f with eigenvalue λ_o , one can find a scalar c with $cf \geq 0$. It is an easy consequence that the multiplicity of λ_o is one.

In the case that X is an abelian covering space of a compact manifold (a direct generalization of periodic Schrödinger operators), we may have more about $\sigma_c(H)$. To state our proposition, we need a concept of "twisted" Laplacians associated with characters of the covering transformation group. Since the concept in more general form turns out to be useful in the later discussion, we shall define here the Laplacians associated with possibly infinite dimensional unitary representations.

Let us fix a normal covering map $\pi: X \to M$ with covering transformation group Γ . Let ρ be a unitary representation of Γ on a Hilbert space V . We denote

$$C^\infty(E_\rho) = \{s: X \to V \text{ smooth; } s(\sigma x) = \rho(\sigma)\, s(x) \text{ for } \sigma \in \Gamma \ , \ x \in X\} ,$$

which is, in a natural manner, regarded as the space of smooth sections of the flat vector bundle $E_\rho = \Gamma \setminus X \times V$. For each smooth $s: X \to V$, one can define $\Delta s = \Delta_X s : X \to V$ in the usual way, and prove that $\Delta(C^\infty(E_\rho)) \subset C^\infty(E_\rho)$. We put $\Delta_\rho = \Delta|C^\infty(E_\rho)$ and $H_\rho = \Delta_\rho + q(x)$. It is shown that Δ_ρ (and H_ρ) has a unique positive self-adjoint extension to $L^2(E_\rho)$, where $L^2(E_\rho)$, the space of square integrable sections of E_ρ , is the completion of

$C^\infty(E_\rho)$ with respect to the norm $\|\cdot\|$ defined by

$$\|s\|^2 = \int_D \|s(x)\|_V^2 \, dx \quad ,$$

D being a fundamental domain.

If ρ is a finite dimensional representation, H_ρ is an elliptic operator in the usual sense, so that $\sigma(H_\rho)$ consists of eigenvalues with finite multiplicity.

Example. Let $\pi: \mathbb{R} \to \mathbb{R}/\mathbb{Z}$ be the standard covering map onto the circle. A one dimensional representation (character) χ of $\mathbb{Z}$ is identified with a real number α with $0 \leqq \alpha < 2\pi$ by the formula

$$\chi(n) = e^{\sqrt{-1}n\alpha} \quad , \qquad n \in \mathbb{Z} \quad ,$$

so that a section of E_χ is identified with a function u on $\mathbb{R}$ satisfying

$$u(x+1) = e^{\sqrt{-1}\alpha} u(x) \ .$$

Therefore the spectral problem of the twisted Schrödinger operator H_χ in this example is equivalent to the periodic boundary value problem for a Hill's operator explained in Example 2.

Proposition 4. Let $\pi: X \to M$ be a normal covering map with abelian covering transformation group Γ. Then we have

(1) $\qquad \sigma(H) = \bigcup_{\chi \in \hat{\Gamma}} \sigma(H_\chi) \quad ,$

where $\hat{\Gamma}$ denotes the character group of Γ. The spectrum $\sigma(H)$ is a disjoint union of closed intervals: $[\lambda_0,\lambda_1] \cup [\lambda_2,\lambda_3] \ \ldots$, where $\lambda_{2i-1} < \lambda_{2i}$ and $\lambda_i \uparrow \infty \ (i \to \infty)$.

(2) $\qquad \lambda_0(H) = \lambda_0(H_M) \qquad (H_M = \Delta_M + q) \quad ,$ and there exists a number $\lambda > \lambda_0(H)$ such that

$$[\lambda_0(H),\lambda] \subset \sigma_c(\Delta_X) \ .$$

An idea of the proof of Proposition 4 is the following (cf. Reed-Simon [15]). We identify the Hilbert space $L^2(X)$ with the direct integral

$$\int_{\hat{\Gamma}}^{\oplus} L^2(E_\chi) d\chi \quad ,$$

where $d\chi$ denotes the normalized Haar measure on the abelian topological group $\hat{\Gamma}$. The identification is given by the extension of the linear map:

$$f \in C_0^\infty(X) \ \to \ s_\chi \in L^2(E_\chi) \quad ,$$

where $s_\chi(x) = \sum_{\sigma\in\Gamma} \chi(\sigma^{-1})f(\sigma x)$. Under this identification, H is also decomposed into a direct integral

$$\int_{\hat{\Gamma}}^{\oplus} H_\chi \, d\chi \quad .$$

Notice that the spectrum of H is characterized in the following way:

$$\sigma(H) = \{ \lambda \ ; \text{ the measure of the set } \{\chi \in \hat{\Gamma} \ ; \ \sigma(H_\chi) \cap (\lambda-\varepsilon, \lambda+\varepsilon) \neq \emptyset \} \text{ is positive for every } \varepsilon > 0 \}$$

$$\sigma_p(H) = \{ \lambda \ ; \text{ the measure of the set } \{\chi \in \hat{\Gamma} \ ; \ \lambda \text{ is an eigenvalue of } H_\chi\} \text{ is positive } \} \quad .$$

Since the eigenvalues of H_χ depend continuously on the character χ, the first assertion follows from the above characterization of $\sigma(H)$.

The second assertion comes from the following observations.

a) $\lambda_o(H_\chi) \geq \lambda_o(H_M)$. The equality holds if and only if $\chi = 1$, the trivial character.

b) $\lambda_o(H_M)$ is a simple eigenvalue of H_M .

These are also consequences of Perron-Frobenius-theorem-arguments. We omit the detail.

One may now ask if H happens to have an eigenvalue. Here are several examples.

Example A. Let G be a simple Lie group of non-compact type with discrete series of irreducible representations, or equivalently G is assumed to have a compact Cartan subgroup. It is easily checked that G has a left invariant metric (obtained by modifying Killing form) with respect to which the Laplacian Δ_G has infinitely many eigenvalues. It is known, on the other hand, that G has a co-compact discrete subgroup, so that G is a Riemannian manifold with compact quotient. In the particular case $G = SL_2(\mathbb{R})$, which is, in a natural manner, identified with the unit tangent sphere bundle over the upper half plane H^2 , we may obtain

$$\sigma(\Delta_G) = [1/4, \infty)$$

$$\sigma_p(\Delta_G) = \{\tfrac{1}{4} (n^2+4mn+2m^2+1); \ n=1,2,3,\ldots, \ m=1,3,5,\ldots\} \quad .$$

Some computations of $\sigma_p(\Delta_G)$ for other simple groups have been carried out by Toshiyuki Kobayashi (see [13]).

Example B (due to Kaoru Ono, see [13]). Let N be a compact manifold with a spin structure $P \to N$. Suppose that the $\hat{A}$-genus of

N is not equal to zero. Then the induced principal bundle $\tilde{P}$ on the universal covering space $\tilde{N}$ has a metric with compact quotient such that $\Delta_{\tilde{P}}$ has an eigenvalue . Using this observation, one may construct a (4-dimensional) compact Riemannian manifold M with free abelian fundamental group such that $\sigma_p(\Delta_{\tilde{M}}) \neq \emptyset$, where $\tilde{M}$ denotes the universal covering manifold of M . This example is interesting from the view of the fact that periodic Schrödinger operators have no point spectre.

Problem. Find an open Riemann surface X with compact quotient such that $\sigma_p(\Delta_X) \neq \emptyset$.

We proceed to the problem which asks relationships between the spectrum of Δ_M and that of Δ_X , X being a covering space of M . Recall the simple relation : $\Delta_X(f \circ \pi) = (\Delta_M f) \circ \pi$, $f \in C^\infty(M)$. One might still think that this relation makes it easy to establish some relations between $\sigma(\Delta_M)$ and $\sigma(\Delta_X)$. Actually this is the case if the covering map has finite fiber, or equivalently if the covering space X is also compact.

Proposition 5. Let $\pi: X \to M$ be a normal finite-fold covering map. Then

$$\sigma(\Delta_X) = \bigcup_{\rho \in \hat{\Gamma}} \sigma(\Delta_\rho) ,$$

where $\hat{\Gamma}$ denotes the set of all the equivalence classes of irreducible representations of the covering transformation group.

Especially, we have that $\sigma(\Delta_M) \subset \sigma(\Delta_X)$, which is a direct consequence of the relation $\Delta_X \circ \pi^* = \pi^* \Delta_M$ (just note that if f is in $L^2(M)$, then $f \circ \pi$ is in $L^2(X)$) . But if Γ is not finite, this argument is not available. In fact, you easily have an example of an infinite-fold covering map with $\sigma(\Delta_M) \not\subset \sigma(\Delta_X)$, say, we put $X = H^2$ and M = a closed surface with constant (-1) negative curvature. In this example, $\sigma(\Delta_M) \ni 0 \notin \sigma(\Delta_{H^2}) = [1/4,\infty)$, and it should be noted that $\pi_1(M)$ contains a free subgroup of rank ≥ 2 .

We have seen in Proposition 4 that the same statement as in Proposition 5 holds for abelian covering case. We should point out that finite groups and abelian groups belong to the class of amenable groups.

A group Γ is called amenable if there exists an invariant mean on Γ . An invariant mean is a continuous functional m on the real Banach space $L^\infty_{\mathbb{R}}(\Gamma)$ of bounded real valued functions on Γ , satisfying the following conditions:

1) $m(1) = 1$,

2) if $f \geq 0$, then $m(f) \geq 0$, and

3) if we put $(\sigma \cdot f)(\mu) = f(\sigma^{-1}\mu)$, then $m(\sigma \cdot f) = m(f)$ for all $\sigma \in \Gamma$ and $f \in L^{\infty}_{\mathbb{R}}(\Gamma)$.

We shall list up several facts on amenability of groups (see F.P.Greenleaf [8] for the detail).

i) Every finite group is amenable. In fact, the ordinary mean

$$m(f) = (\#\Gamma)^{-1} \sum_{\sigma \in \Gamma} f(\sigma)$$

gives an invariant mean.

ii) Every abelian group is amenable. The existence of an invariant mean is a consequence of the Axiom of Choice.

iii) A free group with k generators is non-amenable provided $k > 1$.

iv) If Γ is amenable, then so are all subgroups of Γ .

From iii) and iv) , it follows that the fundamental group of a closed surface with constant negative curvature is not amenable.

We wish to generalize Proposition 4 and 5 (we assume $q(x) \equiv 0$) to the case of amenable coverings. For this, we first look at two examples which illuminate the importance of the concept of twisted Laplacians in the study of the spectrums of Laplacians on covering spaces.

Example 1. Let $\pi: X \to M$ be a normal covering, and $\rho = 1$, the trivial representation of the covering transformation group. Then we easily find

$$C^{\infty}(E_{\rho}) = C^{\infty}(M),$$

$$L^{2}(E_{\rho}) = L^{2}(M) ,$$

$$\Delta_{\rho} = \Delta_{M} .$$

Example 2. Let $\rho = \rho_r$ = the (right) regular representation of the covering transformation group Γ . Namely, Γ acts on $V = L^2(\Gamma)$, the space of square summable functions on Γ , by $(\rho_r(\sigma)f)(\mu) = f(\mu\sigma)$. Then there exists a unitary isomorphism

$$\Phi : L^2(X) \cong L^2(E_{\rho_r})$$

such that $\Phi \circ \Delta_X = \Delta_{\rho_r} \circ \Phi$. Especially one has

$$\sigma(\Delta_X) = \sigma(\Delta_{\rho_r}) .$$

The map Φ is defined by the extension of the linear map

$$C_o^\infty(X) \to L^2(E_{\rho_r})$$

$$f \to s \quad ,$$

where $s(x)(\sigma) = f(\sigma x)$.

Thus the concept of twisted Laplacians includes both the Laplacian Δ_M on the base manifold M and the Laplacian Δ_X on the covering manifold X .

Proposition 6. Let $\pi: X \to M$ be an arbitrary normal covering. Then

$$\sigma(\Delta_X) \subset \overline{\bigcup_{\rho \in \hat{\Gamma}} \sigma(\Delta_\rho)} \ ,$$

where $\hat{\Gamma}$ denotes the set of equivalence classes of unitary representations of the covering transformation group Γ .

This is proved by means of direct integral arguments. Here is a sketch of proof. We decompose the right regular representation of Γ into a direct integral of irreducible unitary representations:

$$L^2(\Gamma) = \int_S^\oplus V_\rho \, d\mu(\rho)$$

(since Γ is, in general, not a group of type I , uniqueness of the decomposition does not hold).
Take now the element δ in $L^2(\Gamma)$ defined by

$$\delta(\sigma) = \begin{cases} 1 & \text{if } \sigma = 1 \\ 0 & \text{otherwise} . \end{cases}$$

Let $v: S \to \bigcup_{\rho \in S} V_\rho$ $(v(\rho) \in V_\rho)$ be the function corresponding to δ ,
Given a function f with compact support, we define $s_\rho: X \to V_\rho$ by setting

$$s_\rho(x) = \sum_{\sigma \in \Gamma} f(\sigma x) \, \rho(\sigma^{-1}) \, v(\rho) \ .$$

It is easy to check that $s_\rho(\sigma x) = \rho(\sigma) \, s_\rho(x)$ for $\sigma \in \Gamma$ and $x \in X$, so that s_ρ is a section of E_ρ . We may conclude, by using rather formal discussion, that the linear map $f \to \{s_\rho ; \rho \in S\}$ extends to a unitary isomorphism

$$L^2(X) \cong \int_S^\oplus L^2(E_\rho) \, d\mu(\rho) \quad ,$$

under which Δ_X is also decomposed into $\int_S^\oplus \Delta_\rho \, d\mu(\rho)$. Therefore

$$\sigma(\Delta_X) \subset \overline{\bigcup_{\rho \in S} \sigma(\Delta_\rho)} \subset \overline{\bigcup_{\rho \in \hat{\Gamma}} \sigma(\Delta_\rho)}$$

Here is an elementary remark. Suppose the group Γ is of finite order. Then the regular representation ρ_r is decomposed into irreducible representations in such a way that

$$\rho_r = \sum_{\rho \in \hat{\Gamma}} \oplus (\dim \rho)\, \rho ,$$

so that

$$\Delta_{\rho_r} = \sum_{\rho \in \hat{\Gamma}} \oplus (\dim \rho)\, \Delta_\rho$$

and

$$\sigma(\Delta_X) = \sigma(\Delta_{\rho_r}) = \bigcup_{\rho \in \hat{\Gamma}} \sigma(\Delta_\rho) ,$$

which proves Proposition 5.

We now get to a generalization of Proposition 4 and 5.

<u>Proposition</u> 7. If $\pi: X \to M$ be a normal covering map with an amenable covering transformation group, then

$$\sigma(\Delta_X) = \overline{\bigcup_{\rho \in \hat{\Gamma}} \sigma(\Delta_\rho)} .$$

In view of Proposition 6, to establish this proposition, it is enough to show

$$\sigma(\Delta_\rho) \subset \sigma(\Delta_X)$$

for all unitary representation ρ. This is, in fact, a corollary of the following theorem in which we do not assume the amenability of Γ.

<u>Theorem</u> I ([25]). Let ρ_1 and ρ be unitary representation of Γ. If ρ_1 is weakly contained in ρ, then

$$\sigma(\Delta_{\rho_1}) \subset \sigma(\Delta_\rho) .$$

The notion of weak containment was introduced by J.M.G.Fell in his article [8]. Recall first the usual notion of containment. Given two unitary representations ρ_1 and ρ of Γ on Hilbert spaces V_1 and V respectively, we say that ρ_1 is contained in ρ if there exists an isometry $V_1 \subset V$ such that $\rho_1 = \rho | V_1$. It is obvious that $\sigma(\Delta_{\rho_1}) \subset \sigma(\Delta_\rho)$ whenever ρ_1 is contained in ρ. The concept of weak containment is a weak version of containment and defined as follows. A representation ρ_1 is said to be weakly contained in ρ if for every positive ε, and every orthonormal set $\{v_1, \ldots, v_n\}$ in V_1, there exists an orthonormal set $\{u_1, \ldots, u_n\}$ in V such that

$$|\langle v_i, \rho_1(\sigma)v_j\rangle_{V_1} - \langle u_i, \rho(\sigma)u_j\rangle_V| < \varepsilon$$

for all $\sigma \in A$ and $i,j=1,\ldots,n$. Here we fix a finite set A of generators of Γ. If ρ_1 is contained in ρ, then ρ_1 is weakly contained in ρ.

If Γ is amenable, then every unitary representation of Γ is weakly contained in the direct sum of copies of the regular representation (see R.J.Zimmer [27]), hence we get Proposition 7.

We may establish the converse of the statement in Proposition 7.

Proposition 8. Let $\pi: X \to M$ be a normal covering. If $\sigma(\Delta_1) \equiv \sigma(\Delta_M) \subset \sigma(\Delta_X)$, then the covering transformation group Γ is amenable.

In fact, if one notes that 0 lies in $\sigma(\Delta_M)$, the proposition comes from the following remarkable theorem by R.Brooks [2].

Theorem. 0 is in $\sigma(\Delta_X)$ (that is, $\lambda_0(\Delta_X) = 0$) if and only if Γ is amenable.

We will observe that the Brooks theorem is also proven by means of twisted Laplacians. For this, we put

$$\lambda_0(\rho) = \lambda_0(\Delta_\rho) \equiv \text{the bottom of the spectrum of } \Delta_\rho .$$

What we are going to do is estimates of $\lambda_0(\rho)$ in purely representation-theoretic terms of ρ. We define $\delta(\rho,1)$, a sort of distance between ρ and the trivial representation 1, by

$$\delta(\rho,1) = \delta_A(\rho,1) = \inf_v \sup_{\sigma\in A} \|\rho(\sigma)v-v\|_V$$

where the supremum is taken over a finite set A of generators of Γ, and v runs over all the unit vectors in V (see R.Brooks [3]).

Theorem II ([25]). There exists positive constants c_1 and c_2 not depending on ρ such that for all ρ,

$$c_1\delta(\rho,1)^2 \leq \lambda_0(\rho) \leq c_2\,\delta(\rho,1)^2 .$$

In particular $\lambda_0(\rho) = 0$ if and only if $\delta(\rho,1) = 0$.

The Brooks theorem is rediscovered from this theorem. In fact, the following lemma is known (see R.J.Zimmer [27]).

Lemma Let ρ_r be the regular representation of Γ. Then $\delta(\rho_r,1) = 0$ if and only if Γ is amenable.

If we restrict ourselves to the one-dimensional representations (characters), we may get more precise estimates for $\lambda_o(\Delta_\rho)$. For this, we identify the component of the character group of $\pi_1(M)$, containing the trivial character, with the Jacobian torus

$$J(M) = H^1(M,\mathbb{R})/H^1(M,\mathbb{Z}).$$

Identifying the 1-cohomology group $H^1(M,\mathbb{R})$ with the space of harmonic one forms on M, we set, for each $\omega \in H^1(M,\mathbb{R})$,

$$\chi_\omega(\sigma) = \exp 2\pi\sqrt{-1} \int_{C(\sigma)}\omega ,$$

where $C(\sigma)$ is a closed curve whose homotopy class corresponds to σ. Then the map $\omega \to \chi_\omega$ yields an isomorphism of $J(M)$ onto the component of the character group. We put

$$\lambda_o(\omega) = \lambda_o(\chi_\omega) ,$$

so that $\lambda_o(\omega)$ is a continuous function on the Euclidean space $H^1(M,\mathbb{R})$, and satisfies $\lambda_o(0) = 0$ and $\lambda_o(\omega) \geqq 0$. Since $\lambda_o(0)$ is a simple eigenvalue of Δ_M, so is $\lambda_o(\omega)$ provided that ω is in a sufficiently small neighborhood of $\omega \equiv 0$. By using a perturbation technique, we obtain

Proposition 9 ([10]). The function $\lambda_o(\omega)$ is smooth in a neighborhood of $\omega \equiv 0$, and satisfies $\lambda_o(-\omega) = \lambda_o(\omega)$ and

$$\lambda_o(\omega) = \frac{4\pi^2}{\mathrm{vol}(M)} \|\omega\|^2 + O(\|\omega\|^4) ,$$

where $\|\omega\|^2 = \int_M|\omega|^2$. Further one has

$$\lambda_o(\omega) \leqq \frac{4\pi^2}{\mathrm{vol}(M)} \|\omega\|^2$$

for every $\omega \in H^1(M,\mathbb{R})$.

The above proposition is useful in establishing a density theorem for closed geodesics in a compact hyperbolic manifold (see [10] and R.Phillips and P.Sarnak [12]).

We conclude this lecture with some remarks in connection with the "Riemann hypothesis" for the Selberg zeta functions. The point is to examine the estimate $\lambda_o(\Delta_{H^2}) \leqq \lambda_1(\Delta_M)$ for a given compact surface with the universal covering H^2. This is generalized to the problem to compare $\lambda_1(\Delta_M)$ with $\lambda_o(\Delta_X)$ when we are given a covering map $\pi: X \to M$.

Proposition 10 ([23]). Let $\pi: X \to M$ be a normal covering map, and let

$$\cdots \to M_{n+1} \to M_n \to \cdots \to M_1 \to M$$

be a tower of subcoverings of π. If $\bigcap_{n=1}^{\infty} \Gamma_n = (1)$ for the covering transformation groups Γ_n of the covering maps $X \to M_n$, then

$$\lim_{n\to\infty} \lambda_1(\Delta_{M_n}) \leqq \lambda_o(X) .$$

In particular, if $X \to M$ is an amenable covering, then

$$\lim_{n\to\infty} \lambda_1(\Delta_{M_n}) = 0 .$$

Applying this to the homology universal covering X of a compact Riemann surface, we could find a compact surface M with $\lambda_1(\Delta_M) < 1/4$. That is, there exists a surface which does not satisfy the "Riemann hypothesis" (see B.Randol [14] and P.Buser [6] for different constructions). This is also the case for the Ihara zeta functions.

Incidentally, it is also known that a series of finite regular graphs and a series of Riemann surfaces constructed in arithmetic manners satisfy "Riemann hypothesis". See [32] for instance.

It might be interesting to mention here the Selberg conjecture which corresponds to "Riemann hypothesis" for some special surfaces with finite volume. Let Γ_N be the principal congruence subgroup of $SL(2,\mathbb{Z})$, that is, a group of two-by-two matrices with integer entries congruent to the identity modulo N. Then $\Gamma_N\backslash H^2$ is a surface of finite volume with cusps. The Laplacian on $\Gamma_N\backslash H^2$ has infinitely many eigenvalues together with continuous spectrum. We denote by $\lambda_1(\Gamma_N\backslash H^2)$ the first positive eigenvalue. The conjecture says: $\lambda_1(\Gamma_N\backslash H^2) \geqq 1/4$. It is known that this is true for several N. A.Selberg [30] established the following estimate:

$$\lambda_1(\Gamma_N\backslash H^2) \geqq 3/16 \qquad (\forall N) .$$

On the other hand, employing a perturbation method of eigenvalues of twisted Laplacian, we may construct a series of subgroups of finite index

$$\cdots \subset \Gamma^k \subset \Gamma^{k-1} \subset \cdots \subset \Gamma^1 \subset SL(2,\mathbb{Z})$$

such that $\lim_{k\to\infty} \lambda_1(\Gamma^k\backslash H^2) = 0$.

The following is interesting from the above view.

<u>Proposition</u> 11 (cf. R.Brooks [3], [4]). Given a normal covering $\pi: X \to M$, we let $\{M_\alpha ; M_\alpha \to M\}$ be the set of finite-fold subcovering spaces of π , and let $\hat{\Gamma}_f$ be the set of all non-trivial finite dimensional irreducible representations of the covering transformation group of π . Then the following two statements are equivalent:

1) $\inf \lambda_1(M_\alpha) > 0$,
2) $\inf \{\delta(\rho, 1);\ \rho \in \hat{\Gamma}_f\} > 0$.

We should point out that the above condition (2) is a weak version of the Kazhdan property (T) (D.A.Kazhdan [11]), that is, if $\hat{\Gamma}_f$ in (2) is replaced by the set of all non-trivial irreducible representations, then the condition is what we call the Kazhdan property (T) .

Appendix 1

The aim of this appendix is to give a formulation of our problems in the category of graphs.

Let M be a graph. We assume that $\# V(x) < \infty$ for every vertex x in M , and put $U(x) = \{x\} \cup V(x)$. We also assume that a non-negative number $p(x,y)$ is assigned to each pair $\{x,y\}$ with $y \in U(x)$ in such a way that

a) $p(x,y) = p(y,x)$, and $p(x,y) > 0$ if $y \in V(x)$.

b) $\sum_{y \in U(x)} p(x,y) = 1$.

Thus $\{p(x,y)\}$ gives transition probabilities of a random walk on the graph M . The transition operator $P = P_M$ of this random walk is defined by

$$(Pf)(x) = \sum_{y \in U(x)} p(x,y)\, f(y) .$$

It is easy to check that P is a bounded self-adjoint operator acting in $L^2(M)$ with $\|P\| \leq 1$.

In the case of regular graphs, putting

$$p(x,y) = \frac{1}{q+1} , \quad y \in V(x) ,$$
$$p(x,x) = 0 ,$$

we have an isotropic random walk. The relation between the adjacency operator and the transition operator is : $P = \frac{1}{q+1} A$.

Let M be a finite graph with transition probabilities $\{p(x,y)\}$, and let $\pi: X \to M$ be a covering map of graphs (that is, π is surjective and gives a bijection of $V(x)$ onto $V(\pi(x))$ for each $x \in X$). Lifting the transition probabilities up to X , we obtain a random walk on X and the relationship

$$P_X(f \circ \pi) = \pi \circ (P_M f) ,$$

f being an arbitrary function on M . Our problem is to seek for relationships between $\sigma(P_X)$ and $\sigma(P_M)$. The reader may find strong resemblance to the problem for Laplacians. Actually, we can establish several results with a similar taste as in the case of Laplacians. The following is one of those results.

Proposition. Let $\pi: X \to M$ be a normal covering map. Then $1 \in \sigma(P_X)$ if and only if the covering transformation group is amenable.

The above proposition is quite a bit related to a criterion on amenability of discrete groups due to H.Kesten [29].

Appendix 2

As is seen in Example 2, the discriminant of a Hill's operator is a useful concept in the study of the spectrum. In the general n-dimensional case, however, we may no longer make use of the concept since the dimension of the space $\{f \in C^\infty(X); Hf = \lambda f\}$ is infinite. We shall seek for something else which compensates the lack.

Let X be a Riemannian manifold with compact quotient $\Gamma\backslash X$, and let $H = \Delta_X + q(x)$ be a Schrödinger operator with a Γ-invariant potential $q(x)$.

We define $End^\Gamma(L^2(X))$ to be the set of bounded operators acting in $L^2(X)$ which commute with the Γ-action on $L^2(X)$. It is known that $End^\Gamma(L^2(X))$ is a von Neumann algebra of type II_∞ , so that it has a trace. We may choose a trace tr_Γ satisfying the formula

$$tr_\Gamma A = \int_D a(x,x)dx \quad ,$$

where A is assumed to have a continuous kernel function $a(x,y)$, and D is a fundamental domain X for the Γ-action (see M.F.Atiyah [1]). What we should notice here is that $e^{-tH} \in End^\Gamma(L^2(X))$ and has the definite trace $tr_\Gamma\, e^{-tH}$.

We now let

$$\int_{-\infty}^{\infty} \lambda \, dE(\lambda)$$

be the spectral decomposition of H. Recall that the family $\{E(\lambda)\}_{\lambda\in\mathbb{R}}$ of projections is assumed to satisfy

a) if $\lambda < \mu$, then $E(\lambda) \leqq E(\mu)$,

b) $E(\infty) = I$, and $E(\lambda+0) = E(\lambda)$.

Notice that

$$\sigma(H) = \{\lambda\in\mathbb{R}\,;\, E(\lambda-\varepsilon) \lneqq E(\lambda+\varepsilon) \text{ for every positive } \varepsilon\},$$

$$\sigma_p(H) = \{\lambda\in\mathbb{R}\,;\, E(\lambda-0) \neq E(\lambda)\} .$$

Since

$$e^{-tH} = \int e^{-t\lambda}\, dE(\lambda) ,$$

putting $\phi(\lambda) = \mathrm{tr}_\Gamma\, E(\lambda)$, we get

$$\mathrm{tr}_\Gamma\, e^{-tH} = \int_{-\infty}^{\infty} e^{-t\lambda}\, d\phi(\lambda) .$$

The function $\phi(\lambda)$ is non-decreasing, and characterizes the spectrum of H in the following way.

$$\sigma(H) = \{\lambda \in \mathbb{R}\,;\, \phi(\lambda+\varepsilon) > \phi(\lambda-\varepsilon) \text{ for all positive } \varepsilon\}$$

$$\sigma_p(H) = \{\lambda \in \mathbb{R}\,;\, \phi(\lambda) > \phi(\lambda-0)\} .$$

<u>Proposition</u>. If Γ acts freely on X, then

$$\phi(\lambda) \sim \frac{\omega_n \,\mathrm{vol}(M)}{(2\pi)^n} \lambda^{n/2} \quad \text{as } \lambda \uparrow +\infty ,$$

where ω_n denotes the volume of the unit ball in $\mathbb{R}^n$, and $M = \Gamma\backslash X$.

It is interesting to note that the asymptotic behaviour of $\phi(\lambda)$ does not depend on the potential, and is the same as that of the counting function of eigenvalues of Δ_M: $\sum_{\lambda_j \leqq \lambda} 1$.

A proof of the above proposition relies on the fact that $\mathrm{tr}_\Gamma\, e^{-tH}$ has the same asymptotics as $\mathrm{tr}\, e^{-t\Delta_M}$ as $t \downarrow 0$, that is,

$$\mathrm{tr}_\Gamma\, e^{-tH} \sim \frac{\mathrm{vol}(M)}{(4\pi t)^{n/2}} \quad (t \downarrow 0) .$$

We omit the detail.

<u>Proposition</u>. Assume that Γ is abelian, and let h be the order of the torsion part of Γ.

a) If the interval $[\lambda_1,\lambda_2]$ is a connected component of $\sigma(H)$, then $\phi(\lambda_2) - \phi(\lambda_1) = \frac{m}{h}$ for some integer $m \geqq 1$.

b) If $\lambda \in \sigma_p(H)$, then $\phi(\lambda) - \phi(\lambda-0) = \frac{m}{h}$ for some integer $m \geqq 1$.

The assertion b) is a special case of H.Donnelly [35] .

Example. In the one dimensional case, the function $\phi(\lambda)$ is connected with the discriminant by the formula

$$\int_{-\infty}^{\infty} \frac{1}{s+\lambda}\, d\phi(\lambda) = \frac{-\Delta'(-s)}{\sqrt{\Delta(-s)^2-4}} .$$

In fact, we established, in [20] , the equality

$$\int_0^{\infty} e^{-st}\, \mathrm{tr}_{\mathbb{Z}}\, e^{-tH} dt = \frac{-\Delta'(-s)}{\sqrt{\Delta(-s)^2-4}}$$

so that, noting

$$\int_0^{\infty} e^{-st} \int_{-\infty}^{\infty} e^{-t\lambda}\, d\phi(\lambda)dt = \int_{-\infty}^{\infty} \frac{1}{s+\lambda}\, d\phi(\lambda) ,$$

we get the desired formula.

The function $\phi(\lambda)$ is related to the Wiener measures on the space of free loops in M , provided that X is the universal covering space of M . We let Ω be the space of continuous maps of the circle S^1 into M . The Wiener measure μ_t, $t>0$, is a measure on Ω characterized by the relationship: For $0 \leqq \tau_1 < \tau_2 < \dots < \tau_N < 1$ and for a continuous function f on the N-ple product $M\times\dots\times M$,

$$\int_\Omega f(c(\tau_1),\dots,c(\tau_N))\, d\mu_t(c)$$

$$= \int_{M\times\dots\times M} k(t(\tau_2-\tau_1),x_1,x_2)\dots k(t(\tau_N-\tau_{N-1}),x_{N-1},x_N)$$

$$\times\, k(t(1+\tau_1-\tau_N),x_N,x_1) f(x_1,\dots,x_N) dx_1\dots dx_N ,$$

where $k(t,x,y)$ denotes the kernel function of the operator $e^{-t\Delta_M}$

Proposition ([21]) . Let Ω_0 be the connected component of consisting of null-homotopic loops. Then

$$\int_{-\infty}^{\infty} e^{-t\lambda}\, d\phi(\lambda) = \int_{\Omega_0} e^{-t\int_0^1 q(c(\tau))d\tau} d\mu_t(c) .$$

This proposition implies that if you know much about (Ω_0, μ_t) , then you may have, at least in principle, some of the information on the spectrum $\sigma(\Delta_X)$.

Appendix 3

For the reader's convenience, we shall give here brief accounts of the Fourier transformations defined on the upper half plane H^2 and regular trees, which give unitary equivalence between those geometric operators (the Laplacian and the adjacency operator) and certain multiplicative operators, thereby leading us to complete descriptions of spectrums.

First consider the case of the upper half plane. We identify, by using the Cayley transformation, H^2 with the unit disc $D = \{ Z \in \mathbb{C} \ ; \ |Z| < 1 \}$ with the metric

$$ds^2 = \frac{4|dZ|^2}{(1-|Z|^2)^2} \quad .$$

Define a function F_ν on $D \times \partial D$ by setting

$$F_\nu(Z, e^{i\theta}) = \left(\frac{1-|Z|^2}{|Z-e^{i\theta}|^2} \right)^{\frac{1}{2}+i\nu}$$

It is easy to see that

$$\Delta_D F_\nu = \left(\frac{1}{4} + \nu^2 \right) F_\nu \ .$$

For f $C_0^\infty(D)$, put

$$(F f)(\nu, e^{i\theta}) = \frac{1}{2\pi} \int_D F_{-\nu}(Z, e^{i\theta}) f(Z) \, dv(Z) \quad ,$$

where $dv(Z)$ is the density associated with the metric ds^2 . Then F is a non-Euclidean version of Fourier transform, and satisfies the following properties

a) F extends to a surjective unitary map

$$F: L^2(D) \to L^2([0,\infty) \times \partial D \ , \ \nu \tanh \pi\nu \, d\nu \, d\theta \ ) \ .$$

b) $(F \Delta_D F^{-1} \phi)(\nu, e^{i\theta}) = (\frac{1}{4} + \nu^2) \, \phi(\nu, e^{i\theta}) \quad .$

We find, therefore, that $\sigma(\Delta_D) = \sigma_c(\Delta_D) = [^1/4, \infty)$.

A regular tree is, in some sense, a discrete model of the unit disc, so that we may imitate the construction to define the Fourier transformation on the tree.

Let X be a regular tree with $\# V(x) \equiv q+1$. Define a distance d on X by setting

$$d(x,x) = 0$$

$$d(x,y) = n \quad \text{if there exists a path } (x_o,\dots,x_n) \text{ with } x_o = x,\ x_n = y, \text{ and } x_{i-1} \neq x_{i+1} \text{ for } i=1,\dots,n-1 .$$

We fix a point 0 in X, and put

$$X_n = \{ x \in X \ ;\ d(0,x) = n \} .$$

For each $x \in X(n)$, there exists a unique vertex $y \in X(n-1)$ with $y \in V(x)$. Let $\pi_n : X(n) \to X(n-1)$ be a map defined by $\pi_n(x) = y$. The boundary of X is then defined as the projective limit

$$\varprojlim\, (X(n), \pi_n) ,$$

and denoted by ∂X. By Kolmogoloff's theorem, there exists a (unique) probability measure μ on ∂X satisfying

$$\mu(\omega_n^{-1}(x)) = (\# X(n))^{-1} = \frac{1}{(q+1)\, q^{n-1}} \quad \text{for } x \in X(n) ,$$

where $\omega_n : \partial X \to X(n)$ is the projection map.

Let $b = (0, x_1, x_2, \dots) \in \partial X$, and put

$$H_n(b) = \{ x \in X \ ;\ d(x, x_k) = k-n \ \text{ for sufficiently large } k \} .$$

One can then easily check that

$$X = \bigcup_{n \in \mathbb{Z}} H_n(b) \qquad \text{(disjoint)}$$

Define a function $\langle x;b \rangle$ on $X \times \partial X$ by putting

$$\langle x;b \rangle = n \qquad \text{when } x \in H_n(b) .$$

We set

$$\lambda_\theta = q^{1/2} e^{i\theta} ,$$

$$\phi(x) = \lambda_\theta^{\langle x;b \rangle} .$$

Then $A\phi = 2\, q^{1/2} \cos\theta\, \phi$.

We can now define the Fourier transformation. Let f be a function on X with finite support. Then the Fourier transform of f is defined by

$$(\mathcal{F} f)(\theta, b) = \sum_{x \in X} f(x)\, \lambda_{-\theta}^{\langle x;b \rangle} .$$

In the same manner as the case of the disc, we may prove the following.

a) $\mathcal{F}$ extends to a surjective isometry

$$\mathcal{F} : L^2(X) \to L^2([0,\pi] \times \partial X\ ,\ \frac{1}{\pi}\, a(\theta)\, d\theta\, d\mu(b)) ,$$

where

$$a(\theta) = \frac{q+1}{2q} \left| \frac{1-e^{2i\theta}}{1-q^{-1}e^{2i\theta}} \right|^2 .$$

b) $(F A_X F^{-1}\phi)(\theta,b) = 2q^{1/2} \cos\theta\, \phi(\theta,b)$.

We find, therefore, that $\sigma(A_X) = \sigma_c(A_X) = [-2q^{1/2}, 2q^{1/2}]$.

References

[1] M.F.Atiyah, Elliptic operators, discrete groups and von Neumann algebra, Astérisque, 32-33 (1976), 43-72.

[2] R.Brooks, The fundamental groups and the spectrum of the Laplacian, Comment. Math. Helvetici 56 (1981), 581-598.

[3] R.Brooks, Combinatorial problems in spectral geometry, in the Proceedings of the Taniguchi Symposium "Curvature and topology of Riemannian manifolds" 1985, Springer Lect. Note 1201, 14-32.

[4] R.Brooks, The spectral geometry of tower of coverings, J.Diff. Geom. 23 (1986), 97-107.

[5] M.Burger, Estimation de petites valeurs propres du Laplacien d'un revetêment de variétés Riemanniennes compactes, C.R.Acad.Sci. Paris 302 (1986), 191-194.

[6] P.Buser, On Cheeger's inequality $\lambda_1 \geqq h^2/4$, in Geometry of the Laplace operator, (Proc.Symp. Pure Math., Hawaii (1979), 29-77.

[7] J.M.G.Fell, Weak containment and induced representations of groups, Canadian J.Math. 14 (1962), 237-268.

[8] F.P.Greenleaf, Invariant Means on Topological Groups and Their Applications, von Nostrand, Reinhald 1969.

[9] Y.Ihara, On discrete subgroups of the two-by-two projective linear group over p-adic field, J.Math.Soc. Japan 18 (1966), 219-235.

[10] A.Katsuda and T.Sunada, Homology and closed geodesics in a compact Riemann surface, to appear in Amer.J.Math.

[11] D.A.Kazhdan, Connection of the dual space of a group with the structure of its closed subgroups, Funct.Anal.Appl. 1 (1967), 63-65.

[12] R.Phillips and P.Sarnak, Geodesics in homology classes, Duke Math.J. 55 (1987), 287-297.

[13] K.Ono, T.Kobayashi and T.Sunada, Spectrum of the Laplacian on a non-compact Riemannian manifold with compact quotient, in preparation.

[14] B.Randol, Small eigenvalues of the Laplace operator on compact Riemann surfaces, Bull.Amer.Math.Soc. 80 (1974), 996-1000.

[15] M.C.Reed and B.Simon, Methods of Modern Mathematical Physics, Vol. IV, Academic Press, 1978.

[16] P.Sarnak, Entropy estimates for geodesic flows, Ergod.Th. and Synam.Sys. 2 (1982), 513-524.

[17] A.Selberg, Harmonic analysis and discontinuous subgroups in weakly symmetric Riemannian spaces with applications to Dirichlet series, J.Indian Math.Soc. 20 (1956), 47-87.

[18] J.P.Serre, Tree, Springer, New York, 1980.

[19] D.Sullivan, Related aspects of positivity in Riemannian geometry, J.Diff.Geom. 25 (1987), 327-351.

[20] T.Sunada, Trace formula for Hill's operators, Duke Math.J. 47 (1980), 529-546.

[21] T.Sunada, Trace formula, Wiener integrals and asymptotics, Proc. Japan-France Seminar (Spectra of Riemannian Manifolds), Kaigai Publ. Tokyo 1983, 159-169.

[22] T.Sunada, Geodesic flows and geodesic random walks, Advanced Studies in Pure Math. (Geometry of Geodesics and Related Topics) Vol.3 (1984), 47-85.

[23] T.Sunada, Riemannian coverings and isospectral manifolds, Ann. of Math. 121 (1985), 169-186.

[24] T.Sunada, L^2-functions in geometry and some applications, Proc. Taniguchi Symp. 1985 (Curvature and Topology of Riemannian Manifolds) 266-284, Springer Lect. Note 1201.

[25] T.Sunada, Unitary representations of fundamental groups and the spectrum of twisted Laplacians, preprint. (to appear in Topology)

[26] T.Sunada, Spectrum of symmetric random walks on a graph, in preparation.

[27] R.J.Zimmer, Ergodic Theory and Semi-simple Groups, Birkhäuser, Boston, 1984.

[28] H.Donnelly and P.Li, Pure point spectrum and negative curvature for non-compact manifolds, Duke Math.J. 46 (1979), 497-503.

[29] H.Kesten, Full Banach mean values of countable groups, Math. Scand. 7 (1959), 146-156.

[30] A.Selberg, On the estimation of Fourier coefficients of modular forms, in Proc.Sym. in Pure Math. Vol.8, A.M.S., Providence, RI 1965.

[31] H.Donnelly, On L^2-Betti numbers for abelian groups, Canad.Math. Bull. 24 (1981), 91-95.

[32] A. Lubotzky, R. Phillips, and P. Sarnak, Ramanujan graphs, preprint.

LECTURE NOTES IN MATHEMATICS

Edited by A. Dold and B. Eckmann

Some general remarks on the publication of proceedings of congresses and symposia

Lecture Notes aim to report new developments - quickly, informally and at a high level. The following describes criteria and procedures which apply to proceedings volumes. The editors of a volume are strongly advised to inform contributors about these points at an early stage.

§1. One (or more) expert participant(s) of the meeting should act as the responsible editor(s) of the proceedings. They select the papers which are suitable (cf. §§ 2, 3) for inclusion in the proceedings, and have them individually refereed (as for a journal). It should not be assumed that the published proceedings must reflect conference events faithfully and in their entirety. Contributions to the meeting which are not included in the proceedings can be listed by title. The series editors will normally not interfere with the editing of a particular proceedings volume - except in fairly obvious cases, or on technical matters, such as described in §§ 2, 3. The names of the responsible editors appear on the title page of the volume.

§2. The proceedings should be reasonably homogeneous (concerned with a limited area). For instance, the proceedings of a congress on "Analysis" or "Mathematics in Wonderland" would normally not be sufficiently homogeneous.

One or two longer survey articles on recent developments in the field are often very useful additions to such proceedings - even if they do not correspond to actual lectures at the congress. An extensive introduction on the subject of the congress would be desirable.

§3. The contributions should be of a high mathematical standard and of current interest. Research articles should present new material and not duplicate other papers already published or due to be published. They should contain sufficient information and motivation and they should present proofs, or at least outlines of such, in sufficient detail to enable an expert to complete them. Thus resumes and mere announcements of papers appearing elsewhere cannot be included, although more detailed versions of a contribution may well be published in other places later.

Surveys, if included, should cover a sufficiently broad topic, and should in general not simply review the author's own recent research. In the case of surveys, exceptionally, proofs of results may not be necessary.

"Mathematical Reviews" and "Zentralblatt für Mathematik" require that papers in proceedings volumes carry an explicit statement that they are in final form and that no similar paper has been or is being submitted elsewhere, if these papers are to be considered for a review. Normally, papers that satisfy the criteria of the Lecture Notes in Mathematics series also satisfy this

.../...

requirement, but we would strongly recommend that the contributing authors be asked to give this guarantee explicitly at the beginning or end of their paper. There will occasionally be cases where this does not apply but where, for special reasons, the paper is still acceptable for LNM.

§4. Proceedings should appear soon after the meeeting. The publisher should, therefore, receive the complete manuscript within nine months of the date of the meeting at the latest.

§5. Plans or proposals for proceedings volumes should be sent to one of the editors of the series or to Springer-Verlag Heidelberg. They should give sufficient information on the conference or symposium, and on the proposed proceedings. In particular, they should contain a list of the expected contributions with their prospective length. Abstracts or early versions (drafts) of some of the contributions are very helpful.

§6. Lecture Notes are printed by photo-offset from camera-ready typed copy provided by the editors. For this purpose Springer-Verlag provides editors with technical instructions for the preparation of manuscripts and these should be distributed to all contributing authors. Springer-Verlag can also, on request, supply stationery on which the prescribed typing area is outlined. Some homogeneity in the presentation of the contributions is desirable.

Careful preparation of manuscripts will help keep production time short and ensure a satisfactory appearance of the finished book. The actual production of a Lecture Notes volume normally takes 6 -8 weeks.

Manuscripts should be at least 100 pages long. The final version should include a table of contents and as far as applicable a subject index.

§7. Editors receive a total of 50 free copies of their volume for distribution to the contributing authors, but no royalties. (Unfortunately, no reprints of individual contributions can be supplied.) They are entitled to purchase further copies of their book for their personal use at a discount of 33.3 %, other Springer mathematics books at a discount of 20 % directly from Springer-Verlag. Contributing authors may purchase the volume in which their article appears at a discount of 33.3 %.

Commitment to publish is made by letter of intent rather than by signing a formal contract. Springer-Verlag secures the copyright for each volume.